W0113214

Marine Macrophytes as Foundation Species

Marine Macrophytes as Foundation Species

Editor

Emil Ólafsson

Senior Researcher – Marine Benthic Ecology
Spanish Institute of Oceanography (IEO)
Centro Oceanográfico de las Baleares
Palma
Spain

CRC Press

Taylor & Francis Group
Boca Raton London New York

CRC Press is an imprint of the
Taylor & Francis Group, an **informa** business

A SCIENCE PUBLISHERS BOOK

Cover illustrations reproduced by kind courtesy of Iván A. Hinojosa, Institute for Marine and Antarctic Studies, University of Tasmania, Australia.

CRC Press
Taylor & Francis Group
6000 Broken Sound Parkway NW, Suite 300
Boca Raton, FL 33487-2742

First issued in paperback 2020

© 2017 by Taylor & Francis Group, LLC
CRC Press is an imprint of Taylor & Francis Group, an Informa business

No claim to original U.S. Government works

ISBN-13: 978-1-4987-2324-4 (hbk)
ISBN-13: 978-0-367-78280-1 (pbk)

This book contains information obtained from authentic and highly regarded sources. Reasonable efforts have been made to publish reliable data and information, but the author and publisher cannot assume responsibility for the validity of all materials or the consequences of their use. The authors and publishers have attempted to trace the copyright holders of all material reproduced in this publication and apologize to copyright holders if permission to publish in this form has not been obtained. If any copyright material has not been acknowledged please write and let us know so we may rectify in any future reprint.

Except as permitted under U.S. Copyright Law, no part of this book may be reprinted, reproduced, transmitted, or utilized in any form by any electronic, mechanical, or other means, now known or hereafter invented, including photocopying, microfilming, and recording, or in any information storage or retrieval system, without written permission from the publishers.

For permission to photocopy or use material electronically from this work, please access www.copyright.com (http://www.copyright.com/) or contact the Copyright Clearance Center, Inc. (CCC), 222 Rosewood Drive, Danvers, MA 01923, 978-750-8400. CCC is a not-for-profit organization that provides licenses and registration for a variety of users. For organizations that have been granted a photocopy license by the CCC, a separate system of payment has been arranged.

Trademark Notice: Product or corporate names may be trademarks or registered trademarks, and are used only for identification and explanation without intent to infringe.

Library of Congress Cataloging-in-Publication Data

Marine macrophytes as foundation species / Emil Olafsson, editor.
  pages cm
 "A CRC title."
 Includes bibliographical references and index.
 ISBN 978-1-4987-2324-4 (hardcover : alk. paper) 1. Aquatic plants. 2. Marine habitats. I. Olafsson, Emil, editor.

QK930.M37 2015
581.7'6--dc23   2015028929

Visit the Taylor & Francis Web site at
http://www.taylorandfrancis.com

and the CRC Press Web site at
http://www.crcpress.com

Preface

Marine macrophytes (macroalgae, seagrasses, and mangroves) comprise thousands of species distributed in shallow water areas along the world's coastlines. They play a key role in marine ecosystems regarding biodiversity and energy flow. A large proportion of macrophyte species can be characterised as ecosystem engineers—organisms that directly or indirectly affect the availability of resources to other species by modifying, maintaining, and creating habitats.

This book is divided into three main themes:

- *Marine macroalgae and seagrasses as sources of biodiversity* gives an overview of the diversity of the main organisms associated with macrophytes, and their functional role and interactions within their hosts.
- *Primary and secondary production of Macrophytes* synthesizes research on food web structures derived from/or associated with, macrophytes and the transfer of macrophytic primary and secondary production from one ecosystem to another.
- *Threats to macrophytic ecosystem engineers* addresses human-induced effects including eutrophication, physical destruction, invasive species, and global warming.

The main features of the book are: Discusses how microalgae and seaweeds contribute to biodiversity; synthesizes research findings on macrophytes as a source of food; covers threats to marine ecosystems caused by human activities; serves as the first book to cover the value of macrophytes for the well-being of marine habitats.

The book is among the first one to concentrate on the value of macrophytes for the well-being of marine habitats. The book is aimed at academics but may also be useful for students, policy makers, and laymen alike.

I dedicate this book to Agnar Ingólfsson and Ragnar Elmgren; both had a strong impact on my career in marine biology.

Emil Ólafsson
Palma de Mallorca, Spain
July 6th, 2016

Contents

I. Macroalgae and Seagrasses as a Source of Biodiversity

1

Microbial Biodiversity Associated with Marine Macroalgae and Seagrasses

Franz Goecke and *Johannes F. Imhoff*[a]

Introduction

Macrophytes are the main primary producers of coastal ecosystems, which include many habitat-forming species with vital ecological importance. Macroalgae and seagrasses as any other macroorganisms in this ecosystem are in permanent contact with microbes. Seawater contains millions of microorganisms per ml of bacteria, cyanobacteria, fungi, microalgae and protists (Goecke et al. 2010), which eventually interact with every single species of macrophytes during the entire lifetime till the next generation. In the marine environment microbes are found in the immediate surroundings, floating in the water or along with currents and tides. The aquatic environment favors the formation of biofilms on surfaces. Besides every mechanical, chemical and physical defense, microorganisms can be found on every living surface in the aquatic environment, and this includes marine macrophytes as well (Wahl et al. 2012).

Such epibiotic biofilms have a huge potential to affect the biology, ecology and fitness of their host (the basibiont). Many direct and indirect effects of epibiotic biofilms have been described, many more can be expected to exist, but the consensus of most investigations is that the presence of biofilms alters the substratum physically and chemically, and that they have the capacity to modulate (reduce, enhance,

Kieler Wirkstoff-Zentrum am GEOMAR Helmholtz-Zentrum für Ozeanforschung, Am Kiel-Kanal 44, Kiel D-24106, Germany.
[a] Email: jimhoff@geomar.de

select) the recruitment of other microbes and macroorganisms (recently reviewed by Wahl et al. 2012).

Macrophytes are especially susceptible to epibiosis and are typically covered by diverse microbial communities that normally include bacteria, microalgae, fungi and diverse protists (Sieburth and Thomas 1973, Raghukumar and Raghukumar 1992, Kohlmeyer and Volkmann-Kohlmeyer 2003, Neuhauser et al. 2011). Uncontrolled biofilm formation produces a permanent threat to macrophytes (see Goecke et al. 2010), although the consequences of epibiotism and endobiosis for the host macrophyte are not always necessarily negative, but instead depend on the community context. Hence, such intimate associations can range from mutualism through commensalism and parasitism in an almost continuous spectrum (Potin 2012).

The colonization of marine macrophytes is quite variable and dense microbial populations (up to 10^6 cells per cm^2) may be found. Density and composition of epibiotic biofilms vary at different scales: Among host species, conspecific host individuals, body regions of a host individual, among habitats, seasons and during the life cycle (Wahl et al. 2012). Microbes and basibionts are not passive players in the colonization process. The macrophyte basibiont can exhibit a variety of defenses against microbial fouling and epibionts. This resistance can be generally mediated by either structural or chemical means, with both mechanisms interfering with the attachment, colonization (and penetration) and dispersion of microorganisms, spores or propagules (Potin 2012). A wide variety of macroalgae and seagrasses have shown to produce diverse antibiotic (antifungal, antibacterial, antiprotozoal secondary metabolites) and antifouling compounds, which may allow the host to modify the composition of the associated microbial populations (Lam et al. 2008, Olson and Kellogg 2010, Goecke et al. 2012a, Grosser et al. 2012). The epibiotic microorganisms can produce biologically active metabolites as well or modify those available and thereby enhance the antifouling protections as suggested by different authors (Armstrong et al. 2001, Goecke et al. 2013a).

Besides some detrimental effects, a wide range of beneficial and neutral interactions have been observed between macrophytes and microorganisms. Many of the microorganisms are specifically associated with their hosts and may enable them to expand physiological capacities (Hollants et al. 2013a), to survive adverse environmental conditions and to overcome competitors.

After different evolutionary and ecological processes, those negative and positive interactions, together with the great diversity of microorganisms and macrophytes may be responsible for differences on microbial associations with different co-occurring hosts (Goecke et al. 2012a, 2013a,b). Macrophytes sustain therefore a huge biodiversity of microbes that need to be explored. In this chapter, we focus on recent advances in this field, especially on interactions between bacterial and fungal communities with macrophytes, and point out some urgent questions for future research. These microbial groups were chosen because they are the most abundant in this habitat and their relation to macrophytes has been studied more intensively than that of other microbes. In addition, many other 'minor' groups of microorganisms which interact with macrophytes, e.g., oomycetes, labyrinthulids, flagellates, heliozoans, ciliates, amoebozoans, thraustochytrids, foraminiferans, and phytomyxids were not considered because of limited space. However, the prokaryotic domain of the Archaea is included.

During the last decades, a diversification and intensification of aquaculture, the sustained nutrient enrichment from land-based activities, pollution, biological invasions, ocean acidification, global warming and climate change, among others, have demonstrated to strongly affect biotic interactions (Eggert et al. 2010, Gachon et al. 2010, Case et al. 2011, Wernberg et al. 2011, Williams 2007, Potin 2012). This points to the necessity to develop an understanding on the microbial communities specifically associated with macrophytes. Which are the ecological processes that are involved? And how can these microbial communities affect and be affected by the habitat-forming and principal primary producers of the coastal ecosystems.

Marine bacteria

Studies of bacteria associated with macrophytes were reported since late 19th century (Hollants et al. 2013a). During the last five years an interesting number of publications have dealt especially with bacteria-macroalgae relationships, trophic and chemical interactions and the associated microbial communities, which are discussed in this chapter. The relationship between macrophyte and bacteria in which the host provides nutrients and habitat, while the bacterial community promotes macrophyte growth and protects the host against pathogens, has been elaborated over the last 20 years (Hollants et al. 2013a). While several bacterial species have been identified as causative agents of diseases of macroalgae, many bacterial associations are clearly beneficial for the algal host (Kurtz et al. 2003, Matsuo et al. 2003, Amin et al. 2009, Goecke et al. 2010, Seyedsayamdost et al. 2011, Grant et al. 2014). Recent comprehensive reviews have examined the body of literature available on bacteria associated with macroalgae and seagrasses, which is largely built on cultivation-based studies (see Duarte et al. 2005, Goecke et al. 2010, Egan et al. 2013, Hollants et al. 2013a). On one hand, these suffer from the known limitations of adequately describing the composition of natural microbial communities by cultivation-dependent approaches (Friedrich 2012, Fig. 1), on the other hand, they offer plenty of possibilities to test ecological hypotheses in controlled experimental setups such as reinfection experiments with defined pathogens.

Biofilms on the surface of marine organisms are usually dominated by prokaryotes (Bacteria). Bacterial biofilm with cell densities of 10^2 to 10^8 cells cm^{-2} have been found on different macroalgae (see Dobretsov et al. 2006, Bengtsson et al. 2010, Wahl et al. 2010); similar values have been determined for roots and leafs of different seagrass species (Kurilenko et al. 2001, García-Martínez et al. 2005). Macrophyte surfaces as a microbial habitat display several advantages and have characteristic properties that influence their association with microbes such as nutrient supply, defense against competitors, protection from damaging ultraviolet radiation and predation, among others (Goecke et al. 2013a); therefore they are highly attractive and competitive marine environments.

The analysis of microbial communities associated with macrophytes is still in its infancy and in particular functional interactions are not known. The molecular microbial ecology of macroalgal or seagrass hosts is a rather young field with only small datasets available but with large potentials to influence theory building in microbial ecology and host–microbe interactions (Friedrich 2012). Nevertheless, in most cases molecular investigations have confirmed the outcome of initial cultivation

Figure 1. Biofilm on the common brown macroalgae *Fucus vesiculosus* in the Baltic Sea. (a) Scanning electron microphotograph (SEM) with details of the microbial biofilm over the surface of the thalli of the alga. (b) Petri dish showing the growth of different colonies of microorganisms after processing a small piece of the alga (indicated by the white box). Clearly just a selection of microorganisms is growing on a particular nutrient medium, which does not reflect the true diversity and real abundances associated with the macroalgae.

studies, i.e., that the attraction of bacteria by macrophytes turns out to be highly specific (Hollants et al. 2013a). While the composition of the bacterial flora can change over seasons, life span and different thallus-parts as a result of biotic and abiotic factors (Crump and Koch 2008, Staufenberger et al. 2008, Bengtsson et al. 2010, Tujula et al. 2010), marine macrophyte generally associate with specific bacterial communities that differ significantly from those occurring in the surrounding seawater or on inanimate (and undefended) substrata in close vicinity (Dobretsov et al. 2006, James et al. 2006, Longford et al. 2007, Lachnit et al. 2009, Bengtsson et al. 2010, Burke et al. 2011b, Hollants et al. 2013a).

Recent research also confirms that different species of marine macrophyte in the same habitat support differently composed bacterial communities (Lachnit et al. 2009, 2011, Trias et al. 2012a), while specimens of the same macrophyte even in different environments tend to be associated with highly similar bacterial communities (Uku et al. 2007, Staufenberger et al. 2008, Lachnit et al. 2009). Even though a consistent bacterial core community at higher taxonomic levels (i.e., of Proteobacteria and Bacteroidetes, Table 1) was observed on different macroalgae (see Staufenberger et al. 2008, Tujula et al. 2010, Burke et al. 2011b), phylogenetically closely related macroalgae may not necessarily harbor the same bacterial taxa (Hollants et al. 2013a). Certain bacterial symbionts are sensitive markers potentially useful to distinguish genetically similar macroalgae as suggested by Balakirev et al. (2012).

Table 1. Bacterial phyla associated with macrophytes as determined by culture-independent methods. A consistent bacterial core community formed by Proteobacteria and Bacteroidetes was observed on different macrophyte samples (based on Cifuentes et al. 2000, Weidner et al. 2000, Jensen et al. 2007, Uku et al. 2007, Crump and Koch 2008, García-Martínez et al. 2009, Balakirev et al. 2012, Bengtsson et al. 2012, de Oliveira et al. 2012, Garcias-Bonet et al. 2012, Fernandes et al. 2012, Hollants et al. 2013a,b, Wahl et al. 2012, Aires et al. 2013, Miranda et al. 2013, Sweet et al. 2013, Wang et al. 2013, Stratil et al. 2014).

	Dominant	Common	Rare
Macroalgae			
	Alphaproteobacteria	Firmicutes	Chloroflexi
	Gammaproteobacteria	Actinobacteria	Chlorobi
	Bacteroidetes	Betaproteobacteria	Lentisphaerae
		Deltaproteobacteria	Epsilonproteobacteria
		Planctomycetes	Deinococcus-Thermus
		Cyanobacteria	Spirochaetes
			Verrucomicrobia
			Fusobacteria
			Tenericutes
			Acidobacteria
Seagrasses			
	Gammaproteobacteria	Planctomycetes	Chloroflexi
	Alphaproteobacteria	Betaproteobacteria	Holophaga-Acidobacterium
		Bacteroidetes	Spirochaetes
		Epsilonproteobacteria	Actinobacteria
		Verrucomicrobia	Cyanobacteria
		Firmicutes	Deferribacter
		Deltaproteobacteria	Nitrospira

Studies of seagrasses have particularly been concerned with specific functional groups of bacteria (Table 1), such as sulfate-reducing bacteria (Küsel et al. 1999, Nielsen et al. 1999, Finster et al. 2001), nitrogen-fixing bacteria (Kirshtein et al. 1991, Hansen et al. 2000, Bagwell et al. 2002) and acetogenic bacteria (Küsel et al. 1999), whereas only in a few studies on the general bacterial diversity has been considered by using molecular analyses (see Cifuentes et al. 2000, Weidner et al. 2000, Jensen et al. 2007, Uku et al. 2007, Crump and Koch 2008, Garcias-Bonet et al. 2012). The large similarities observed at higher taxonomic ranks among the associated-microbial communities of macroalgae and seagrasses decrease notoriously at lower ranks of bacteria (for details see Table 1 and references therein) and point out more specific associations at the lower ranks.

In terms of biodiversity and because of this specificity, macrophytes are an interesting environment for discovery of new bacterial taxa. In the case of algae, several new bacterial species (101 spp.) and new genera (36 genera) have been isolated (reviewed by Goecke et al. 2013b). Furthermore, the authors conducted a phylogenetic study (supported by phenotypic information) of those new species based on 16S rRNA gene sequences and obtained a similar phylogenetic distribution as found in cultivation-independent molecular studies (as in Table 1). Also seagrasses were a source of a number of new bacteria, e.g., *Desulfomusa hansenii* (Finster et al. 2001),

Desulfovibrio zosterae (Nielsen et al. 1999), *Granulosicoccus coccoides* (Kurilenko et al. 2010), and diverse *Marinomonas* species (Lucas-Elío et al. 2011).

It feels like home: The phyto(phyco)-sphere

As mentioned before, the establishment of epibiosis is not a simple process, and various physical and chemical properties of the host surface, as well as interactions among the settlers are determinants of the formation of specific communities (e.g., García-Martínez et al. 2005, Wahl et al. 2010, 2012). The ecological function of different macrophytes can be similar as they are primary producers that provide physical habitat and trophic support to other species which in turn are important food items for other organisms (Thomsen 2010). However, species-specific attributes with respect to the phenology, morphology and biology of each taxon can influence the structure of associated communities differently (Drouin et al. 2011).

The phytosphere is a unique and attractive environment, which consists of the phyllosphere, the endosphere and the rhizosphere (the last one normally absent in macroalgae with exception of, for example, some Bryopsidales). Each of these three habitats provides a considerably diverse physical, chemical and biological environment, and as a consequence can support a wide range of different microbial groups (Saito et al. 2007).

Thallus morphology and microtopography of the macrophyte surface play an important initial role in the colonization and association with the microbiota and invertebrates (Wahl et al. 2012) as does the surface chemistry of a macrophyte, which is the sum of exuded secondary metabolites and extracellular exopolymeric substances present on the thallus (Grosser et al. 2012). This complex mixture of compounds, presumably in equilibrium with the ambient water body, is a primary source of nutrients for those bacteria attached to the surface (Lachnit et al. 2010, Salaün et al. 2012). Rhizoplane bacteria obtained from the roots of *Zostera marina* demonstrated chemotaxis towards root exudates, in particular amino acids (Wood and Hayasaka 1981). Bacteria may use the host not only as a surface for settlement but also as a source of substrates for their own nutrition. The utilization of substrates produced or released by macrophytes, including (structural) polysaccharide components of cell walls, e.g., alginate, agar, carrageenan, cellulose, fucoidan, laminaran, porphyran, ulvan, is an important aspect of surface colonization by bacteria (Bengtsson et al. 2011, Salaün et al. 2012). Such macromolecular composition is characteristically different among the major evolutionary lineages, linking specific life style or nutritional habits to specifically encountered biopolymers (Goecke et al. 2013b). Hence, each algal/seagrass species can be considered to be a unique micro-environment.

It is important to consider that macrophyte structural complexity is an important factor influencing the abundance and taxon richness of microfauna, e.g., rotifers, copepods, cladocerans (Lucena-Moya and Duggan 2011), macrofauna, e.g., amphipods, mollusks, annelids and even fishes (Tuya et al. 2009). Macrophytes with a more complex morphology such as those presenting diverse ramifications as *Ceramium*, *Cystoseira* and *Corallina* spp., support a greater diversity and abundance of associated macroorganisms in comparison to macrophytes with flat and simple fronds such as

Chondrus, *Porphyra* and *Palmaria* spp. (Chemello and Milazzo 2002). The reasons suggested are a reduction of mortality by predation, reduction of physical stress (wave action, desiccation, UV irradiation), by accumulating species transported passively by currents; and by proportioning a major surface for perifiton and other small food items (Dean and Connell 1987, Warfe and Barmuta 2004). Surprisingly, its effect on microbial communities remains uninvestigated. The accumulation of potential predators (microfauna) could negatively affect the abundance of microorganisms, but also the facilitation of settlement, accumulation of microorganisms and organic matter, along with the presence of a variety of microhabitats could instead promote diversity and abundance of microbes.

One hand clean the other: Bacterial interrelationships in the seagrass rhizosphere

Bacterial abundance is generally higher in sediments from seagrass meadows than in adjacent unvegetated sediments (Duarte et al. 2005, and references therein). Seagrasses develop a complex rhizosphere, typically deploying several hundred meters of root material per square meter of seagrass meadow and occupying up to 80% of the sediment space (García-Martínez et al. 2005). These sediments contain active bacterial communities that consume molecules excreted by plant roots such as organic molecules (amino acids, sugars and organic acids) and gases (O_2, N_2 and CO_2). In return, microbial processes, including organic matter mineralization, phosphorous solubilization and nitrogen fixation, provide nutrients for the macrophytes (Hansen et al. 2000, Crump and Koch 2008).

Inside the intricate seagrass meadow, the increase in physical structure complexity affect sedimentation, turbulence and water flow, trapping sediments and stimulating the accumulation of organic matter below the meadows (Fig. 2). This organic matter is available for microbial processes and can easily develop anoxic conditions. As a reflection of this, sulfate-reducing bacteria, the predominant bacteria involved in anaerobic degradation of organic matter, proliferate and consequently sulfate-reduction is increased with the posterior accumulation of toxic sulfide (Duarte et al. 2005, Jensen et al. 2007, and references therein). Anoxic and toxic conditions can be detrimental for seagrass growth and survival. However, these plants release oxygen through the roots and thereby (partially) prevent anoxic conditions. This oxygen may be crucial for host-associated microbial processes and for aerobic bacterial metabolism (Duarte et al. 2005). Some microbes may have beneficial effects on seagrasses by metabolizing toxic substances in the rhizoplane, e.g., sulfide-oxidizing bacteria using nitrate as electron acceptor (Küsel et al. 2006) or ammonia-oxidizing bacteria (Ando et al. 2009). The presence of bacteria related to well-known nitrogen fixers is especially interesting because these may have a beneficial role in the establishment, growth and survival of macrophytes (such as *Posidonia oceanica*) in oligotrophic environments (Kirshtein et al. 1991, Bagwell et al. 2002, Garcias-Bonet et al. 2012); with a resulting mosaic of microhabitats and diverse coexistent phytosphere-microbial communities.

The effects of seagrasses on sediment organic matter and oxygen inputs modify the bulk biogeochemical redox conditions in the sediments. Plant root exudates can

Figure 2. Habitat-forming seagrasses in the marine environment. (a) A patch of *Posidonia oceanica* growing in the Mediterranean Sea. Sometimes these macrophytes can form 'island' between clear sandbanks which provide shelter and refuge for diverse macro- and microorganisms. (b) Detail into a seagrass meadow of *Zostera marina* in Germany, Baltic Sea. The intricate meadow increases in physical structure complexity, trap sediments and stimulate the accumulation of organic matter below the meadows, reduce detection of preys by bigger predators and proportionate a major surface for perifiton. (c) Once dislodged (e.g., by storms, herbivores or death) piles of the seagrasses may accumulate on the beach and form very special conditions in support of a large number of macro- and microorganisms.

selectively stimulate and inhibit microbial growth and thus foster the establishment of specific microbial communities (Jensen et al. 2007). Bacterial rhizosphere communities can vary in composition according to plant species, growth stage, root exudates, and available carbon source (Kurtz et al. 2003, and references therein). Nevertheless, the role of bacteria (as in sulfur cycling) can also determine the health and therefore the growth rates of marine angiosperms (Garcias-Bonet et al. 2012). Recent declines in seagrass distribution underscore the importance of understanding microbial community structure-function relationships in the seagrass rhizosphere that might affect the viability of these macrophytes (Küsel et al. 1999), and may be vital for restoration programs. Moreover, van der Heide et al. (2012) hypothesized that seagrass meadows may provide an optimal habitat for certain bivalves (Lucinidae, Bivalvia) and their bacterial symbionts by indirectly stimulating sulfide production by increasing organic matter content and by providing oxygen through radial oxygen release from the roots. In turn, those lucinids remove sulfide, which could relieve any stress caused to seagrass growth by sulfide. In this way a three-stage symbiosis is formed supporting the success of seagrass meadows.

A new perspective: Specific function instead of species-specific

In order to differentiate between community structure and function in the bacterial assemblages associated with the green alga *Ulva australis*, a new perspective came from Burke et al. (2011a), who analyzed metagenomic sequences. Despite high phylogenetic variability in the microbial species composition associated with that alga (see Tujula et al. 2010), the authors discovered little functional variability in the functional composition (measured as presence of functional gene clusters). Phylogenetically different bacterial species (or strains)—able to colonize one particular host species—that can carry out similar metabolic and other functions apparently compete with each other in the colonization of particular algal surfaces (Burke et al. 2011a). This suggests that the assembly, structure and ecological role of bacterial communities in the future should be described preferentially at the functional level. In this sense, different bacterial species and strains that carry out similar metabolic functions were found to colonize similar algal taxa or algal groups (Goecke et al. 2013b). Also, studies on different seagrass species revealed that a variety of microorganisms able to fix di-nitrogen are associated with the phyllosphere and rhizosphere of seagrasses (Kirshtein et al. 1991, Cifuentes et al. 2000, Weidner et al. 2000, Uku et al. 2007). Since for the host (and its interactions with the environment) microbial function matters more than phylogenetic composition, investigations at the functional level based on genomic or metabolomic information should become more prominent in the future (Wahl et al. 2012).

Traditionally, classification of bacteria has relied on cultivation. Since only a selected fraction of bacteria will grow on a specific medium, the cultivated fraction of bacteria may not reflect the abundant/diverse bacterial populations of their habitat (Friedrich 2012, Fig. 1). Metagenomic approaches on the other hand can reveal the complex biodiversity of the bacterial assemblages, but give information on the functional aspect only to a rather limited extent. There is no doubt that we are only starting to discover the specific and complex relationships of macrophytes and bacteria in nature. Interactions between bacteria and algae (macrophytes) are thought to be important in controlling the dynamics of both communities and yet are barely understood at the species and functional level (Grossart et al. 2005). Moreover, it is likely that macrophyte-bacterial relationships will change as the host ages (Seyedsayamdost et al. 2011). Recent technical advances in environmental microbiology have enabled the evaluation of microbial diversity and/or genomic diversity using rapid, simple and less biased culture-independent molecular techniques (Saito et al. 2007). These changes include typically commensal epiphytic microorganisms that are common on the surface but could become detrimental if entering into the macrophyte tissue (Goecke et al. 2010) or if the host/epibiotic microbial community are strongly affected by environmental factors such as ocean acidification or warming (Eggert et al. 2010, Campbell et al. 2011, Case et al. 2011, Fernandes et al. 2012). A better insight into these mutualistic and pathogenic interactions is necessary for understanding and predicting algal bloom formations, disease outbreaks, and the response of populations of macrophytes and bacteria to changes in their environment (Amin et al. 2009).

Marine archaea

The dogma until around 1992 was that Archaea inhabit mainly 'extreme' environments, inhospitable to most other forms, but later, based on cultivation-independent studies, they were surprisingly found to be abundant in coastal and open-ocean waters (DeLong 2007). Archaea were detected in association with different marine organisms, sponges and corals (Olson and Kellogg 2010) and also with the rhizosphere of different macrophytes, e.g., *Ruppia maritima* (Trias et al. 2012b).

With respect to seagrasses, Cifuentes et al. (2000) identified members of Euryarchaeota and Crenarchaeota in sediments taken from a *Zostera noltii* meadow, and later the authors were able to grow (although not in pure culture) one marine non-extremophile Crenarchaeota (Cifuentes et al. 2003). In a recent study, seasonal changes in abundance of ammonia-oxidizing archaea were observed with quantitative PCR within sand of a *Zostera marina* meadow (Ando et al. 2009). The biogeochemical implications can be essential: Nitrification is important to sustain and control the nitrogen cycle in eelgrass zones and ammonium can be toxic to the seagrass species. Archaea have been detected on macroalgae only recently, in mesophotic depth in the Mediterranean Sea (Trias et al. 2012a). But this may be only as a result of lack of research in this field. Only few investigations have been made on macrophyte-associated archaea and information on community compositions (abundance and diversity) of archaea is lacking. Besides the ammonia-oxidation, no other potential beneficial or detrimental effects of archaea for the macrophytes have been postulated. One of the likely reasons for this is the fact that most microbial ecology studies of macrophyte surfaces have been culture-based and proper culture conditions for these archaea are lacking (Olson and Kellogg 2010). It is evident, however, that archaea other than methanogens may be an important part of the prokaryotic community in the marine phytosphere (Cifuentes et al. 2000), though there is a massive gap of knowledge on archaea-macrophyte interactions.

Marine fungi: Filamentous fungi and yeasts

The occurrence of fungi in the marine environment was first reported late in the 19th century, therefore marine mycology is considered as a recent science field (Shearer et al. 2007). Fungal species in the marine environment include those that are adapted to complete their life cycles in marine aquatic habitats and are not found outside of the aquatic environment <'residents', 'obligate marine'>, those that occur in water fortuitously by being washed or blown in <'transients'>, and those 'facultative' marine species, which may grow in marine as well as in freshwater or terrestrial habitats (Kohlmeyer and Kohlmeyer 1979, Shearer et al. 2007).

Fungi from coastal and marine ecosystems are often neglected. Although marine fungi grow on most organic substrates occurring in oceans and estuaries, the fraction of cultivable isolates is very low (1% or less) with regard to the overall estimated biodiversity (Hawksworth and Rossman 1997, Rateb and Ebel 2011). Nevertheless, fungi have key roles in the marine environment and act as decomposers in nutrient turnover of organic matter (Sridhar et al. 2012). They are of particular interest due to their ecological significance and interactions (Apt 1988, Zuccaro et al. 2003). Pathogens, parasites, saprobes, and mycobionts are the predominant fungi of marine macrophyte communities. However, there is little data available about the ecology of these organisms (Zuccaro et al. 2008). It is often difficult to define the ecological role of associated-fungi in coastal habitats, especially because of unstable relationships (symbiotic, to parasitic and saprophytic) over time (Kohlmeyer and Kohlmeyer 1979).

It has been suggested that macroalgae harbor the highest diversity of aquatic fungi, closely followed by sponges and mangroves as habitats (Rateb and Ebel 2011). Reviews of the literature on algicolous fungi have been published by Andrews (1976), Kohlmeyer and coworkers (1973, 1979, 1981, 1991 and 2003) and Zuccaro and Mitchell (2006).

Of particular interest are permanent symbiotic associations between systemic marine fungi and macroalgae in which the habit of the alga dominates <mycophycobioses>, because most probably they are considered as primitive lichenizations (Kohlmeyer 1973). In this sense, the interactions between the brown macroalgae *Ascophyllum nodosum* and *Pelvetia canaliculata* and their obligate symbiont *Mycophycias ascophylli* (Ascomycetes) are of importance (Deckert and Garbary 2005, Xu et al. 2008). Their studies suggest that *M. ascophylli* may have a protecting role by increasing the growth of algal zygotes and protecting them from desiccation. Although *A. nodosum* and *M. ascophylli* have each been cultured in isolation, both grow very slowly without their symbiont and neither *A. nodosum* nor *P. canaliculata* is found in nature without the fungus (Toxopeus et al. 2011). The ecological implications of the faster growth rate are important in that this might provide an escape mechanism from herbivory (Deckert and Garbary 2005).

In comparison with bacteria, cultivation-independent studies on the diversity of fungal species associated with macrophytes are limited to only a couple of investigations (see Zuccaro et al. 2003, 2008), which do not allow comparisons or general conclusions. Most information on the fungal community 'diversity' on macrophytes is obtained from cultivation-dependent studies (Meyers et al. 1965, Miller and Whitney 1981, Newell 1981, Phillips 1982, Cuomo et al. 1985, 1988, Sathe and Raghukumar 1991, Almaraz et al. 1994, Genilloud et al. 1994, Wilson 1998, Alva et al. 2002, Devarajan et al. 2002, Loque et al. 2010, Panno et al. 2010, Sakayaroj et al. 2010, Sridhar et al. 2012, Suryanarayanam et al. 2010, Godinho et al. 2013, Mata and Cebrián 2013, Shoemaker and Wyllie-Echeverria 2013), which have gained valuable knowledge in this field, but suffered from the mentioned biases.

The most commonly encountered fungi belongs to the Ascomycota, in agreement with cultivation-based studies. Basidiomycetes, Chytridiomycota and Zygomycota are also present normally on algal samples (Eggert et al. 2010, Gachon et al. 2010, Gleason et al. 2011), but they have not usually been detected in these molecular studies. Phylogenetic analysis of *Fucus serratus* of the Atlantic coast of Germany, revealed the presence of four main ascomycete orders: The Halosphaeriales, Hypocreales, Lulworthiales and the Pleosporales (Zuccaro et al. 2003), predominantly the *Lindra, Lulworthia, Engyodontium, Sigmoidea/Corollospora* complex and *Emericellopsis/Acremonium* types (Zuccaro et al. 2008).

While an increasing number of studies have identified fungi as important agents in the pathology of marine plants, very little is known about the mechanism of pathogenesis. Studies of these fungi have been undertaken mostly from a taxonomic approach, but they provide only few data on the host-parasite interaction, pathogenicity, predisposition and epidemiology (Andrews 1976, Kohlmeyer and Kohlmeyer 1979). Also, there are only a few studies on the asymptomatic fungal endosymbionts (Schulz et al. 2002, Suryanarayanan 2012). Recognizing the enormous undiscovered diversity in fungi, especially those associated with macrophytes, is both an opportunity and a challenge (Hawksworth and Rossman 1997). While a lot of information has accumulated on the roles of fungi in structuring and maintaining terrestrial ecosystems like forests, their role in orchestrating an important part of the marine ecosystem consisting of macrophytes is not clear (Suryanarayanan 2012). Supported by the high diversity of hosts and fungi, the extended evidence on symbiotic relationships with (terrestrial) algae (as lichens) and land plants (in mycorrhyzal symbiosis), we can expect that many close and specific associations between macrophytes and fungi still remain to be discovered.

Marine fungi as pharmacy for macrophytes

Algicolous fungi are proposed to have important ecological functions for the host that includes providing increased resistance against biotic stresses. It is known that a number of secondary metabolites play a major role in the symbiotic life forms and that natural products derived from associated microorganisms may function as a chemical defense for the host (Armstrong et al. 2001, Goecke et al. 2013a). In this sense, protective effects are thought to be mediated by fungal-derived natural products which also makes algicolous fungi a valuable resource for new bioactive compounds (Schulz et al. 2002, Miao et al. 2012). In comparison with seagrasses, those living in association with marine macroalgae are a particularly promising source of novel natural products due to the special and diverse ecological niche in which they exist (Kohlmeyer and Kohlmeyer 1979, Flewelling et al. 2013, Godinho et al. 2013). Macroalgae are a very prolific source of bioactive secondary metabolites themselves (Goecke et al. 2010), and certain microorganisms associated with them are able to metabolize these compounds (as nutrients), detoxify them or even modify them for their own purposes.

There are only a few examples in this case. In one, bromosesquiterpenes like aplysistatin, palisading A and 12-hydroxypalisadin B (isolated from the red alga *Laurencia luzonensis*) were biotransformed by the algicolous fungus *Rhinocladiella* sp. K-001, yielding two new compounds: 3,4-dihydroaplysistatin and 9,10-dehydrobromopalisadin A (Koshimura et al. 2009). Similar experiments were made with other algicolous fungal strains of *Aspergillus, Chrysosporium* and *Hypocrea* sp. (for more information see Leutou et al. 2009, Ramesh and Kalaiselvam 2011, Yun et al. 2011). Although the specific meaning of these transformations has not been ecologically tested, biotransformation can be applied to generate new active or less toxic derivatives of bioactive natural products (Leutou et al. 2009). It offers not only possible physiological and ecological benefits to the host (i.e., more powerful chemical defenses or detoxification), but also interesting biotechnological applications. Unfortunately, most of the reports on secondary metabolites have been directed toward the examination of fungal metabolites for biomedically relevant activity, largely ignoring the ecological roles of fungal secondary metabolites in the marine ecosystem (Jenkins et al. 1998). Clearly further research at the physiological, metabolic and the molecular level is necessary to obtain better insight into the chemical ecology of host-fungal relationships (König et al. 2006).

Marine yeast: Important agents on macrophyte-degradation processes

Yeasts (a polyphyletic group of fungi) are ubiquitous in their distribution and their populations mainly depend on the type and concentration of organic materials. Their environmental role is similar to many other fungi, acting as saprophytes and pathogens of plants and animals, and they participate in a range of ecologically significant processes in the sea (e.g., decomposition of vegetal substrates, nutrient-recycling), especially in estuarine and near-shore environments (Kutty and Philip 2008).

Yeast can be commonly found in large numbers associated with marine macrophytes (1 to 10^5 yeast cells gram of the host, see Suehiro 1960, van Uden and Castelo-Branco 1963, Sieburth and Jensen 1967, Seshadri and Sieburth 1971). Many reports describe the abundant development of yeasts, especially on decomposing macroalgae (Bunt 1955, Suehiro and Tomiyazu 1962, Patel 1975, Seshadri and Sieburth 1975, and references therein). Yeasts are truly versatile agents of biodegradation and may utilize exudates of their living (and dead) hosts. For example, different species of *Candida* assimilate a variety of macrophyte-derived compounds including alginic acid, fucoidin, fucose, galactose, glucose, mannitol and phloroglucinol (Seshadri and Sieburth 1975). In this way, macrophytes can act as an important reservoir for yeasts. For example, huge heaps of kelps (like *Macrocystis pyrifera*) when deposited above the high tide line may remain landlocked for several days and undergo decomposition. Later, when eventually washed back into the sea, the yeast population can be released into the seawater (van Uten and Castelo Branco 1963).

A literature survey revealed that investigations on marine yeasts are comparatively few and that this group of marine mycota is still poorly understood (Kutty and Philip 2008). Most of the studies on yeast associated to macrophytes are concerned with isolation of strains from fresh and natural or decomposing samples (Table 2). There are numerous reports on seasonal and host variation (Suehiro 1960, Seshadri and Sieburth 1971, Patel 1975, Summerbell 1983, Wilson 1998), but there are no cultivation-independent studies on the distribution of yeast on macrophytes. Though diverse organic extracts and compounds of macrophytes exhibited anti-yeast bioactivities which could be useful as host defense (e.g., Ballesteros et al. 1992, Choi et al. 2009), no field test have been made of ecological associations. There is currently no evidence for symbiotic relationships between yeast and macroalgae. Only recently, the yeast *Metschnikowia australis* was found in high densities in the intra-vesicular fluid of the brown Antarctic macroalga *Adenocystis utricularis*. This is an interesting habitat since the yeast is protected inside the alga against the stressful environmental conditions and probably is able to utilize photosynthetic nutrients released by the macroalga (Loque et al. 2010), but the possible ecological association between *M. australis* and *A. utricularis* and it prevalence needs further investigation with molecular methods.

A comment on marine protozoa

Protists are microscopic eukaryotic microbes that are ubiquitous, diverse, and major participants in oceanic food webs and in marine biogeochemical cycles (Sherr et al. 2007). Several studies have documented the presence of heterotrophic protists (Armstrong et al. 2000), naked amoebae (Rogerson 1991), foraminiferans (Semeniuk 2001, Debanay and Payri 2010), labyrinthulids (Bergmann et al. 2011, Garcias-Bonet et al. 2011), phytomyxids (den Hartog 1989, Neuhauser et al. 2011, Goecke et al. 2012b), thraustochytrids (Phillips 1982, Raghukumar and Raghukumar 1992), and diatoms (Sieburth and Thomas 1973, Siqueiros-Beltrones and Ibarra-Obando 1987,

Table 2. Yeast genera associated with marine macrophytes in different localities around the world. The yeast genera are: *Aureobasidium* (Au), *Candida* (Ca), *Cryptococcus* (Cr), *Debaryomyces* (De), *Kluyveromyces* (Kl), *Leucosporidiurn* (Le), *Metschnikowia* (Me), *Meyerozyma* (My), *Pichia* (Pi), *Rhodotorula* (Rh), *Saccharomyces* (Sa), *Sporobolomyces* (Sp), *Torulopsis* (To), *Trichosporon* (Tr), *Yamadazyma* (Ya), and undetermined genera are represented by (un). The numbers of species are in parenthesis. Also, the states (St) of the sources are established as decomposing sample (d), fermented sample (f), fresh from the natural environment (n) and processed sample (p).

Yeast genus	Source	St	Locality	Reference
Le	Algae (5/7), seagrass (1)	n	Canada	Summerbell 1983
un	Alga (1)	n/p	Norway	Sieburth and Jensen 1967
un	Alga (1)	d	Antarctica	Bunt 1955
un(4)	Seagrass (3)	n	USA	Shoemaker and Wyllie-Echeverria 2013
un(6)	Macroalgae (11/25)	n	India	Suryanarayanan et al. 2010
Me, Rh	Macroalgae (2/9)	n	Shetland	Flewelling et al. 2013
Ca, De	Macroalgae (un)	f	Japan	Uchida and Murata 2004
Ca, Me	Alga (1)	d	USA	van Uden and Castelo-Branco 1963
Ca(2), Rh(2)	Macroalgae (9)	n	USA	Seshadri and Sieburth 1971
Ca(2), Sp	Seagrass (3)	n	Bermuda	Wilson 1998
Ca, Cr, Rh(2)	Algae (8), seagrass (2)	n	USA	Roth et al. 1962
Ca(19), Pi, Rh(6)	Macroalgae (9)	n	USA	Seshadri and Sieburth 1975
Ca(2), Sa, Pi, To(2)	Phycosphere water (un)	n	USA	van Uden and Zobell 1962
Au?, Ca, Rh, Tr?	Macroalgae (2)	n	USA	Phillips 1982
Au, Cr, Me, Rh, un	Macroalgae (3)	n	Antarctica	Loque et al. 2010
Ca, Cr, Rh, Sa(2),To(2)	Macroalgae (5)	n	India	Patel 1975
Ca(2), Rh, To, Tr(3)	Macroalgae (9/24)	d	Japan	Suehiro 1960
Ca(5), Cr(4), Rh, To, Tr(2)	Macroalgae (7)	d	Japan	Suehiro and Tomiyazu 1962
Ca, Cr(2), De, Me(2), My, Rh, Ya	Macroalgae (8)	d	Antarctica	Godinho et al. 2013

Chung and Lee 2008, Lam et al. 2008) on macrophytes. The extraordinary species diversity and variety of interactions of protists in the sea are only slowly being appreciated (Sherr et al. 2007).

Therefore, there is a lack of information on distribution of many of those protists associated with marine hosts. It is suspected that some of these organisms may be common parasites in many marine ecosystems worldwide; however, at present there are only anecdotal data available to support this hypothesis (Neuhauser et al. 2011). It is surprising that only few studies have considered the abundance of heterotrophic protists on macrophytes as macroalgae (Armstrong et al. 2000). Molecular approaches have demonstrated that poorly recognized groups can be important ecosystem components (Richards et al. 2012), and parasites are increasingly being considered to be equally important to predators for the functioning and the stability of these important coastal ecosystems (Gachon et al. 2010, Li et al. 2010). Beyond characterizing the diversity and distribution of protists in the ocean, major lines of research continue to elucidate the ecological roles of protists in marine ecosystems, i.e., food webs, pathogens, degraders (Sherr et al. 2007). More research on these microbes, their life cycle, abundance and distribution is needed to reliably estimate the impacts that these associations might have (Goecke et al. 2012b).

Conclusions and future perspectives

Macrophytes provide suitable microniches for microbes more favorable than in free water. They can act as an important reservoir for diverse microbes during adverse conditions (van Uten and Castelo Branco 1963, Englebert et al. 2008, Barott et al. 2011). The pathogenic role of marine microbes, although dispersed, has been corroborated (e.g., Andrews 1976, Apt 1988, Gachon et al. 2010) however, the presence and relevance of symbiotic microbial communities in marine macrophytes remain unexplored. Particularly understudied is the role of epibiotic biofilms for infection and disease of the host. Do biofilms repel host-pathogens and parasites and if so, when and how? What are the conditions under which biofilms switch from beneficial or neutral to adverse or even toxic effects upon the host (Wahl et al. 2012). It remains to be seen how changes in environmental conditions such as increased eutrophication and elevated sea surface temperature influence the microbial communities associated with macrophytes (as Case et al. 2011, Fernandes et al. 2012) and how these changes affect the physiology and success of macrophytes around the world (Barott et al. 2011). Characterization of the microbiota closely associated with macrophytes is therefore a first step that may provide further insights into the complex interactions between microbes and macrophytes (Garcias-Bonet et al. 2012). We have to extend our knowledge of 'normal' microbial communities associated with macrophytes, what are the reasons for such different associations, and which ecological functions are these microbes responsible for. For example, bacterial endophytes are crucial for the survival of many terrestrial plants, but little is known about the presence and importance of endophytes of marine plants (Garcias-Bonet et al. 2012). Moreover, coastal ecosystems and in particular seagrass meadows, are currently declining at an alarming rate worldwide, leading to loss of biodiversity. Unfortunately, extensive and extremely high costing restoration efforts have had little success so far. Recent findings indicate that restoration efforts should not only focus on environmental stressors (e.g., eutrophication, sediment runoff and high salinity) as a cause of decline but should also consider internal ecological interactions, such as the presence and vigor of symbiotic or mutualistic relations (Campbell et al. 2011, van der Heide et al. 2012).

Marine macrophyte survival, growth, and reproduction are known to vary with numerous climatically-sensitive environmental variables, and there is mounting evidence that acidification or warming will negatively impact macrophytes by facilitating microbial infections (Campbell et al. 2011, Case et al. 2011, Fernandes et al. 2012). Although seaweeds are known to be vulnerable to physical and chemical changes in the marine environment, the impacts of ongoing and future climate change in seaweed-dominated ecosystems remain poorly understood (Harley et al. 2012). Schulz et al. (2002) hypothesized that the fungal endophyte-plant host interaction is characterized by a finely tuned equilibrium between fungal virulence and plant defense, and if this balance is disturbed by either a decrease in plant defense or an increase in fungal virulence, disease can develop. Not only is the physiological state of the host/microbes affected but also connectivity among host and potential vectors. Today there is a consensus that the recent shift towards intensive algal aquaculture and production methods correlates with more damaging disease outbreaks (Gachon et al. 2010). For example, macroalgae with buoyant structures which have a high floating potential, continue to grow after detachment and persist in a floating condition for long time periods. Depending on the prevailing winds and currents, macrophytes can be common dispersal vehicles for associated benthic invertebrates (Wichmann et al. 2012, and references therein). Microbes are transported as well (Thiel and Gutow 2005). Thus, this has been postulated as one route to transfer pathogens and symbionts among macrophytes (Meusnier et al. 2001, Hollants et al. 2013a, Aires et al. 2013). For example, roots of *Zostera* sp. infected with the phytomyxid *Plasmodiophora bicaudata* produce poorly developed roots and uprooting takes place relatively easily, releasing many floating infected *Zostera* plants (den Hartog 1989). This theory can be corroborated with molecular methods and experiments on survival and infective potential.

Kelp forests and seagrass meadows worldwide are known as hotspots for macroscopic biodiversity and primary production, yet very little is known about the biodiversity and roles of microorganisms in these ecosystems (Bengtsson et al. 2012). Macrophytes are important foundation species as they can modify their abiotic environment, e.g., reduce hydrodynamic energy from currents and waves, increase sediment accretion, alter sediment quality and stabilize sediments (Bos et al. 2007, Venier et al. 2012, and references therein). Such foundation species may enhance diversity by facilitating the presence of other organisms or communities which may eventually lead to succession (Bouma et al. 2009). Macrophytes serve as feeding, breeding and nursery grounds for economically important marine organisms including endangered species such as some fishes, mollusks and marine turtles (Sakayaroj et al. 2010). Microbes are benefited and also sustained by macrophytes. Through full and partial degradation of dissolved and particulate organic carbon, microbes make macrophyte primary production available to many animal consumers (Bengtsson et al. 2012). Microorganisms are an intrinsic part of those biotic environments, and in fact those microbes can be the real reason of ecological success of macrophytes in certain environments. Marine macrophytes have been challenged throughout their evolution by microorganisms and have developed in a world of microbes (Goecke et al. 2010), and during the last two decades exciting new studies have opened the door to understand such invisible but often vital connections in coastal ecosystems.

References

Aires, T., E.A. Serrao, G. Kendrick, C.M. Duarte and S. Arnaud-Haond. 2013. Invasion is a community affair: Clandestine followers in the bacterial community associated to green algae, *Caulerpa racemosa*, track the invasion source. PLoS ONE 8: e68429.

Almaraz, T., J. Checa and T. Gallardo. 1994. Ascomycetes asociados a *Pelvetia canaliculata* Decne. and Thur. (Fucales, Phaeophyta). Anales Jard. Bot. Madrid 52: 3–6.

Alva, P., E.H.C. McKenzie, S.B. Pointing, R. Pena-Muralla and K.D. Hyde. 2002. Do sea grasses harbour endophytes? Fungal Divers 7: 167–178.

Amin, S.A., D.H. Green, M.C. Hart, F.C. Kupper, W.G. Sunda and C.J. Carrano. 2009. Photolysis of iron-siderophore chelates promotes bacterial-algal mutualism. Proc. Natl. Acad. Sci. U.S.A. 106: 17071–17076.

Ando, Y., T. Nakagawa, R. Takahashi, K. Yoshihara and T. Tokuyama. 2009. Seasonal changes in abundance of ammonia-oxidizing Archaea and ammonia-oxidizing bacteria and their nitrification in sand of an eelgrass zone. Microbes Environ. 24: 21–27.

Andrews, J.H. 1976. The pathology of marine algae. Biol. Rev. Camb. Philos. Soc. 51: 211–253.

Apt, K. 1988. Galls and tumor-like growths on marine macroalgae. Dis. Aquat. Organ. 4: 211–217.

Armstrong, E., A. Rogerson and J.W. Leftley. 2000. The abundance of heterotrophic protists associated with intertidal seaweeds. Estuar. Coast Shelf Sci. 50: 415–424.

Armstrong, E., L.M. Yan, K.G. Boyd, P.C. Wright and J.G. Burgess. 2001. The symbiotic role of marine microbes on living surfaces. Hydrobiologia 461: 37–40.

Bagwell, C.E., J.R. La Rocque, G.W. Smith, S.W. Polson, M.J. Friez, J.W. Longshore and C.R. Lovell. 2002. Molecular diversity of diazotrophs in oligotrophic tropical seagrass bed communities. FEMS Microb. Ecol. 39: 113–119.

Balakirev, E.S., T.N. Krupnova and F.J. Ayala. 2012. Symbiotic associations in the phenotypically-diverse brown alga *Saccharina japonica* Plos One 7: e39587.

Ballesteros, E., D. Martin and M.J. Uriz. 1992. Biological activity of extracts from some Mediterranean macrophytes. Bot. Mar. 35: 481–485.

Barott, K.L., B. Rodriguez-Brito, J. Janouskovec, K.L. Marhaver, J.E. Smith, P. Keeling and F.L. Rohwer. 2011. Microbial diversity associated with four functional groups of benthic reef algae and the reef-building coral *Montastraea annularis*. Environ. Microbiol. 13: 1192–1204.

Bengtsson, M.M., K. Sjøtun and L. Øvreås. 2010. Seasonal dynamics of bacterial biofilms on kelp (*Laminaria hyperborea*). Aquat. Microb. Ecol. 60: 71–83.

Bengtsson, M.M., K. Sjøtun, J.E. Storesund and L. Øvreås. 2011. Utilization of kelp-derived carbon sources by kelp surface associated bacteria. Aquat. Microb. Ecol. 62: 191–199.

Bengtsson, M.M., K. Sjøtun, A. Lanzén and L. Øvreås. 2012. Bacterial diversity in relation to secondary production and succession on surfaces of the kelp *Laminaria hyperborea*. ISME J. 6: 2188–2198.

Bergmann, N., B. Fricke, M.C. Schmidt, V. Tams, K. Beining, H. Schwitte, A.A. Boettcher, D.L. Martin, A.C. Bockelmann, T.B. Reusch and G. Rauch. 2011. A quantitative real-time polymerase chain reaction assay for the seagrass pathogen *Labyrinthula zosterae*. Mol. Ecol. Resour. 11: 1076–1081.

Bos, A.R., T.J. Bouma, G.L.J. de Kort and M.M. van Katwijk. 2007. Ecosystem engineering by annual intertidal seagrass beds: Sediment accretion and modification. Estuar. Coast. Shelf Sci. 74: 344–348.

Bouma, T.J., V. Ortells and T. Ysebaert. 2009. Comparing biodiversity effects among ecosystem engineers of contrasting strength: Macrofauna diversity in *Zostera noltii* and *Spartina anglica* vegetations. Helgol. Mar. Res. 63: 3–18.

Bunt, J.S. 1955. The importance of bacteria and other microorganisms in the seawater at Macquarie Island. Aust. J. Mar. Fresh. Res. 6: 60–65.

Burke, C., P. Steinberg, D. Rusch, S. Kjelleberg and T. Thomas. 2011a. Bacterial community assembly based on functional genes rather than species. Proc. Natl. Acad. Sci. U.S.A. 108: 14288–14293.

Burke, C., T. Thomas, M. Lewis, P. Steinberg and S. Kjelleberg. 2011b. Composition, uniqueness and variability of the epiphytic bacterial community of the green alga *Ulva australis*. ISME J. 5: 590–600.

Campbell, A.H., T. Harder, S. Nielsen, S. Kjelleberg and P.D. Steinberg. 2011. Climate change and disease: Bleaching of a chemically defended seaweed. Glob. Chang. Biol. 17: 2958–2970.

Case, R.J., S.R. Longford, A.H. Campbell, A. Low, N.A. Tujula, P.D. Steinberg and S. Kjelleberg. 2011. Temperature induced bacterial virulence and bleaching disease in a chemically defended marine macroalga. Environ. Microbiol. 13: 529–537.

Chemello, R. and M. Milazzo. 2002. Effect of algal architecture on associated fauna: Some evidence from phytal molluscs. Mar. Biol. 140: 981–990.

Choi, H.G., J.H. Lee, H.H. Park and F.A.Q. Sayegh. 2009. Antioxidant and antimicrobial activity of *Zostera marina* L. extract. Algae 24: 179–184.

Chung, M.H. and K.S. Lee. 2008. Species composition of the epiphytic diatoms on the leaf tissues of three *Zostera* species distributed on the Southern Coast of Korea. Algae 23: 75–81.

Cifuentes, A., J. Antón, S. Benlloch, A. Donnelly, R.A. Herbert and F. Rodríguez-Valera. 2000. Prokaryotic diversity in *Zostera noltii*-colonized marine sediments. Appl. Environ. Microbiol. 66: 1715–1719.

Cifuentes, A., J. Antón, R. De Wit and F. Rodríguez-Valera. 2003. Diversity of Bacteria and Archaea in sulphate-reducing enrichment cultures inoculated from serial dilution of *Zostera noltii* rhizosphere samples. Environ. Microbiol. 5: 754–764.

Crump, B.C. and E.W. Koch. 2008. Attached bacterial populations shared by four species of aquatic angiosperms. Appl. Environ. Microbiol. 74: 5948–5957.

Cuomo, V., F. Vanzanella, F. Fresi, F. Cinelli and L. Mazzella. 1985. Fungal flora of *Posidonia oceanica* and its ecological significance. Trans. Br. Mycol. Soc. 84: 35–40.

Cuomo, V., E.B.G. Jones and S. Grasso. 1988. Occurrence and distribution of marine fungi along the Coast of the Mediterranean Sea. Prog. Oceanog. 21: 189–200.

de Oliveira, L.S., G.B. Gregoracci, G.G.Z. Silva, L.T. Salgado, G.A. Filho, M.A. Alves-Ferreira, R.C. Pereira and F.L. Thompson. 2012. Transcriptomic analysis of the red seaweed *Laurencia dendroidea* (Florideophyceae, Rhodophyta) and its microbiome. BMC Genomics 13: 487.

Dean, R.L. and J.H. Connell. 1987. Marine invertebrates in algal succession. III. Mechanisms linking habitat complexity with diversity. J. Exp. Mar. Biol. Ecol. 109: 249–273.

Debenay, J.P. and C.E. Payri. 2010. Epiphytic foraminiferal assemblages on macroalgae in reefal environments of New Caledonia. J. Foramin. Res. 40: 36–60.

Deckert, R.J. and D.J. Garbary. 2005. *Ascophyllum* and its symbionts. VI. Microscopic characterization of the symbiotum. Algae 20: 225–232.

DeLong, E.F. 2007. Microbial domains in the ocean: A lesson from the Archaea. Oceanography 20: 124–129.

Den Hartog, C. 1989. Distribution of *Plasmodiophora bicaudata*, a parasitic fungi on small *Zostera* species. Dis. Aquat. Org. 6: 227–229.

Devarajan, P.T., T.S. Suryanarayanan and V. Geetha. 2002. Endophytic fungi associated with the tropical seagrass *Halophila ovalis* (Hydrocharitaceae). Indian J. Mar. Sci. 31: 73–74.

Dobretsov, S., H.U. Dahms, T. Harder and P.Y. Qian. 2006. Allelochemical defense against epibiosis in the macroalga *Caulerpa racemosa* var. *turbinata*. Mar. Ecol. Prog. Ser. 318: 165–175.

Drouin, A., C.W. McKindsey and L.E. Johnson. 2011. Higher abundance and diversity in faunal assemblages with the invasion of *Codium fragile* ssp. *fragile* in eelgrass meadows. Mar. Ecol. Prog. Ser. 424: 105–117.

Duarte, C.M., M. Holmer and N. Marbá. 2005. Plant microbe-interactions in seagrass meadows. pp. 31–60. *In*: Kristensen, E., R. Haese and J. Kotska (eds.). Macro- and Microorganisms in Marine Sediments. Coastal and Estuarine Studies 60, American Geophysical Union, Washington.

Egan, S., T. Harder, C. Burke, P. Steinberg, S. Kjelleberg and T. Thomas. 2013. The seaweed holobiont: Understanding seaweed-bacteria interactions. FEMS Microbiol. Rev. 37: 462–476.

Eggert, A., A.F. Peters and F.C. Küpper. 2010. The potential impact of climate change on endophyte infections in kelp sporophytes. pp. 139–154. *In*: Israel, A., R. Einav and J. Seckbach (eds.). Seaweeds and their Role in Globally Changing Environments. Cellular Origin, Life in Extreme Habitats and Astrobiology 15. Springer.

Englebert, E.T., C. McDermott and G.T. Kleinheinz. 2008. Effects of the nuisance algae, *Cladophora*, on *Escherichia coli* at recreational beaches in Wisconsin. Sci. Total Environ. 404: 10–17.

Fernandes, N., P. Steinberg, D. Rusch, S. Kjelleberg and T. Thomas. 2012. Community structure and functional gene profile of bacteria on healthy and diseased thalli of the red seaweed *Delisea pulchra*. PLoS ONE 7: e50854.

Finster, K., T.R. Thomsen and N.B. Ramsing. 2001. *Desulfomusa hansenii* gen. nov., sp. nov., a novel marine propionate-degrading, sulfate-reducing bacterium isolated from *Zostera marina* roots. Int. J. Syst. Evol. Microbiol. 51: 2055–2061.

Flewelling, A.J., J.A. Johnson and C.A. Gray. 2013. Isolation and bioassay screening of fungal endophytes from North Atlantic marine macroalgae. Bot. Mar. 56: 287–297.

Friedrich, M.W. 2012. Bacterial communities on macroalgae. pp. 189–201. *In*: Wiencke, C. and K. Bischof (eds.). Seaweed Biology, Ecological Studies 219. Springer-Verlag, Berlin Heidelberg.

Gachon, C.M.M., T. Sime-Ngando, M. Strittmater, A. Chambouvet and G.H. Kim. 2010. Algal diseases: Spotlight on a black box. Trends Plant Sci. 15: 633–640.

Garcias-Bonet, N., T.D. Sherman, C.M. Duarte and N. Marbà. 2011. Distribution and pathogenicity of the protist *Labyrinthula* sp. in Western Mediterranean seagrass meadows. Estuar. Coast. 34: 1161–1168.

Garcias-Bonet, N., J.M. Arrieta, C.N. de Santana, C.M. Duarte and N. Marbà. 2012. Endophytic bacterial community of a Mediterranean marine angiosperm (*Posidonia oceanica*). Front. Microbio. 3: 342.

García-Martínez, M., J. Kuo, K. Kilminster, D. Walker, R. Rosselló-Mora and C.M. Duarte. 2005. Microbial colonization in the seagrass *Posidonia* spp. roots. Mar. Biol. Res. 1: 388–395.

García-Martínez, M., A. López-López, M.L. Calleja, N. Marbà and C.M. Duarte. 2009. Bacterial community dynamics in a seagrass (*Posidonia oceanica*) meadow sediment. Estuar. Coast. 32: 276–286.

Genilloud, O., F. Pelaez, I. Gonzalez and M.T. Diez. 1994. Diversity of actinomycetes and fungi on seaweeds from the Iberian coasts. Microbiologia SEM 10: 413–422.

Gleason, F.H., F.C. Küpper, J.P. Amon, K. Picard, C.M.M. Gachon, A.V. Marano, T. Sime-Ngando and O. Lilje. 2011. Zoosporic true fungi in marine ecosystems: A review. Mar. Freshwater Res. 62: 383–393.

Goecke, F., A. Labes, J. Wiese and J.F. Imhoff. 2010. Chemical interactions between marine macroalgae and bacteria. Mar. Ecol. Prog. Ser. 409: 267–300.

Goecke, F., A. Labes, J. Wiese and J.F. Imhoff. 2012a. Dual effect of macroalgal extracts on growth of bacteria in Western Baltic Sea. Rev. Biol. Mar. Oceanogr. 47: 75–86.

Goecke, F., J. Wiese, A. Núñez, A. Labes, J.F. Imhoff and S. Neuhauser. 2012b. A novel phytomyxean parasite associated with galls on the bull-kelp *Durvillaea antarctica* (Chamisso) Hariot. Plos One 7: e45358.

Goecke, F., A. Labes, J. Wiese and J.F. Imhoff. 2013a. Phylogenetic analysis and antibiotic activity of bacteria isolated from the surface of two co-occurring macroalgae from the Baltic Sea. Eur. J. Phycol. 48: 47–60.

Goecke, F., V. Thiel, J. Wiese, A. Labes and J.F. Imhoff. 2013b. Algae as an important environment for bacteria–phylogenetic relationships among bacterial species isolated from algae. Phycologia 52: 14–24.

Godinho, V.M., L.E. Furbino, I.F. Santiago, F.M. Pellizzari, N.S. Yokoya, D. Pupo, T.M.A. Alves, P.A.S. Junior, A.J. Romanha, C.L. Zani, C.L. Cantrell, C.A. Rosa and L.H. Rosa. 2013. Diversity and bioprospecting of fungal communities associated with endemic and cold-adapted macroalgae in Antarctica. ISME J. 7: 1434–1451.

Grant, M.A.A., E. Kazamia, P. Cicuta and A.G. Smith. 2014. Direct exchange of vitamin B12 is demonstrated by modelling the growth dynamics of algal–bacterial cocultures. ISME J. Pub. ahead doi: 10.1038/ismej.2014.9.

Grossart, H.P., F. Levold, M. Allgaier, M. Simon and T. Brinkhoff. 2005. Marine diatom species harbour distinct bacterial communities. Environ. Microbiol. 7: 860–873.

Grosser, K., L. Zedler, M. Schmitt, B. Dietzek, J. Popp and G. Pohnert. 2012. Disruption-free imaging by Raman spectroscopy reveals a chemical sphere with antifouling metabolites around macroalgae. Biofouling 28: 687–696.

Hansen, J.W., J.W. Udy, C.J. Perry, W.C. Dennison and B.Aa. Lomstein. 2000. Effect of the seagrass *Zostera capricorni* on sediment microbial processes. Mar. Ecol. Prog. Ser. 199: 83–96.

Harley, C.D.G., K.M. Anderson, K.W. Demes, J.P. Jorve, R.L. Kordas, T.A. Coyle and M.H. Graham. 2012. Effects of climate change on global seaweed communities. J. Phycol. 48: 1064–1078.

Hawksworth, D.L. and A.Y. Rossman. 1997. Where are all the undescribed fungi? Phytopathology 87: 888–891.

Hollants, J., F. Leliaert, O. de Clerck and A. Willems. 2013a. What we can learn from sushi: A review on seaweed-bacterial associations. FEMS Microbiol. Ecol. 83: 1–16.

Hollants, J., F. Leliaert, H. Verbruggen, A. Willems and O. de Clerck. 2013b. Permanent residents or temporary lodgers: Characterizing intracellular bacterial communities in the siphonous green alga Bryopsis. Proceedings of the Royal Society B-Biological Sciences 280: 20122659.

James, J.B., T.D. Sherman and R. Devereux. 2006. Analysis of bacterial communities in seagrass bed sediments by double-gradient denaturing gradient gel electrophoresis of PCR-amplified 16S rRNA genes. Microb. Ecol. 52: 655–661.

Jenkins, K.M., S.G. Toske, P.R. Jensen and W. Fenical. 1998. Solanapyrones E-G, antialgal metabolites produced by a marine fungus. Phytochemistry 49: 2299–2304.

Jensen, S.I., M. Kühl and A. Priemé. 2007. Different bacterial communities associated with the roots and bulk sediment of the seagrass *Zostera marina*. FEMS Microbiol. Ecol. 62: 108–117.

Kirshtein, J.D., H.W. Paerl and J. Zehr. 1991. Amplification, cloning and sequencing of a *nifH* segment from aquatic microorganisms and natural communities. Appl. Environ. Microbiol. 57: 2645–2650.

Kohlmeyer, J. 1973. Fungi on marine algae. Bot. Mar. 16: 201–215.

Kohlmeyer, J. and V. Demoulin. 1981. Parastic and symbiotic fungi on marine algae. Bot. Mar. 24: 9–18.

Kohlmeyer, J. and E. Kohlmeyer (eds.). 1979. Marine Mycology. The Higher Fungi. Academic Press, New York.

Kohlmeyer, J. and B. Volkmann-Kohlmeyer. 1991. Illustrated key to the filamentous marine fungi. Bot. Mar. 34: 1–61.

Kohlmeyer, J. and B. Volkmann-Kohlmeyer. 2003. Marine Ascomycetes from algae and animal hosts. Bot. Mar. 46: 285–306.

König, G.M., S. Kehraus, S.F. Seibert, A. Abdel-Lateff and D. Müller. 2006. Natural products from marine organisms and their associated microbes. ChemBioChem 7: 229–238.

Kurilenko, V.V., R. Christen, N.V. Zhukova, N.I. Kalinovskaya, V.V. Mikhailov, R.J. Crawford and E.P. Ivanova. 2010. *Granulosicoccus coccoides* sp. nov., isolated from leaves of seagrass (*Zostera marina*). Int. J. Syst. Evol. Microbiol. 60: 972–976.

Kurilenko, V.V., E.P. Ivanova and V.V. Mikhaĭlov. 2001. Zonal distribution of epiphytic microorganisms on the sea grass *Zostera marina*. Microbiol. (Moscou) 70: 427–428.

Kurtz, J.C., D.F. Yates, J.M. Macauley, R.L. Quarles, F.J. Genthner, C.A. Chancy and R. Devereux. 2003. Effects of light reduction on growth of the submerged macrophyte *Vallisneria americana* and the community of root-associated heterotrophic bacteria. J. Exp. Mar. Biol. Ecol. 291: 199–218.

Kutty, S.N. and R. Philip. 2008. Marine yeasts—a review. Yeast 25: 465–483.

Küsel, K., H.C. Pinkart, H.L. Drake and R. Devereux. 1999. Acetogenic and sulfate-reducing bacteria inhabiting the rhizoplane and deeper cortex cells of the sea grass *Halodule wrightii*. Appl. Environ. Microbiol. 65: 5117–5123.

Lachnit, T., M. Blümel, J.F. Imhoff and M. Wahl. 2009. Specific epibacterial communities on macroalgae: Phylogeny matters more than habitat. Aquat. Biol. 5: 181–186.

Lachnit, T., M. Wahl and T. Harder. 2010. Isolated thallus-associated compounds from the macroalga *Fucus vesiculosus* mediate bacterial surface colonization in the field similar to that on the natural alga. Biofouling 26: 247–255.

Lachnit, T., D. Meske, M. Wahl, T. Harder and R. Schmitz. 2011. Epibacterial community patterns on marine macroalgae are host-specific but temporally variable. Environ. Microbiol. 13: 655–665.

Lam, C., A. Grage, D. Schulz, A. Schulte and T. Harder. 2008. Extracts of North Sea macroalgae reveal specific activity patterns against attachment and proliferation of benthic diatoms: A laboratory study. Biofouling 24: 59–66.

Leutou, A.S., G. Yang, V.N. Nenkep, X.N. Siwe, Z. Feng, T.T. Khong, H.D. Choi, J.S. Kang and B.W. Son. 2009. Microbial transformation of a monoterpene, geraniol, by the marine-derived fungus *Hypocrea* sp. J. Microbiol. Biotechnol. 19: 1150–1152.

Li, W., T. Zhang, X. Tang and B. Wang. 2010. Oomycetes and fungi: Important parasites of marine algae. Acta Oceanol. Sin. 29: 74–81.

Longford, S.R., N.A. Tujula, G.R. Crocetti, A.J. Holmes, C. Holmström, S. Kjelleberg, P.D. Steinberg and M.W. Taylor. 2007. Comparisons of diversity of bacterial communities associated with three sessile marine eukaryotes. Aquat. Microb. Ecol. 48: 217–229.

Loque, C.P., A.O. Medeiros, F.M. Pellizzari, E.C. Oliveira, C.A. Rosa and L.H. Rosa. 2010. Fungal community associated with marine macroalgae from Antarctica. Polar Biol. 33: 641–648.

Lucas-Elío, P., E. Marco-Noales, E. Espinosa, M. Ordax, M.M. López, N. Garcías-Bonet, N. Marbà, C.M. Duarte and A. Sanchez-Amat. 2011. *Marinomonas alcarazii* sp. nov., *M. rhizomae* sp. nov., *M. foliarum* sp. nov., *M. posidonica* sp. nov. and *M. aquiplantarum* sp. nov., isolated from the microbiota of the seagrass *Posidonia oceanica*. Int. J. Syst. Evol. Microbiol. 61: 2191–2196.

Lucena-Moya, P. and I.C. Duggan. 2011. Macrophyte architecture affects the abundance and diversity of littoral microfauna. Aquat. Ecol. 45: 279–287.

Mata, J.L. and J. Cebrián. 2013. Fungal endophytes of the seagrasses *Halodule wrightii* and *Thalassia testudinum* in the north-central Gulf of Mexico. Bot. Mar. 56: 541–545.

Matsuo, Y., M. Suzuki, H. Kasai, Y. Shizuri and S. Harayama. 2003. Isolation and phylogenetic characterization of bacteria capable of inducing differentiation in the green alga *Monostroma oxyspermum*. Environ. Microbiol. 5: 25–35.

Meusnier, I., J.L. Olsen, W.T. Stam, C. Destombe and M. Valero. 2001. Phylogenetic analyses of *Caulerpa taxifolia* (Chlorophyta) and of its associated bacterial microflora provide clues to the origin of the Mediterranean introduction. Mol. Ecol. 10: 931–946.

Meyers, S.P., P.A. Orpurt, J. Simms and L.L. Boral. 1965. Thalassiomycetes VII. Observations on fungal infestations of turtle grass, *Thalassia testudinum* König. Bull. Mar. Sci. Gulf. Caribb. 15: 548–564.

Miao, F.P., X.D. Li, X.H. Liu, R.H. Cichewicz and N.Y. Ji. 2012. Secondary metabolites from an algicolous *Aspergillus versicolor* strain. Mar. Drugs 10: 131–139.

Miller, J.D. and N.J. Whitney. 1981. Fungi from the Bay of Fundy II. Observations on fungi from riving and cast seaweeds. Bot. Mar. 24: 405–411.

Miranda, L.N., K. Hutchison, A.R. Grossman and S.H. Brawley. 2013. Diversity and abundance of the bacterial community of the red macroalga *Porphyra umbilicalis*: Did bacterial farmers produce macroalgae? PLoS ONE 8: e58269.

Neuhauser, S., M. Kirchmair and F.H. Gleason. 2011. Ecological roles of the parasitic phytomyxids (plasmodiophorids) in marine ecosystems—A review. Mar. Freshwater Res. 62: 365–371.

Newell, S.Y. 1981. Fungi and bacteria in or on leaves of eelgrass (*Zostera marina* L.) from Chesapeake Bay. Appl. Environ. Microbiol. 41: 1219–1224.

Nielsen, J.T., W. Liesack and K. Finster. 1999. *Desulfovibrio zosterae* sp. nov., a new sulfate reducer isolated from surface-sterilized roots of the seagrass *Zostera marina*. Int. J. Syst. Bacteriol. 49: 859–865.

Olson, J.B. and C.A. Kellogg. 2010. Microbial ecology of corals, sponges, and algae in mesophotic coral environments. FEMS Microbiol. Ecol. 73: 17–30.

Panno, L., S. Voyron, A. Anastasi and G.C. Varese. 2010. Marine fungi associated with the sea grass *Posidonia oceanica* L.: A potential source of novel metabolites and enzymes. J. Biotechnol. 150: 383–384.

Patel, K.S. 1975. The relationship between yeasts and marine algae. Proc. Ind. Acad. Sci. B 82: 25–28.

Phillips, R.E. 1982. Observations of higher fungi and protists associated with the marine red algae *Rhodoglossum affine* and *Gelidium coultri*. M.S. Thesis, California Polytechnic State University.

Potin, P. 2012. Intimate associations between epiphytes, endophytes, and parasites of seaweeds. pp. 203–234. *In*: Wiencke, C. and K. Bischof (eds.). Seaweed Biology, Ecological Studies 219. Springer-Verlag, Berlin Heidelberg.

Raghukumar, C. and S. Raghukumar. 1992. Association of thraustochytrids and fungi with living marine algae. Mycol. Res. 96: 542–546.

Ramesh, T. and M. Kalaiselvam. 2011. An experimental study on citric acid production by *Aspergillus niger* using *Gelidiella acerosa* as a substrate. Indian J. Microbiol. 51: 289–293.

Rateb, M.E. and R. Ebel. 2011. Secondary metabolites of fungi from marine habitats. Nat. Prod. Rep. 28: 290–344.

Richards, T.A., M.D.M. Jones, G. Leonard and D. Bass. 2012. Marine fungi: Their ecology and molecular diversity. Annu. Rev. Mar. Sci. 4: 495–522.

Rogerson, A. 1991. On the abundance of marine naked amoebae on the surfaces of five species of macroalgae. FEMS Microbiol. Ecol. 85: 301–312.

Roth, F.J., D.G. Ahearn, J.W. Fell, S.P. Meyers and S.A. Meyer. 1962. Ecology and taxonomy of yeasts isolated from various marine substrates. Limnol. Oceanogr. 7: 178–185.

Saito, A., S. Ikeda, H. Ezura and K. Minamizawa. 2007. Microbial community analysis of the phytosphere using culture-independent methodologies. Microbes Environ. 22: 93–105.

Salaün, S., S. la Barre, M. dos Santos-Goncalvez, P. Potin, D. Haras and A. Bazire. 2012. Influence of exudates of the kelp *Laminaria digitata* on biofilm formation of associated and exogenous bacterial epiphytes. Microbial Ecol. 64: 359–369.

Sathe, V. and S. Raghukumar. 1991. Fungi and their biomass in detritus of the seagrass *Thalassia hemprichii* (Ehrenberg) Ascherson. Bot. Mar. 34: 271–277.

Sakayaroj, J., S. Preedanon, O. Supaphon, E.B.G. Jones and S. Phongpaichit. 2010. Phylogenetic diversity of endophyte assemblages associated with the tropical seagrass *Enhalus acoroides* in Thailand. Fungal Divers 42: 27–45.

Schulz, B., C. Boyle, S. Draeger, A.K. Römmert and K. Krohn. 2002. Endophytic fungi: A source of novel biologically active secondary metabolites. Mycol. Res. 106: 996–1004.

Semeniuk, T.A. 2001. Epiphytic foraminifera along a climatic gradient, Western Australia. J. Foramin. Res. 31: 191–200.

Seshadri, R. and J.M. Sieburth. 1975. Seaweeds as a reservoir of *Candida* yeasts in inshore waters. Mar. Biol. 30: 105–117.

Seyedsayamdost, M.R., R.J. Case, R. Kolter and J. Clardy. 2011. The Jekyll- and Hyde chemistry of *Phaeobacter gallaeciensis*. Nat. Chem. 3: 331–335.

Shearer, C.A., E. Descals, B. Kohlmeyer, J. Kohlmeyer, L. Marvanova, D. Padgett, D. Porter, H.A. Raja, J.P. Schmit, H.A. Thorton and H. Voglymayr. 2007. Fungal biodiversity in aquatic habitats. Biodivers. Conserv. 16: 49–67.

Sherr, B.F., E.B. Sherr, D.A. Caron, D. Vaulot and A.Z. Worden. 2007. Oceanic protists. Oceanography 20: 130–134.

Shoemaker, G. and S. Wyllie-Echeverria. 2013. Occurrence of rhizomal endophytes in three temperate northeast pacific seagrasses. Aquat. Bot. 111: 71–73.

Sieburth, J.M. and A. Jensen. 1967. Effect of processing on the microflora of Norwegian seaweed meal, with observations on *Sporendonema minutum* (Hoye) Frank and Hess. Appl. Microbiol. 15: 830–838.

Sieburth, J.M. and C.D. Thomas. 1973. Fouling on eelgrass (*Zostera marina* L.). J. Phycol. 9: 46–50.

Siqueiros-Beltrones, D.A. and S.E. Ibarra-Obando. 1987. Lista florística de las diatomeas epifitas de *Zostera marina* en Bahía Falsa, San Quintín. Cienc. Mar. 11: 21–67.

Sridhar, K.R., K.S. Karamchand, C. Pascoal and F. Cássio. 2012. Assemblage and diversity of fungi on wood and seaweed litter of seven northwest Portuguese beaches. *In*: Raghukumar, C. (ed.). Biology of Marine Fungi. Progress in Molecular and Subcellular Biology 53: 209–228.

Staufenberger, T., V. Thiel, J. Wiese and J.F. Imhoff. 2008. Phylogenetic analysis of bacteria associated with *Laminaria saccharina*. FEMS Microbiol. Ecol. 64: 65–77.

Stratil, S.B., S.C. Neulinger, H. Knecht, A.K. Friedrichs and M. Wahl. 2014. Salinity affects compositional traits of epibacterial communities on the brown macroalga *Fucus vesiculosus*. FEMS Microbiol. Ecol. Pub. ahead doi: 10.1111/1574-6941.12292.

Suehiro, S. 1960. Studies on the yeasts developing in the putrefied marine algae. Sci. Bull. Fac. Agric. Kyushu Univ. 17: 443–449.

Suehiro, S. and Y. Tomiyasu. 1962. Studies on the marine yeasts. V. Yeasts isolated from seaweeds. J. Fac. Agric. Kyushu Univ. 12: 163–169.

Summerbell, R.C. 1983. The heterobasidiomycetous yeast genus *Leucosporidium* in an area of temperate climate. Can. J. Bot. 61: 1402–1410.

Suryanarayanan, T.S., A. Venkatachalam, N. Thirunavukkarasu, J.P. Ravishankar, M. Doble and V. Geetha. 2010. Internal mycobiota of marine macroalgae from the Tamilnadu coast: Distribution, diversity and biotechnological potential. Bot. Mar. 53: 457–468.

Suryanarayanan, T.S. 2012. Fungal endosymbionts of seaweeds. *In*: Raghukumar, C. (ed.). Biology of Marine Fungi. Progress in Molecular and Subcellular Biology 53: 53–69.

Sweet, M.J., J.C. Bythell and M.M. Nugues. 2013. Algae as reservoirs for coral pathogens. PLoS One 8: e69717.

Thiel, M. and L. Gutow. 2005. The ecology of rafting in the marine environment. II. The rafting organisms and community. *In*: Gibson, R.N., R.J.A. Atkinson and J.D.M. Gordon (eds.). Oceanography and Marine Biology: An Annual Review 43: 279–418. Taylor & Francis.

Thomsen, M.S. 2010. Experimental evidence for positive effects of invasive seaweed on native invertebrates via habitat-formation in a seagrass bed. Aquat. Inv. 5: 341–346.

Toxopeus, J., C.J. Kozera, S.J.B. O'Leary and D.J. Garbary. 2011. A reclassification of *Mycophycias ascophylli* (Ascomycota) based on nuclear large ribosomal subunit DNA sequences. Bot. Mar. 54: 325–334.

Trias, R., A. García-Lledó, N. Sánchez, J.L. López-Jurado, S. Hallin and L. Bañeras. 2012a. Abundance and composition of epiphytic bacterial and archaeal ammonia oxidizers of marine red and brown macroalgae. Appl. Environ. Microbiol. 78: 318–325.

Trias, R., O. Ruiz-Rueda, A. García-Lledó, A. Vilar-Sanz, R. López-Flores, X.D. Quintana, S. Hallin and L. Bañeras. 2012b. Emergent macrophytes act selectively on ammonia oxidizing bacteria and archaea. Appl. Environ. Microbiol. Pub. ahead. doi: 10.1128/AEM.00919-12.

Tujula, N.A., G.R. Crocetti, C. Burke, T. Thomas, C. Holmström and S. Kjelleberg. 2010. Variability and abundance of the epiphytic bacterial community associated with a green marine Ulvacean alga. ISME J. 4: 301–311.

Tuya, F., T. Wernberg and M.S. Thomsen. 2009. Habitat structure affect abundances of labrid fishes across temperate reefs in south-western Australia. Environ. Biol. Fish. 86: 311–319.

Uchida, M. and M. Murata. 2004. Isolation of a lactic acid bacterium and yeast consortium from a fermented material of *Ulva* spp. (Chlorophyta). J. Appl. Microbiol. 97: 1297–1310.

Uku, J., M. Bjork, B. Bergman and B. Diez. 2007. Characterization and comparison of prokaryotic epiphytes associated with three east African seagrasses. J. Phycol. 43: 768–779.

van der Heide, T., L.L. Govers, J. de Fouw, H. Olff, M. van der Geest, M.M. van Katwijk, T. Piersma, J. van de Koppel, B.R. Silliman, A.J.P. Smolders and J.A. van Gils. 2012. A three-stage symbiosis forms the foundation of seagrass ecosystems. Science 336: 1432–1434.

van Uden, N. and C.E. Zobell. 1962. *Candida marina* nov. sp., *Torulopsis torresii* nov. sp. and *T. maris* nov. sp., three yeasts from the Torres Strait. Antonie van Leeuwenhoek 28: 275–283.

van Uden, N. and R. Castelo-Branco. 1963. Distribution and population densities of yeast species in Pacific water, air, animals and kelp off southern California. Limnol. Oceanogr. 8: 323–329.

Venier, C., J. Figueiredo da Silva, S.J. McLelland, R.W. Duck and S. Lanzoni. 2012. Experimental investigation of the impact of macroalgal mats on flow dynamics and sediment stability in shallow tidal areas. Estuar. Coast. Shelf Sci. 112: 52–60.

Wahl, M., L. Shahnaz, S. Dobretsov, M. Saha, F. Symanowski, K. David, T. Lachnit, M. Vasel and F. Weinberger. 2010. Ecology of antifouling resistance in the bladder wrack *Fucus vesiculosus*: Patterns of microfouling and antimicrobial protection. Mar. Ecol. Prog. Ser. 411: 33–48.

Wahl, M., F. Goecke, A. Labes, S. Dobretsov and F. Weinberger. 2012. The second skin: Ecological role of epibiotic biofilms on marine organisms. Front. Microbio. 3: 292.

Wang, X., X. Liu, S. Kono and G. Wang. 2013. The ecological perspective of microbial communities in two pairs of competitive Hawaiian native and invasive macroalgae. Microb. Ecol. 65: 361–370.

Warfe, D.M. and L.A. Barmuta. 2004. Habitat structural complexity mediates the foraging success of multiple predator species. Oecologia 141: 171–178.

Weidner, S., W. Arnold, E. Stackebrandt and A. Pühler. 2000. Phylogenetic analysis of bacterial communities associated with leaves of the seagrass *Halophila stipulacea* by a culture-independent small-subunit rRNA gene approach. Microbial Ecol. 39: 22–31.

Wernberg, T., B.D. Russell, P.J. Moore, S. Ling, D.A. Smale, A.H. Campbell, M.A. Coleman, P.D. Steinberg, G.A. Kendrick and S.D. Connell. 2011. Impacts of climate change in a global hotspot for temperate marine biodiversity and ocean warming, J. Exp. Mar. Biol. Ecol. 400: 7–16.

Wichmann, C.S., I.A. Hinojosa and M. Thiel. 2012. Floating kelps in Patagonian Fjords: An important vehicle for rafting invertebrates and its relevance for biogeography. Mar. Biol. 159: 2035–2049.

Williams, S.L. 2007. Introduced species in seagrass ecosystems: Status and concerns. J. Exp. Mar. Biol. Ecol. 350: 89–110.

Wilson, W.L. 1998. Isolation of endophytes from seagrasses from Bermuda. M.S. Thesis. University of New Brunswick, Canada.

Wood, D.C. and S.S. Hayasaka. 1981. Chemotaxis of rhizoplane bacteria to amino acids comprising eelgrass (*Zostera marina* L.) root exudate. J. Exp. Mar. Biol. Ecol. 50: 153–161.

Xu, H., R.J. Deckert and D.J. Garbary. 2008. *Ascophyllum* and its symbionts. X. Ultrastructure of the interaction between *A. nodosum* (Phaeophyceae) and *Mycophycias ascophylli* (Ascomycetes). Botany 86: 185–193.

Yun, K., C.M. Kondempudi, H.D. Choi, J.S. Kang and B.W. Son. 2011. Microbial mannosidation of bioactive chlorogentisyl alcohol by the marine-derived fungus *Chrysosporium synchronum*. Chem. Pharm. Bull. 59: 499–501.

Zuccaro, A. and J.I. Mitchell. 2006. Fungal communities of seaweeds. pp. 533–579. *In*: Dighton, J., J.F. White and P. Oudemans (eds.). The Fungal Community. CRC, Taylor and Francis, New York.

Zuccaro, A., B. Schulz and J.I. Mitchell. 2003. Molecular detection of Ascomycetes associated with *Fucus serratus*. Mycol. Res. 107: 1451–1466.

Zuccaro, A., C.L. Schoch, S. Draeger, J.W. Spatafora, J. Kohlmeyer and J.I. Mitchell. 2008. Detection and identification of fungi intimately associated with the brown seaweed *Fucus serratus*. Appl. Environ. Microbiol. 74: 931–941.

2

The Role of Chemically Defended Seaweeds as Biodiversity Sources

Renato Crespo Pereira,* Bernardo Antonio Perez da Gama
and *Daniela Bueno Sudatti*

Introduction

Seaweeds are ecologically important components of coastal areas as primary producers, competitors, and ecosystem engineers that play a central role in coastal habitats such as rhodolith beds, kelp forests and coral reefs. These organisms have been recognized for long time as fundamentally important worldwide to benthic marine environments, in polar (Wiencke and Amsler 2012), temperate (Huovinen and Gómez 2012) and tropical (Mejia et al. 2012) regions.

As primary producers, seaweeds are the basis of most marine food webs (with the exception of chemosynthetic, deep sea ecosystems), and provide direct sources of nutrients and energy for consumers, such as food for herbivores (Aquilino and Stachowicz 2012), dissolved particles for filter feeders (Wilson 2002) and available substrata for bacteria (Paradas et al. 2010) and macrobenthic colonization (da Gama et al. 2008a). Besides these basic ecological roles, many seaweeds provide shelter and habitat for a rich and varied fauna of small invertebrates (Christie et al. 2009) and epibiotic communities (Egan et al. 2013), promote sediment stabilization (Madsen et al. 2001) and supply atmospheric oxygen.

Departamento de Biologia Marinha, Instituto de Biologia, Universidade Federal Fluminense, C.P. 100.644, CEP 24001-970, Niterói, RJ, Brazil.
* Corresponding author: rcrespo@id.uff.br

This is a typical marine scenario in which the seaweeds exert a bottom-up control, and a number of counter adaptations are supposed to have occurred in organisms associated to them, which represent potential steps toward a top-down control of seaweed communities (Pereira and da Gama 2008). All the above aspects are the reasons why seaweeds are characterized as interactive organisms, influencing directly or indirectly the surrounding environment by controlling the availability of resources to other species, and by modifying, maintaining and creating habitats. Imagine for a second a huge kelp forest and all associated organisms that depend on them. It is easy to understand the importance of seaweeds to the maintenance of both biodiversity and natural environments on benthic marine ecosystems. However, many links between seaweeds and the surrounding diversity are less obvious and more intricate.

For instance, no reason should be given for not recognizing the paramount role played by trees as biodiversity sources in terrestrial ecosystems. However, seaweeds are not trees, and as thallophytes, they do not have bodies split into roots, trunk, leaves, fruits and flowers as they bear relatively undifferentiated tissues in their thalli. But instead of flowers, fruits and seeds, they offer unique and diversified compounds: to date, several seaweeds continue to be sources of a variety of new secondary metabolites (Maschek and Baker 2008, Blunt et al. 2013). These chemical signals, once in contact with the aquatic medium, may be perceived by conspecific algae under attack ("talking trees"), or by other algal species ("eavesdropping"), detected by symbionts, avoided by pathogens, as well as deceive and deter competitors and herbivores. Once in the marine food web, seaweed natural products may be modified, sequestered, used as defensive shields, chemical camouflage, weapons, and cues for settlement by the most diversified community ever: marine organisms. This chapter explores the importance of chemicals produced by seaweeds to modulate biological interactions, influence the structure of benthic marine communities and maintain biodiversity.

Secondary metabolites or natural products from seaweeds

Over 3,000 secondary metabolites from different biosynthetic pathways are known for red, brown and green seaweeds (Maschek and Baker 2008), the majority of which produced by tropical species. Although these compounds generally occur in low concentrations (0.5–5.0 percent of algal dry mass, e.g., terpenoids), some of them such as the polyphenolics in brown algae can occur at concentrations as high as 15.0 percent of algal dry mass. Most of the seaweed compounds are terpenoids, especially sesqui- and diterpenoids, acetogenins (acetate-derived compounds), amino acid derivates, and polyphenolics. Contrary to terrestrial plants, secondary metabolites from seaweeds are scarce in nitrogen but possess a higher proportion of halogenated compounds as occurred in red species, probably due to the relative differences in availability of nitrogen and halides such as bromine and chlorine in terrestrial versus marine ecosystems (Fenical 1975, Cabrita et al. 2010).

Green seaweeds, mainly those species belonging to the order Bryopsidales, such as species of *Chlorodesmis*, *Caulerpa*, *Halimeda*, *Penicillus*, and *Udotea*, and others, are abundant and widely distributed in tropical seas, and mainly known by production of sesquiterpenoid and diterpenoid compounds (Blunt et al. 2013), e.g., halimedatrial and caulerpenin.

Brown seaweeds are prolific in the chemistry of terpenoids, acetogenins, and terpenoid-aromatic compounds of mixed biosynthetic origin or not as their most common secondary metabolites (Blunt et al. 2013), e.g., dictyopterene C, pachydictyol A, fucodiphlorethol and epitaondiol. In addition, these macroalgal types also produce phlorotannins, but higher concentrations of these chemicals are storage in species of temperate (Toth and Pavia 2006) and polar regions (Fairhead et al. 2006).

Red species are the richest among most seaweeds in terms of secondary metabolite diversity and abundance (Blunt et al. 2013), encompassing more than 1,500 different compounds belonging to all major classes of secondary metabolites (Maschek and Baker 2008), e.g., lanosol and elatol. The family Rhodomelaceae alone produces ca. 60 percent of all secondary metabolites described for red seaweeds, and the genus *Laurencia* is responsible for 85 percent of these, providing a good example of the potential of this group for the production of halogenated terpenes and acetogenins.

Ecological roles of secondary metabolites from seaweeds

Seaweeds are prolific producers of natural products or secondary metabolites responsible for the chemical mediation of several biological interactions. In general, the ecological roles of seaweed secondary metabolites include sexual pheromones (Pohnert and Boland 2002), chemicals against consumers, both constitutive (Amsler 2012) and induced defenses (Haavisto et al. 2010), also defense against pathogens (Águila-Ramírez et al. 2012), epibionts (Da Gama et al. 2014), and competitors (Morrow et al. 2011). These chemicals constitute valuable phenotypic expression mediating crucial biological interactions in the sea, from molecular to community levels of organization. Understanding the different effects of seaweed chemicals is also important aspect to recognize the ecological roles of these organisms such as structuring species, especially in the context of chemically defended species.

Although we do not fully understand the mechanisms underlying the genetic expression of seaweed secondary metabolites, new studies are trying to cast some light upon this aspect. A molecular approach has been performed in the analysis of the transcriptomic profile of the red seaweed *Laurencia dendroidea* to access the genes involved on the biosynthesis of terpenoid compounds, an aspect that may be further explored in ecological and evolutionary approaches (Oliveira et al. 2012).

Chemical speciation due to interactions between genetic and environmental conditions allows the investigation of biogeographic patterns from a molecular standpoint. The geographic distribution and structural variations of the diterpenes found in Dictyotaceae, brown seaweed species, from the Tropical Atlantic American seem to reflect the past distribution, spread and speciation of this group (Vallim et al. 2005), and the fact that molecules have been conserved highlights their adaptive importance in face of environmental and ecological challenges.

At the cellular scale, elucidating the underlying mechanisms that regulate the production, storage and release of secondary metabolites is paramount to understand how these chemicals interact with other organisms and the environment. Secondary metabolites from seaweeds may be stored in specialized structures, including gland cells (or vesicle cells), in which the vesicle occupies practically the entire cellular space (Young and West 1979, Dworjanyn et al. 1999), or in cells with refractile inclusions, such as *corps en cerise* among red species (Salgado et al. 2008). Brown algal phlorotannins may also be stored inside specialized structures, the so-called physodes (Schoenwaelder 2002). Recent evidences showed that the exudation of the chemicals from the storage structures to the surface of the seaweeds is an important and crucial event to propitiate mediation with bacteria (Paradas et al. 2010) and other epibionts (Sudatti et al. 2008).

Several populations of the same species of green, brown or red seaweeds are known as producers of distinct types and concentrations of chemical defenses against consumers when living in different habitats (Paul and Van Alstyne 1988c, Matlock et al. 1999, Pereira et al. 2004). In general, populations of seaweeds from habitats with high herbivory pressure tend to contain higher levels of potent defense agents than in populations from areas with low herbivory (Paul and Van Alstyne 1988c). In addition, intra-populational variability in the amount of defensive chemicals may be very expressive among individuals of the same populations of *Laurencia obtusa* (Sudatti et al. 2006), but also among individuals of *Laurencia dendroidea* (previously named *L. obtusa*) from distinct populations living along a broad biogeographic scale (Oliveira et al. 2013).

At the community level, one of the most expressive effects of chemically defended seaweeds on the benthic community structure resulted from the establishment of the introduced green seaweed *Caulerpa taxifolia* in the Mediterranean. Crude extracts of *C. taxifolia* exert strong negative effects on productivity of the natives, the seaweeds *Gracilaria bursa-pastoris* and *Cystoseria barbata* (Ferrer et al. 1997). In addition, the mere presence of *C. taxifolia* is a sufficient condition to cause changes in structure and biodiversity of the benthic community (Francour et al. 1995, Bellan-Santini et al. 1996). Other introduced species living in the Mediterranean, such as *Caulerpa racemosa* also produce chemicals that exert allelochemical effects on the native seagrass *Cymodea nodosa* (Raniello et al. 2007), and have a deep impact on the local biodiversity (Piazzi et al. 2001).

Other examples of broad magnitude effects on marine communities include chemicals from the surface of the red alga *Galaxaura filamentosa* and green alga *Chlorodesmis fastigiata* that may kill or inhibit the growth of reef-building corals, supporting the hypothesis that competition with these algal species could be a factor in the worldwide decline of coral reefs (Rasher et al. 2011). Chemicals from *Dictyota* may cause physiological alterations in the reef-building green seaweed *Halimeda tuna* (Beach et al. 2003). *Halimeda* species are known to possess structural and induced chemical defenses (diterpenes) against herbivores, but are also reported to attract non-herbivore amphipods, which build domiciles with these algae to avoid predation. Species-specific interactions can therefore regulate the dynamics of the involved populations, thus affecting the community as a whole. From molecules to communities, chemical mediation by seaweeds seem to have cascading effects in the marine environment.

Chemically defended seaweeds as engineering species

Some species of chemically defended seaweeds possess characteristics or produce known significant ecological impacts to be considered as engineering species. For example, some Laminariales, including species of *Macrocystis*, *Laminaria*, *Nereocystis*, *Ecklonia*, *Egregia*, *Eisenia*, and *Pterygophora* with large thalli may form huge kelp forests, and are also known to produce phlorotannins as chemical defense against consumers. The laminarian species *Ecklonia stolonifera* and *Eisenia bicyclis* produce chemical defense against the abalone *Haliotis discus* (Taniguchi et al. 1991, 1992), and *Laminaria digitata* exhibits induced defenses against the gastropod *Helcion pellucidum* (Leblanc et al. 2011).

Kelps usually contain phlorotannins, phloroglucinol-based polymers present in most brown seaweeds (Ragan and Glombitza 1986, Amsler and Fairhead 2006), and this is especially true and ecologically important for species of Laminariales from temperate Australia and New Zealand that typically contain very high levels of phlorotannins (Steinberg 1989). In general, higher quantities of polyphenols are suggested to be active against herbivory (Steinberg and Van Altena 1992). In this way, kelps are potentially chemically defended against consumers and can indirectly extend this protection umbrella over all associated organisms. Kelps are important components of intertidal and subtidal benthic marine communities in temperate seas (e.g., Schiel and Foster 1986), and this protective effect to associated organisms could be also thought to occur in cold regions of the world. In fact, due to global distribution of kelp forests (Mann 1973), we can hypothesize a broad, yet fairly unknown impact of chemical mediation in community structure.

Crustose coralline algae (CCA) may also be important ecosystem engineers in benthic marine systems worldwide (Foster et al. 2007), forming structurally complex habitats, which shelter a high diversity of associated organisms (e.g., Amado-Filho et al. 2010, Peña and Barbara 2008), aside their contribution for calcium carbonate production and dynamising in the marine environment (Pereira-Filho et al. 2012). Chemicals from CCAs such as species of *Hydrolython*, *Lithophyllum*, *Neogoniolithon*, *Mesophyllum*, and *Peysonnelia* may induce metamorphosis in marine invertebrates, such as corals (Morse et al. 1988, Morse and Morse 1991, Heyward and Negri 1999) or even increase their survival (e.g., *Titanoderma prototypum*, Price 2010). Metamorphose of commercially important species, such as the abalone *Haliotis*, occur in response to chemicals from the CCA *Phymatolithon*, *Lithothamnium*, and *Lithophyllum* (Morse and Morse 1984, Roberts et al. 2010). The CCAs from the *Lithophyllum* genus also produce allelopathic compounds that destroy or inhibit spores of brown seaweeds (Suzuki et al. 1998), and inhibit the germination of spores from 13 species of green, brown and red seaweeds (Kim et al. 2004). These chemical signals encompassing settlement cues and inhibitory effects highlight the potential that CCAs produce a deep impact on the structure and dynamics of marine benthic communities, in particular, considering the wide rodolith beds recently discovered along the Brazilian coast (e.g., Amado-Filho et al. 2010) and also known elsewhere (e.g., Foster et al. 2007).

Green seaweeds of the genus *Halimeda* (Lamouroux), are important components of coral reef communities. The segmented, calcified thalli of *Halimeda* are abundant around the world, in a range of habitats (Hillis-Colinvaux 1980, Hine et al. 1988,

Tsuda and Kamura 1991, Hillis 2001), and have long been described as contributors to sand- to mud-size carbonate sediment from the mid-Jurassic to the Holocene (Elliot 1960, 1965), taking part in reef-building and sequestration of carbon from the atmosphere (Rees 2007). The capacity of several species of *Halimeda* to produce chemical defenses and effectively protect themselves against herbivory (Paul and Van Alstyne 1988a) undoubtedly also contribute to the maintenance of biodiversity in tropical reef habitats.

Other structuring importance of seaweeds

Chemically defended seaweeds as protection for more susceptible species

Benthic marine environments under high herbivore pressure are often dominated by unpalatable, frequently chemically defended seaweeds, which constitute "associational refuges" for more susceptible species, becoming natural high diversity spots. Several associational defenses between marine organisms have been hypothesised to be chemically mediated, but few examples have been experimentally investigated.

Fishes easily consume the palatable red seaweed *Hypnea musciformis* when living alone, but it is protected when in contact with the brown alga *Sargassum filipendula* (Hay 1986). The palatable seaweed species would completely depend on chemically defended ones to provide a space of reduced herbivory, therefore escaping possible local extinction.

On coral reef habitats, several palatable seaweed species would be removed by fishes when living alone, but they are protected due to their occurrence close to chemically defended brown alga *Stypopodium zonale* (Littler et al. 1986).

The brown alga *Dictyota* sp. protects *Sargassum furcatum* living as epiphyte on this brown alga. Experimental evidence revealed that when offered separately to the herbivorous sea urchin *Lytechinus variegatus*, *S. furcatum* was highly consumed compared to *Dictyota* sp., but it was less consumed when offered to this sea-urchin associated with *Dictyota* sp. Furthermore, *Dictyota* chemicals provided protection to *Sargassum,* while *Dictyota* mimics did not (Pereira et al. 2010).

The scope of this protection can be very broad, since many of these palatable seaweeds harbour a rich and varied associated fauna that would be maintained by protection afforded by chemically defended seaweeds. Considering that species of *Hypnea*, *Stypopodium* and *Dictyota* are typical from subtropical and tropical environments, associational defenses involving these seaweeds may be frequent and important mechanisms to maintain biodiversity in a wide geographic scale.

Seaweeds as shelter, camouflage and chemical sources for marine invertebrates

In many cases, the compounds that deter fishes are used as cues by small specialist herbivores to locate or recognize their host seaweeds. In numerous cases, it appears that "feeding" specialization is really "habitat" specialization, and is driven by the need for small sedentary herbivores to escape predation rather than by nutritional traits of the host (Sotka et al. 1999). These seaweeds that are chemically defended against fishes and sea-urchins would provide an evolutionary opportunity for small sedentary herbivores to escape or deter their own consumers (Hay 1992).

The amphipod *Pseudoamphithoides incurvaria* constructs their domiciles using preferentially the brown seaweed *Dictyota bartayresii*, which is chemically defended against fishes by production of the diterpene pachydictyol A (Hay et al. 1990). An experimental approach revealed that amphipods in this domiciles are rejected as food by fishes, but they are eaten when this protection is removed (Hay et al. 1990). Another example of seaweed that is chemically defended against fish is *Chlorodesmis fastigiata* a green species from the Great Barrier Reef that provides habitat for crabs, amphipods and molluscs (Hay et al. 1989).

Chemically defended seaweeds are commonly used by a variety of marine invertebrates to live in association with or to cover their bodies, both strategies providing protection against predation on them. For majid crabs, the strategy to avoid predation is a key factor driving variation in camouflage "decoration" (Hultgren and Stachowicz 2011), providing different relationships between these small invertebrates and the host seaweeds. The species *Libinia dubia* is ornamented almost exclusively with the brown seaweed *Dictyota menstrualis* (Stachowicz and Hay 1999), which produces several diterpenes that make it unpalatable to fishes (Hay et al. 1987), a chemical camouflage that minimizes predation. There are other examples of this type of camouflage by majid crabs using noxious seaweeds (and even chemically defended invertebrates, such as sponges) and probably this behaviour is widespread (Hultgren and Stachowicz 2011).

The association between invertebrates and chemically defended seaweeds may be a more specialized relationship, as observed among molluscs. The sacoglossan *Elysia halimedae* prefers to feed on the seaweed *Halimeda macroloba* from which it obtains the diterpene halimedateraacetate, modified to produce halimedatrial for its own defense against predators (Paul and Van Alstyne 1988b). Other species of molluscs around the world sequester and store secondary metabolites from seaweeds-diet and use them for its own defense (Cimino and Ghiselin 2009). Some herbivores specialized in feeding on chemically defended seaweeds (e.g., opistobranchs from the genus *Aplysia*) not only tolerate algal defenses, but can sequester, and even transform or synthesise (i.e., *de novo* synthesis) new compounds to deter their own predators (e.g., Pennings and Paul 1993, Ginsburg and Paul 2001). In fact, *Aplysia* chemical defenses are so effective in preventing predation that it has been suggested they tend to lose their shells as a consequence of being chemically defended (Cimino and Ghiselin 2009). Some molluscs can therefore be considered as promoters of chemical diversity, since they can perform structural changes in seaweed compounds, thus increasing the overall molecular diversity in the communities where they thrive (Pereira and Teixeira 1999).

Talking trees and eavesdropping among seaweeds?

Terrestrial plant individuals have long been known to exhibit the capacity to increase their levels of chemical defense as a response to chemical cues from neighbours affected by herbivores or pathogens—a "talking trees" or "listening trees" effect (e.g., Baldwin et al. 2006). Few studies thoroughly investigate the existence of this interaction among seaweeds, and evidence so far is somewhat contradictory, despite the fact that the marine environment could propitiate the dispersion of this kind of intraspecific signaling. When *Littorina obtusata*, a herbivorous gastropod, feeds on the brown seaweed *Ascophyllum nodosum*, waterborne chemical cues are released, inducing the production of defensive chemicals (phlorotannins) on neighbour thalli (Toth and Pavia 2000). Response to grazing by the isopod *Idotea granulosa* by neighbour seaweeds was also reported in the red alga *Polyioides rotundus* and two green seaweeds, *Caulerpa rupestris* and *Ulva lactuca* (Toth 2007). Induced resistance preventing future attack by herbivores on *Fucus vesiculosus* was detected when neighbour individuals of this brown seaweed were grazed by *Idotea baltica* (Haavisto et al. 2010). Chemical cues also appear to be inducible of chemical defenses in *Fucus vesiculosus* by action of *Littorina littorea* (Rohde et al. 2004). In some cases, the herbivore's saliva has been shown to be responsible for defense induction (e.g., Pavia and Toth 2000), a striking difference when compared to true terrestrial "talking trees" effects, where plant-to-plant signaling is the rule.

Although examples of inducible defenses in seaweeds by waterborne cues ("talking trees" effect) related to intraspecific interaction abound Yun et al. (2012) took it to another level, identifying interspecific capture of waterborne grazing signals or "eavesdropping", resulting in an increase in chemical defenses by seaweeds native from Portugal and Germany in response to grazing signals from other seaweed species. Interestingly, while native species responded to interspecific grazing signals, the invasive *Sargassum muticum* did not.

How widespread waterborne signaling occurs among seaweeds remains controversial. Negative results in this field may be absent from the literature due to publication bias. Weidner et al. (2004) investigated inducible defenses among 9 southwestern Atlantic seaweeds, as well as the effects of waterborne cues from herbivores in the absence of consumption or from conspecific seaweeds (green algae *Codium decorticatum*) under grazing by the amphipod *Elasmopus brasiliensis*, and did not find any "talking tree" effect from waterborne cues, although two other seaweeds exhibited induced defense in response to direct grazing.

If, however, the talking trees effect and in particular eavesdropping are pervasive in the marine environment, then the community effects of seaweed chemical defense may indeed be huge, since one chemically defended seaweed species can deter the effects of grazing by increasing the resistance of the whole seaweed community, with potential consequences throughout the whole food web, beyond the obvious direct predator-prey effect.

Seaweeds as biodiversity regulators through settlement cues and antifoulants

Seaweeds have traditionally been studied as sources of positive cues for settlement, i.e., as sources of settlement or metamorphosis inducers (Crisp 1974). In the last decades they have emerged as major sources of chemical signals or cues for the settlement and/or metamorphosis of planktonic life stages of a wide variety of benthic marine organisms, varying from bacteria (Águila-Ramírez et al. 2012) to a plethora of marine invertebrates (reviewed by Pawlik 1992, Hadfield and Paul 2001). Although relatively inconspicuous, this seems to be a major venue by which seaweeds promote the diversity of their habitats.

The vast majority of benthic marine organisms have a life history comprising a planktonic life stage that lasts from minutes to months, sometimes dispersing over great distances before settling and metamorphosing on suitable substrata. Although widespread studies are still lacking, a large body of literature, in particular encompassing commercially important (such as barnacles, major biofoulers) or cultivable species (such as abalone—*Haliotis* spp.), exists on the physical and chemical cues involved in larval settlement and metamorphosis (Pawlik 1992), a major critical step in the life history of benthic species, as survival is dependent on larvae choosing an appropriate habitat in which to settle.

Natural chemical cues for settlement are usually extremely difficult to fully identify (Hadfield and Paul 2001), but known examples vary from fatty acids to small peptides and larger proteins, as well as carbohydrates. Settlement cues originate from a wide variety of sources, from biofilms (e.g., Shin 2008) to adult conspecifics (therefore called cues for gregarious settlement) (Toonen and Pawlik 1996), including prey organisms or organisms that are used as substratum or habitat by settlers—hence called associative settlement (Crisp 1974).

Seaweed chemical cues may signal a suitable substrate for settlement of larvae from different phyla of marine benthic invertebrates. Among red seaweeds, a compound produced by *Delesseria sanguinea* induces the settlement of the scallop *Pecten maximus* (Chevolot et al. 1991, Nicolas et al. 1998). Metamorphosis of the sea urchin *Holopneustes purpurascens* is strongly induced by *Delisea pulchra* chemicals, as well as by cues from other red seaweeds (Williamson et al. 2000).

However, CCAs seen to be the most prolific in settlement and/or metamorphosis cues for marine invertebrates, such as the sea urchin *Strongylocentrotus droebachiensis* (Pearce and Scheibling 1990), the chiton *Tonicella lineate* (Barnes and Gonor 1973), the serpulid worm *Spirorbis rupestris* (Gee 1965), the sea hare *Aplysia californica* (Nadeau et al. 1989), and the gastropod *Strombus gigas* (Boettcher and Targett 1996).

Among brown seaweeds, the compound tocopherol epoxide induces the settlement of the hydroid *Coryne uchidai on Sargassum tortile* (Kato et al. 1975). Many molluscs are found settling on macroalgae, although adults are usually found in different habitats. Epitaondiol, a meroditerpene isolated from *Stypopodium zonale*, promoted the attachment of brown mussel juveniles (*Perna perna*) in laboratory assays (Soares et al. 2008). Interestingly, this seaweed has long been postulated to double seaweed diversity in its close neighbourhood (Littler et al. 1986).

Even though seaweeds can serve as habitat, shelter, substrata and food to an enormous variety of marine benthic organisms, known positive or negative settlement cues from seaweeds only recently began to be targeted by scientists within a more ecologically relevant context. Although it has long been hypothesised that these compounds play a role at the algal surface (Steinberg and de Nys 2002), the dynamics of storage, transport and release of compounds to the thallus surface have only recently been addressed (Salgado et al. 2008, Paradas et al. 2010).

Seaweeds are known to make use of a variety of strategies to cope with epibionts, from epithallus sloughing (Keats et al. 2007) to chemical defenses (e.g., da Gama et al. 2002, 2003, 2014). These defense strategies against epibiosis are usually called antifouling defenses (Hellio et al. 2009) due to the possible applicability of these mechanisms in future commercial technologies, such as marine natural product-based or other bioinspired antifouling coatings (Clare and Aldred 2009).

Most marine benthic organisms undergo a shift from a planktonic, free-living life stage (e.g., larvae, spores, etc.) prior to settling and metamorphosing into benthic juveniles. However, settlement of competent larvae is constrained by the availability of a proper substratum, either biological (i.e., a basibiont) or not, and further survival depends on successful settlement and metamorphosis into the juvenile benthic form. In addition, larvae of many invertebrate species and seaweed propagules are influenced by specific chemical cues to settle and metamorphose in appropriate sites (Pawlik 1992).

Chemicals from seaweeds were already known to prevent fouling or epibiont settlement in the 1950's (e.g., Sieburth and Conover 1965). Recent studies have revealed how widespread chemical defenses against fouling are among seaweeds, including species of Chlorophyta, Phaeophyta and Rhodophyta (e.g., da Gama et al. 2008a, Águila-Ramírez et al. 2012), with red seaweeds emerging as the best defended division, perhaps due to their unique ability to synthesise halogenated chemical defenses (de Nys et al. 1995, da Gama et al. 2002, Dworjanyn et al. 2006). One classical example is that from the Australian red alga *Plocamium hamatum*. Through the production of a halogenated compound, the monoterpene chloromertensene, this alga causes contact necrosis on the sympatric, competing soft coral *Sinularia cruciata*. In addition, this compound exhibited antifouling properties (de Nys et al. 1991).

Coral reefs are, indeed, the most studied ecosystems in what concerns marine chemical ecology, and it has been widely assumed that tropical seaweeds are better defended than their higher latitude counterparts (but see Pereira and da Gama 2008). However, the diversity of marine flora does not peak in tropical regions, with higher diversity actually at subtropical sites (Kerswell 2006). Therefore, the role of chemically defended seaweeds in shaping community diversity may be well underestimated, since the real diversity hotspots remain to be thoroughly investigated.

Another example of how biofouling prevention by seaweeds may affect community structure comes from studies carried out in Brazilian tropical ecosystems: the common red seaweed *Laurencia dendroidea* (= *L. obtusa*) was shown to inhibit biofouling in laboratory (da Gama et al. 2002) as well as in field assays (da Gama et al. 2003). Interestingly, *L. dendroidea* also inhibited the predation by the fish *Kyphosus sectatrix* on gel surfaces used to embed seaweed extracts in antifouling field assays (Pereira et al. 2003). Evidence that fouling prevention can also influence herbivory and predation is therefore accumulating, with possible consequences for marine food webs (Wahl and Hay 1995, Karez et al. 2000, da Gama et al. 2008b).

On the other hand, settlement inhibition by seaweed chemical defenses does not necessarily imply in a biodiversity reduction at the community level, since settling larvae can usually select other substrata on which to settle. However, given the competition for space in hard-bottom communities (see Witman and Dayton 2001), it is possible that, in benthic communities dominated by chemically defended seaweeds, settling organisms can be constrained by defenses against epibionts, but this hypothesis remains to be tested experimentally.

By preventing settlement of benthic organisms, in addition to defenses against herbivores, secondary metabolites from seaweeds can supposedly shape the diversity of marine communities, in a bottom-up, although subtle, manner. Various aspects of molecular structures and biological (processes) of seaweeds inducers of settlement and metamorphosis of marine invertebrates remain unknown, but the existing examples are sufficient to assume the importance of them in structuring marine benthic communities.

Concluding remarks

Despite the fact that chemically defended seaweeds may influence or even constitute important drivers of ecosystem processes in the sea, their role in the maintenance of marine biodiversity seems to have been largely underestimated, with the possible exception of trophic interactions (Paul et al. 2007). The wide variety of secondary metabolites produced by seaweeds constitute *per se* an important phenotypic expression of diversity in the marine environment, and their ecological roles may well produce cascading effects that impact the ecology and evolution of marine systems as a whole. The importance of these molecules is presumably extensive since they are present in seaweeds living in all climate zones, from polar (Amsler et al. 2008) to temperate (Jormalainen and Honkanen 2008) and tropical regions (Pereira and da Gama 2008). One neglected aspect that deserves further investigation concerns the possibly extensive effects of natural products from kelp beds, drift algae (see Chapter by Arroyo and Bonsdorff), and beached algae in general, which transfer not only biomass, but also huge amounts of seaweed chemicals with unknown ecological consequences (Dugan et al. 2003, Andrades et al. 2014).

Since seaweed chemicals take part in many processes, from reproduction, attraction of associated species, inhibition of competitors, epibionts and consumers, seaweeds may play a key role in structuring marine ecosystems across the evolutionary time scale. Chemical cues therefore represent an invaluable adaptation to seaweeds, and can be transported, transferred, and modified by ecological interactions, in a true chemical dialog in the sea that has probably shaped marine diversity through time.

Acknowledgments

The authors thank Conselho Nacional de Desenvolvimento Científico e Tecnológico (CNPq), Fundação de Amparo à Pesquisa do Estado do Rio de Janeiro (FAPERJ), Financiadora de Estudos e Projetos (FINEP) and Coordenação de Aperfeiçoamento de Pessoal de Nível Superior (CAPES), who have long been sponsors of our work on marine chemical ecology and bioactivity of marine natural products.

References

Águila-Ramírez, R.N., A. Arenas-González, C.J. Hernández-Guerrero, B. González-Acosta, J.M. Borges-Souza, B. Véron, J. Pope and C. Hellio. 2012. Antimicrobial and antifouling activities achieved by extracts of seaweeds from Gulf of California, Mexico. Hidrobiológica 22: 8–15.

Amado-Filho, G.M., G. Maneveldt, G.H. Pereira-Filho, R.C.C. Manso, R.G. Bahia, M.B. Barros-Barreto and S.M.P.B. Guimarães. 2010. Seaweed diversity associated with a Brazilian tropical rhodolith bed. Cien. Mar. 36: 371–391.

Amsler, C.D. 2012. Chemical ecology of seaweeds. pp. 177–188. *In*: Wiencke, C. and K. Bischof (eds.). Seaweed Biology: Novel Insights into Ecophysiology, Ecology and Utilization. Springer-Verlag, Berlin.

Amsler, C.D. and V.A. Fairhead. 2006. Defensive and sensory chemical ecology of brown algae. Adv. Bot. Res. 43: 1–91.

Amsler, C.D., J.B. McClintock and B.J. Baker. 2008. Macroalgal chemical defenses in polar marine communities. pp. 91–104. *In*: Amsler, C.D. (ed.). Algal Chemical Ecology. Springer, Heidelberg, Germany.

Andrades, R., M.P. Gomes, G.H. Pereira-Filho, J.F. Souza-Filho, C. Queiroz de Albuquerque and A.S. Martins. 2014. The influence of allochthonous macroalgae on the fish communities of tropical sandy beaches. Est. Coast. Shelf Sci. 75–81.

Aquilino, K.M. and J.J. Stachowicz. 2012. Seaweed richness and herbivory increase rate of community recovery from disturbance. Ecology 93: 879–890.

Baldwin, I.T., R. Halitschke, A. Paschold, C.C. von Dahl and C.A. Preston. 2006. Volatile signaling in plant-plant interactions: "talking trees" in the genomics era. Science 311: 812–815.

Barnes, J.R. and J.J. Gonor J. 1973. The larval settling response of the lined chiton *Tonicella lineata*. Mar. Biol. 201: 259–264.

Beach, K., L. Walters, H. Borgeas, C. Smith, J. Coyer and P. Vroom. 2003. The impact of *Dictyota* spp. on *Halimeda* populations of Conch Reef, Florida Keys. J. Exp. Mar. Biol. Ecol. 297: 141–159.

Bellan-Santini, D., P.M. Arnaud, G. Bellan and M. Verlaque. 1996. The influence of the introduced tropical alga *Caulerpa taxifolia* on the biodiversity on the Mediterranean marine biota. J. Mar. Biol. Assoc. UK 76: 235–237.

Blunt, J.W., B.R. Copp, R.A. Keyzers, M.H.G. Munro and M.R. Prinsep. 2013. Marine natural products. Nat. Prod. Rep. 30: 237–323.

Boettcher, A.A. and N.M. Targett. 1996. Induction of metamorphosis in queen conch, *Strombus gigas* L., larvae by cues associated with red algae from their nursery grounds. J. Exp. Mar. Biol. Ecol. 196: 29–52.

Cabrita, M.T., C. Vale and A.P. Rauter. 2010. Halogenated compounds from marine algae. Mar. Drugs 8: 2301–2317.

Chevolot, L., J.C. Cochard and J.C. Yvin. 1991. Chemical induction of larval metamorphosis of *Pecten maximus* with a note on the nature of naturally occurring triggering substances. Mar. Ecol. Prog. Ser. 74: 83–89.

Christie, H., K.M. Norderhaug and S. Fredriksen. 2009. Macrophytes as habitat for fauna. Mar. Ecol. Prog. Ser. 396: 221–233.

Cimino, G. and M.T. Ghiselin. 2009. Chemical defense and the evolution of opisthobranch gastropods. Proc. Calif. Acad. Sci. 60: 175–422.

Clare, A.S. and N. Aldred. 2009. Surface colonisation by marine organisms and its impact on antifouling research. pp. 46–79. *In*: Hellio, C. and D.M. Yebra (eds.). Advances in Marine Antifouling Coatings and Technologies. Woodhead Publishing, Cambridge, UK.

Crisp, D.J. 1974. Factors influencing the settlement of marine invertebrate larvae. pp. 177–185. *In*: Grant, P.T. and A.M. Mackie (eds.). Chemoreception in Marine Organisms. Academic Press, New York, USA.

Da Gama, B.A.P., R.C. Pereira, A.G.V. Carvalho, R. Coutinho and Y. Yoneshigue-Valentin. 2002. The effects of seaweed secondary metabolites on biofouling. Biofouling 18: 13–20.

Da Gama, B.A.P., R.C. Pereira, A.R. Soares, V.L. Teixeira and Y. Yoneshigue-Valentin. 2003. Is the mussel test a good indicator of antifouling activity? A comparison between laboratory and field assays. Biofouling 19: 161–169.

Da Gama, B.A.P., A.G.V. Carvalho, K. Weidner, A.R. Soares, R. Coutinho, B.G. Fleury, V.L. Teixeira and R.C. Pereira. 2008a. Antifouling activity of natural products from Brazilian seaweeds. Bot. Mar. 51: 191–201.

Da Gama, B.A.P., R.P.A. Santos and R.C. Pereira. 2008b. The effect of epibiosis on the susceptibility of the red seaweed *Cryptonemia seminervis* to herbivory and fouling. Biofouling 24: 209–218.

Da Gama, B.A.P., E. Plougerné and R.C. Pereira. 2014. The antifouling defence mechanisms of marine macroalgae. Adv. Bot. Res. 71: 413–440.

de Nys, R., J.C. Coll and I.R. Price. 1991. Chemically mediated interactions between the red alga *Plocamium hamatum* (Rhodophyta) and the octocoral *Sinularia cruciata* (Alcyonacea). Mar. Biol. 108: 315–320.

de Nys, R., P.D. Steinberg, P. Willemsen, S.A. Dworjanyn, C.L. Gabelish and R.J. King. 1995. Broad spectrum effects of secondary metabolites from the red alga *Delisea pulchra* in antifouling assays. Biofouling 8: 259–271.

Dugan, J.E., D.M. Hubbard, M.D. McCrary and M.O. Pierson. 2003. The response of macrofauna communities and shorebirds to macrophyte wrack subsidies on exposed sandy beaches of southern California. Est. Coast. Shelf Sci. 58S: 25–40.

Dworjanyn, S.A., R. de Nys and P.D. Steinberg. 1999. Localization and surface quantification of secondary metabolites in the red alga *Delisea pulchra*. Mar. Biol. 133: 727–736.

Dworjanyn, S.A., R. de Nys and P.D. Steinberg. 2006. Chemically mediated antifouling in the red alga *Delisea pulchra*. Mar. Ecol. Prog. Ser. 318: 153–163.

Egan, S., T. Harder, C. Burke, P.D. Steinberg, S. Kjelleberg and T. Thomas. 2013. The seaweed holobiont: understanding seaweed-bacteria interactions. FEMS Microbiol. Rev. 37: 462–476.

Elliot, G.F. 1960. Fossil calcareous algal floras of the Middle East with a note on a Cretaceous problematicum, *Hensonella cylindrica*. Quart. J. Geol. Soc. London 15: 217–232.

Elliot, G.F. 1965. The interrelationships of some Cretaceous Codiaceae (calcareous algae). Paleontology 8: 199–203.

Fairhead, V.A., C.D. Amsler, J.B. McClintock and B.J. Baker. 2006. Lack of defense or phlorotannins induction by UV radiation or mesograzers in *Desmarestia anceps* and *D. menziesii* (Phaeophyceae). J. Phycol. 42: 1174–1183.

Fenical, W. 1975. Halogenation in the Rhodophyta: a review. J. Phycol. 11: 245–259.

Ferrer, E., A.G. Garreta and M.A. Ribera. 1997. Effect of *Caulerpa taxifolia* on the productivity of two Mediterranean macrophytes. Mar. Ecol. Prog. Ser. 149: 279–287.

Foster, M.S., L.M. McConnico, L. Lundsten, T. Wadsworth, T. Kimball, L.B. Brooks, M. Medina-López, R. Riosmena-Rodríguez, G. Hernández-Carmona, R.M. Vásquez-Elizondo, S. Johnson and D.L. Steller. 2007. Diversity and natural history of a *Lithothamnion muelleri-Sargassum horridum* community in the Gulf of California. Cien. Mar. 33: 367–384.

Francour, P., M. Harmelin-Vivien, J.G. Harmelin and J. Duclerc. 1995. Impact of *Caulerpa taxifolia* colonization on the littoral ichthyofauna of North-western Mediterranean: preliminary results. Hydrobiologia 300-301: 345–353.

Gee, J.M. 1965. Chemical stimulation of settlement of larvae of *Spirorbis rupestris* (Serpulidae). Anim. Behav. 13: 181–186.

Ginsburg, D.W. and V.J. Paul. 2001. Chemical defenses in the sea hare *Aplysia parvula*: importance of diet and sequestration of algal secondary metabolites. Mar. Ecol. Prog. Ser. 215: 261–274.

Haavisto, F., T. Valikangas and V.J. Jormalainen. 2010. Induced resistance in a brown alga: phlorotannins, genotypic variation and fitness costs for the crustacean herbivore. Oecologia 162: 685–695.

Hadfield, M.G. and V.J. Paul. 2001. Natural chemical cues for settlement and metamorphosis of marine-invertebrate larvae. pp. 431–461. *In*: McClintock, J.B. and B.J. Baler (eds.). Marine Chemical Ecology. CRC Press, Boca Raton, Florida, USA.

Hay, M.E. 1986. Associational plant defenses and the maintenance of species diversity: turning competitors into accomplices. Am. Nat. 128: 617–641.

Hay, M.E. 1992. The role of seaweed chemical defenses in the evolution of feeding specialization and in the mediation of complex interactions. pp. 93–118. *In*: Paul, V.J. (ed.). Ecological Roles for Marine Natural Products. Comstock Press, Ithaca, NY, USA.

Hay, M.E., J.E. Duffy, C. Pfister and W. Fenical. 1987. Chemical defense against different marine herbivores: are amphipods insect equivalents? Ecology 68: 1567–1580.

Hay, M.E., J.R. Pawlik, J.E. Duffy and W. Fenical. 1989. Seaweed-herbivore-predator interactions: host-plant specialization reduces predation on small herbivores. Oecologia 81: 418–427.

Hay, M.E., J.E. Duffy and W. Fenical. 1990. Host-plant specialization decreases predation on a marine amphipod: an herbivore in plant's clothing. Ecology 71: 733–743.

Hellio, C., J.P. Maréchal, B.A.P. da Gama, R.C. Pereira and A. Clare. 2009. Marine natural products with antifouling activities. pp. 572–622. *In*: Hellio, C. and D.M. Yebra (eds.). Advances in Marine Antifouling Coatings and Technologies. Woodhead Publishing, Cambridge, UK.

Heyward, A.J. and A.P. Negri. 1999. Natural inducers for coral larval metamorphosis. Coral reefs 18: 273–279.

Hillis-Colinvaux, L. 1980. Ecology and taxonomy of *Halimeda*: primary producer of coral reefs. Adv. Mar. Biol. 17: 1–327.

Hillis, L.W. 2001. The calcareous reef alga *Halimeda* (Chlorophyta, Bryopsidales): a cretaceous genus that diversified in the Cenozoic. Palaeogeogr. Palaeoclimat. Palaeoecol. 166: 89–100.

Hine, A.C., P. Hallock, M.W. Harris, H.T. Mullins, D.F. Belknap and W.C. Jaap. 1988. *Halimeda* bioherms along an open seaway: Miskito Channel, Nicaraguan Rise, SW Caribbean Sea. Coral Reefs 6: 173–178.

Hultgren, K. and J.J. Stachowicz. 2011. Camouflage in decorator crabs: Integrating ecological, behavioural and evolutionary approaches. pp. 214–229. *In*: Stevens, M. and S. Merlaita (eds.). Animal Camouflage. Cambridge University Press, UK.

Huovinen, P. and I. Gómez. 2012. Cold-temperate seaweed communities of the southern hemisphere. pp. 293–313. *In*: Wiencke, C. and K. Bischof (eds.). Seaweed Biology, Ecological Studies 219. Springer-Verlag, Heidelberg, Berlin.

Jormalainen, V. and T. Honkanen. 2008. Macroalgal chemical defenses and their roles in structuring temperate marine communities. pp. 57–90. *In*: Amsler,C.D. (ed.). Algal Chemical Ecology. Springer, Heidelberg, Germany.

Karez, R., S. Engelbert and U. Sommer. 2000. 'Co-consumption' and 'protective coating': two new proposed effects of epiphytes on their macroalgal hosts in mesograzer epiphyte-host interactions. Mar. Ecol. Prog. Ser. 205: 85–93.

Kato, T., A.A. Kumanireng, I. Ichinose, Y. Kitahara, Y. Kakinuma, M. Nishihara and M. Kato. 1975. Active components of *Sargassum tortile* affecting the settlement of swimming larvae of *Coryne uchidai*. Experientia 31: 433–434.

Keats, D.W., M.A. Knight and C.M. Pueschel. 1997. Antifouling effects of epithallial shedding in three crustose coralline algae (Rhodophyta, Corallinales) on a coral reef. J. Exp. Mar. Biol. Ecol. 213: 281–293.

Kerswell, A.P. 2006. Global biodiversity patterns of benthic marine algae. Ecology 87: 2479–2488.

Kim, M.J., J.S. Choi, S.E. Kang, J.Y. Cho, H.J. Jin, B.S. Chun and Y.K. Hong. 2004. Multiple allelopathic activity of the crustose coralline alga *Lithophyllum yessoense* against settlement and germination of seaweed spores. J. Appl. Phycol. 16: 175–179.

Leblanc, C., G. Schaal, A. Cosse, C. Destombe, M. Valero, P. Riera and P. Pottn. 2011. Trophic and biotic interactions in *Laminaria digitata* beds: which factors could influence the persistence of marine kelp forests in northern Brittany? Cah. Biol. Mar. 52: 415–427.

Littler, M.M., P.R. Taylor and D.S. Littler. 1986. Plant defense associations in the marine environment. Coral Reefs 5: 63–71.

Madsen, J.D., P.A. Chambers, W.F. James, E.W. Koch and D.F. Westlake. 2001. The interaction between water movement, sediment dynamics and submersed macrophytes. Hydrobiologia 444: 71–84.

Mann, K.H. 1973. Seaweeds: Their productivity and strategy for growth. Science 182: 975–981.

Maschek, J.A. and B.J. Baker. 2008. The chemistry of algal secondary metabolism. pp. 1–24. *In*: Amsler, C.D. (ed.). Algal Chemical Ecology. Springer-Verlag, Berlin.

Matlock, D.B., D.W. Ginsburg and V.J. Paul. 1999. Spatial variability in secondary metabolite production by the tropical red alga *Portieria hornemannii*. Hydrobiologia 399: 267–273.

Mejia, A.Y., G.N. Puncher and A.H. Engelen. 2012. Macroalgae in tropical marine coastal systems. pp. 329–357. *In*: Wiencke, C. and K. Bischof (eds.). Seaweed Biology, Ecological Studies 219. Springer-Verlag, Heidelberg, Berlin.

Morrow, K.M., V.J. Paul, M.R. Liles and N.E. Chadwick. 2011. Allelochemicals produced by Caribbean macroalgae and cyanobacteria have species-specific effects on reef coral microorganisms. Coral Reefs 30: 309–320.

Morse, A.N.C. and D.E. Morse. 1984. Recruitment and metamorphosis of *Haliotis* larvae are induced by molecules uniquely available at the surfaces of crustose red algae. J. Exp. Mar. Biol. Ecol. 75: 191–215.

Morse, D.E. and A.N.C. Morse. 1991. Enzymatic characterization of the inducer recognized by *Agaricia humilis* (scleractinian coral) larvae. Biol. Bull. 181: 104–122.

Morse, D.E., N. Hooker, A.N.C. Morse and R.A. Jensen. 1988. Control of larval metamorphosis and recruitment in sympatric agariciid coral. J. Exp. Mar. Biol. Ecol. 116: 193–217.

Nadeau, L., J.A. Paige, V. Starczak, T. Capo, J. Lafler and J.P. Bidwell. 1989. Metamorphic competence in *Aplysia californica* Cooper. J. Exp. Mar. Biol. Ecol. 131: 171–193.

Nicolas, L., R. Robert and L. Chevolot. 1998. Comparative effects of inducers on metamorphosis of the Japanese oyster *Crassostrea gigas* and the great scallop *Pecten maximus*. Biofouling 12: 189–203.

Oliveira, A.S., D.B. Sudatti, M.T. Fujii, S.V. Rodrigues and R.C. Pereira. 2013. Inter- and intrapopulation variation in the defensive chemistry of the red seaweed (Ceramiales, Rhodophyta). Phycologia 52: 130–136.

Oliveira, L.S., G.B. Gregoracci, G.G.Z. Silva, L.T. Salgado, G.M. Amado Filho, M.A. Ferreira, R.C. Pereira and F.L. Thompson. 2012. Transcriptomic analysis of the red seaweed *Laurencia dendroidea* (Florideophyceae, Rhodophyta) and its microbiome. BMC Genomics 13: 487.

Paradas, W.C., L.T. Salgado, D.B. Sudatti, M.A.C. Crapez, M.T. Fujii, R. Coutinho, R.C. Pereira and G.M. Amado-Filho. 2010. Induction of halogenated vesicle transport in cells of the red seaweed *Laurencia obtusa*. Biofouling 26: 277–286.

Paul, V.J. and K.L. Van Alstyne. 1988a. Antiherbivore defenses in *Halimeda*. Proc. 6th Intl. Coral Reef Symp. 3: 133–138.

Paul, V.J. and K.L. Van Alstyne. 1988b. Use of ingested algal diterpenoids by *Elysia halimedae* Macnae (Opistobranchia: Ascoglossa) as antipredator defenses. J. Exp. Mar. Biol. Ecol. 119: 15–29.

Paul, V.J. and K.L. Van Alstyne. 1988c. Chemical defense and chemical variation in some tropical Pacific species of *Halimeda* (Halimedaceae, Chlorophyta). Coral Reefs 6: 263–269.

Paul, V.J., K.E. Arthur, R. Ritson-Williams, R. Ross and K. Sharp. 2007. Chemical defenses: from compounds to communities. Biol. Bull. 213: 226–51.

Pavia, H. and G.B. Toth. 2000. Inducible chemical resistance in the brown seaweed *Ascophyllum nodosum*. Ecology 81: 3212–3225.

Pawlik, J.R. 1992. Chemical ecology of the settlement of benthic marine invertebrates. Oceanog. Mar. Biol. Ann. Rev. 30: 273–335.

Pearce, C.M. and R.E. Scheibling. 1990. Induction of metamorphosis of larvae of the green sea-urchin, *Strongylocentrotus droebachiensis*, by coralline red algae. Biol. Bull. 179: 304–311.

Peña, V. and I. Bárbara. 2008. Maërl community in the northwestern Iberian Peninsula: a review of floristic studies and long-term changes. Aq. Conserv. Mar. Freshw. Ecosyst. 18: 339–366.

Pennings, S.C. and V.J. Paul. 1993. Secondary chemistry does not limit dietary range of the specialist sea hare *Stylocheilus longicauda* (Quoy et Gaimard 1824). J. Exp. Mar. Biol. Ecol. 174: 97–113.

Pereira, R.C. and V.L. Teixeira. 1999. Sesquiterpenos das algas marinhas *Laurencia* Lamouroux (Ceramiales, Rhodophyta). 1. Significado ecológico. Quím. Nova 22: 369–374.

Pereira, R.C. and B.A.P. da Gama. 2008. Macroalgal chemical defenses and their roles in structuring tropical marine communities. pp. 25–56. *In*: Amsler, C.D. (ed.). Algal Chemical Ecology. Springer, Heidelberg, Germany.

Pereira, R.C., B.A.P. da Gama, V.L. Teixeira and Y. Yoneshigue-Valentin. 2003. Ecological roles of natural products of the Brazilian red seaweed *Laurencia obtusa*. Braz. J. Biol. 63: 665–672.

Pereira, R.C., A.R. Soares, V.L. Teixeira, B.A.P. da Gama and R. Villaça. 2004. Variation in chemical defenses against herbivory in Southwestern Atlantic *Stypopodium zonale* (Phaeophyta). Bot. Mar. 47: 202–208.

Pereira, R.C., E.M. Bianco, L.B. Bueno, M.A.L. Oliveira, O.S. Pamplona and B.A.P. da Gama. 2010. Associational defense against herbivory between brown seaweeds. Phycologia 49: 424–428.

Pereira-Filho, G.H., G.M. Amado-Filho, R.L. de Moura, A.C. Bastos, S.M.P.B. Guimarães, L.T. Salgado, R.B. Francini-Filho, R.G. Bahia, D.P. Abrantes, A.Z. Guth and P.S. Brasileiro. 2012. Extensive rhodolith beds cover the summits southwestern Atlantic Ocean seamounts. J. Coast. Res. 28: 261–269.

Piazzi, L., G. Ceccherelli and F. Cinelli. 2001. Threat to macroalgal diversity: effects of the introduced green alga *Caulerpa racemosa* in the Mediterranean. Mar. Ecol. Prog. Ser. 210: 149–159.

Pohnert, G. and W. Boland. 2002. The oxylipin chemistry of attraction and defense in brown algae and diatoms. Nat. Prod. Rep. 19: 108–122.

Price, N. 2010. Habitat selection, facilitation, and biotic settlement cues affect distribution and performance of coral recruits in French Polynesia. Oecologia 163: 747–748.

Ragan, M.A. and K.W. Glombitza. 1986. Phlorotannins, brown algal polyphenolics. pp. 129–241. *In*: Round, F.E. and D.J. Chapman (eds.). Progress in Phycological Research. Biopress, Bristol.

Raniello, R., E. Mollo, M. Lorenti, M. Gavagnin and M.C. Buia. 2006. Phytotoxic activity of caulerpenyne from the Mediterranean invasive variety of *Caulerpa racemosa*: a potential allelochemical. Biol. Inv. 9: 361–368.

Rasher, D.B., E.P. Stout, S. Engel, J. Kubanek and M.E. Hay. 2011. Macroalgal terpenes function as allelopathic agents against reef corals. Proc. Natl. Acad. Sci. 108: 17726–17731.

Rees, S.A. 2007. Significance of *Halimeda* bioherms to the global carbonate budget based on a geological sediment budget for the Northern Great Barrier Reef, Australia. Coral Reefs 26: 177–188.

Roberts, R.D., M.F. Barker and P. Mladenov. 2010. Is settlement of *Haliotis iris* larvae on coralline algae triggered by the alga or its surface biofilm? J. Shellf. Res. 29: 671–678.

Rohde, S., M. Molis and M. Wahl. 2004. Regulation of anti-herbivore defence by *Fucus vesiculosus* in response to various cues. J. Ecol. 92: 1011–1018.

Salgado, L.T., N.B. Viana, L.R. Andrade, R.N. Leal, B.A.P. da Gama, M. Attias, R.C. Pereira and G.M. Amado-Filho. 2008. Intra-cellular storage, transport and exocytosis of halogenated compounds in marine red alga *Laurencia obtusa*. J. Struct. Biol. 162: 345–355.

Schiel, D.R. and M.S. Foster. 1986. Structure of subtidal algal stands in temperate waters. Oceanogr. Mar. Biol. Ann. Rev. 24: 265–308.

Schoenwaelder, M.E.A. 2002. The occurrence and cellular significance of physodes in brown algae. Phycologia 41: 125–139.

Shin, H.W. 2008. Rapid attachment of spores of the fouling alga *Ulva fasciata* on biofilms. J. Environ. Biol. 29: 613–619.

Sieburth, J.M. and J.T. Conover. 1965. *Sargassum* tannin, an antibiotic which retards fouling. Nature 208: 52–53.

Soares, A.R., B.A.P. da Gama, A.P. Cunha, V.L. Teixeira and R.C. Pereira. 2008. Induction of attachment of the mussel *Perna perna* by natural products from the brown seaweed *Stypopodium zonale*. Mar. Biotechnol. 10: 158–165.

Sotka, E.E., M.E. Hay and J.D. Thomas. 1999. Host-plant specialization by a non-herbivorous amphipod: advantages for the amphipod and costs for the seaweed. Oecologia 118: 471–482.

Stachowicz, J.J. and M.E. Hay. 1999. Reducing predation through chemically mediated camouflage: indirect effects of plant defenses on herbivores. Ecology 80: 495–509.

Steinberg, P.D. 1989. Biogeographical variation in brown algal polyphenolics and other secondary metabolites: comparison between temperate Australasia and North America. Oecologia 78: 373–382.

Steinberg, P.D. and I. Van Altena. 1992. Tolerance of marine invertebrate herbivores to brown algal phlorotannins in temperate Australasia. Ecol. Monogr. 62: 189–222.

Steinberg, P.D. and R. de Nys. 2002. Chemical mediation of colonization of seaweed surface. J. Phycology 38: 621–629.

Sudatti, D.B., S.V. Rodrigues and R.C. Pereira. 2006. Quantitative GC-ECD analysis of halogenated metabolites: determination of elatol on surface and within-thallus of *Laurencia obtusa*. J. Chem. Ecol. 32: 835–843.

Sudatti, D.B., S.V. Rodrigues, R. Coutinho, B.A.P. da Gama, L.T. Salgado, G.M. Amado Filho and R.C. Pereira. 2008. Transport and defensive role of natural products at the surface of the red seaweed *Laurencia obtusa*. J. Phycol. 44: 584–591.

Suzuki, Y., T. Takabayashi, T. Kawaguchi and K. Matsunaga. 1998. Isolation of an allelophatic substance from the crustose coralline algae, *Lithophyllum* spp., and its effect of on the brown alga, *Laminaria religiosa* Miyabe (Phaeophyta). J. Exp. Mar. Biol. Ecol. 225: 66–77.

Taniguchi, K., M. Shiogama, K. Kurata and M. Suzuki. 1991. *Ecklonia stolonifera* against the abalone *Haliotis discus hannai*. Bull. Jpn. Soc. Sci. Fish. 57: 2065–2071.

Taniguchi, K., Y. Akimoto, K. Kurata and M. Suzuki. 1992. Chemical defense mechanism of the brown alga *Eisenia bicyclis* against marine herbivores. Bull. Jpn. Soc. Sci. Fish. 58: 571–575.

Toonen, R.J. and J.R. Pawlik. 1996. Settlement of the tube worm *Hydroides dianthus* (Polychaeta: Serpulidae): cues for gregarious settlement. Mar. Biol. 126: 725–733.

Toth, G.B. 2007. Screening for induced herbivore resistance in Swedish intertidal seaweeds. Mar. Biol. 151: 1597–1604.

Toth, G.B. and H. Pavia. 2000. Water-borne cues induce chemical defense in a marine alga (*Ascophyllum nodosum*). Proc. Natl. Acad. Sci. 97: 14418–14420.

Toth, G.B. and H. Pavia. 2006. Artificial wounding decreases plant biomass and shoot strength of the brown seaweed *Ascophyllum nodosum* (Fucales, Phaeophyceae). Mar. Biol. 148: 1193–1199.

Tsuda, R.T. and S. Kamura. 1991. Floristics and geographic distribution of *Halimeda* (Chlorophyta) in the Ryukyu Islands. Jpn. J. Phycol. 39: 57–76.

Vallim, M.A., R.C. Pereira, J.C. De-Paula and V.L. Teixeira. 2005. The diterpenes from Dictyotacean marine brown algae in the tropical Atlantic America regions. Biochem. Syst. Ecol. 33: 1–16.

Wahl, M. and M.E. Hay. 1995. Associational resistance and shared doom: effects of epibiosis on herbivory. Oecologia 102: 329–340.

Weidner, K., B.G. Lages, B.A.P. da Gama, M. Molis, M. Wahl and R.C. Pereira. 2004. Effects of mesograzers and nutrient levels on the induction of defenses in several Brazilian macroalgae. Mar. Ecol. Prog. Ser. 283: 113–125.

Wiencke, C. and C.D. Amsler. 2012. Seaweeds and their communities in polar regions. pp. 265–291. *In*: Wiencke, C. and K. Bischof (eds.). Seaweed Biology, Ecological Studies 219. Springer-Verlag, Heidelberg, Berlin.

Williamson, J.E., R. de Nys, N. Kumar, D.G. Carson and P.D. Steinberg. 2000. Induction of metamorphosis of a marine invertebrate by the metabolites floridoside and isethionic acid from its algal host. Biol. Bull. 198: 332–345.

Wilson, W. 2002. Nutritional value of detritus and algae in blenny territories on the Great Barrier Reef. J. Exp. Mar. Biol. Ecol. 271: 155–169.

Witman, J.D. and P.K. Dayton. 2001. Rocky subtidal communities. pp. 339–366. *In*: Bertness, M.D., S.D. Gaines and M.E. Hay (eds.). Marine Community Ecology. Sinauer Associates, Sunderland, Massachusetts, USA.

Young, D.N. and J.A. West. 1979. Fine structure and histochemistry of vesicle cells of the red alga *Antithamnion defectum* (Ceramiaceae). J. Phycol. 15: 49–57.

Yun, H.Y., A.H. Engelen, R.O. Santos and M. Molis. 2012. Water-borne cues of a non-indigenous seaweed mediate grazer-deterrent responses in native seaweeds, but not vice versa. Plos One 7: e38804.

Epibiont-Marine Macrophyte Assemblages

Carol S. Thornber,[1],* *Emily Jones*[2] and *Mads S. Thomsen*[3,4]

Introduction: Ecological importance of basiphyte-epibiont interactions

What is epibiont ecology?

The biology and ecology of marine seagrasses and macroalgae have likely been influenced by the presence of epibionts for millions of years, given the evolutionary history of both groups, and the potential for co-evolution (Taylor and Wilson 2003). Epibionts are ubiquitous in marine environments, span numerous taxonomic divisions and phyla, occur on a wide taxonomic diversity of basiphytes (hosts), and can either be host-specific (obligate) or host non-specific (facultative). Their importance in ecosystem functioning has been well documented in systems ranging from estuaries to subtidal rocky reefs, as well as from tropical to polar regions (Thomsen et al. 2010).

As widely recognized ecosystem engineers and foundation species, seagrasses and their associated epibionts have been widely studied (e.g., Tomas et al. 2005, Cook et al. 2011, York et al. 2012, Lobelle et al. 2013; Fig. 1). While the majority of algal epiphytes are located on older regions of seagrass leaves, epibionts can also occur

[1] Dept. of Biological Sciences, University of Rhode Island, 120 Flagg Rd, Kingston, RI 02881 USA.
[2] Department of Biology, San Diego State University, 5500 Campanile Dr., San Diego, CA 92182 USA.
Email: emjones@gmail.com
[3] Marine Ecology Research Group, School of Biological Sciences, University of Canterbury, Private Bag 4800, Christchurch, New Zealand.
[4] UWA Oceans Institute and School of Plant Biology, University of Western Australia, Hackett Drive, Crawley 6009 WA, Australia.
Email: mads.solgaard.thomsen@ gmail.com
* Corresponding author: thornber@uri.edu

Figure 1. Photos of typical subtidal seagrass beds at Rottnest Island, Western Australia, heavily covered by epibionts. (A) *Amphibolis antarctica* (Labillardière) Sonder & Ascherson ex Ascherson with *Laurencia* sp. epiphyte; (B) *Amphibolis antarctica* with *Metagoniolithon stelliferum* (Lamarck) Ducker (calcareous epiphyte); (C) *Posidonia australis* J.D. Hooker with several epiphytes (including filamentous red algae and *Jania* sp.). *Amphibolis* stems live for over 2 years and are heavily epiphytized, while *Posidonia* leaves are shed in < 100 days.

on seagrass stems and/or rhizomes (Borowitzka et al. 2006). Epibionts can enhance seagrass leaf turnover rates (Cook et al. 2011), occur in dense patches on seagrasses, and have a greater biomass than the seagrasses to which they are attached (Cook et al. 2011). Algal epiphytes can also contribute up to 50–60% of the primary productivity in seagrass meadows (Borowitzka et al. 2006, Cebrian et al. 2013), and play important roles in nutrient cycling (e.g., Pereg-Gerk et al. 2002) and controlling biodiversity of invertebrates (Edgar 1990).

Macroalgal hosts include taxa spanning a wide range of morphological, ecological, and evolutionary diversity (Chlorophyta, Rhodophyta, and Phaeophyceae; Fig. 2). While reports of epibiota on macroalgal taxa are widespread and common (e.g., Jones and Thornber 2010, Kersen et al. 2011, Rohr et al. 2011, Ávila et al. 2012, Engelen et al. 2013), their ecological impacts and interactions are less thoroughly studied than for epibiota on seagrasses (Potin 2012). Like on seagrasses, epibionts on macroalgae are typically more common on older tissues (Arrontes 1990, Pearson and Evans 1990), and can vary in species identity and community across algal thalli (Fricke et al. 2011).

Definitions

For the purposes of this review, we define a basiphyte as a living organism (host) that is typically anchored directly to a (non-living) substrate and is either a seagrass or a macroalga, that has one or more living, attached epibiont species. Non-living substrates can include a variety of habitat types, including sand, shells, pebbles, cobbles, boulders, and rocky reefs. Throughout this review, we follow Wahl (1989) for consistency of epibiont terminology. We therefore use the term epibiosis to refer to a relationship between two organisms, one of which (epibiont) lives on the other (basiphyte), in an interaction that is neither parasitic nor symbiotic. Here, we discuss epibionts (epibiota) that are sessile animals (called epibiotic fauna or epizoans), plants and algae (called epiphytes), while recognizing that epibionts can also include marine bacteria and fungi (e.g., Zhang et al. 2009, Burke et al. 2011, Lachnit et al. 2011) and mobile animals (Wernberg et al. 2004). Epibionts can also live attached to other epibionts, in secondary or tertiary relationships (Wahl 1989, Thomsen et al. 2010). We use the terms epibiotic to describe an epibiont-basiphyte, and epiphytic to describe an epiphyte-basiphyte, relationship. While there is a well-documented gradient across different taxa from purely epiphytic interactions between sessile epibiota and basiphytes to completely parasitic interactions between parasite and basiphytes (e.g., Potin 2012), the focus of this review is on non-parasitic interactions, which may be negative, positive, or neutral.

Objectives of this chapter

In this chapter, we explore the associations between epibionts and their basiphytes through numerous ecological and taxonomic perspectives. We first describe the diversity of key groups and species of epiphytes and epizoans. Next, we explore a suite of documented ecological interactions between epibionts and their hosts, ranging from negative to positive to neutral (e.g., van Montfrans et al. 1984, Leonardi et al.

Figure 2. Photos of macroalgae in RI, USA, heavily covered by epibionts. (A) shallow subtidal, Brenton Point, Newport RI algal community including *Fucus vesiculosus* with epiphytic *Ectocarpus* sp., *Codium fragile* subsp. *fragile* with the epiphytes *Ectocarpus, Ceramium*, and *Neosiphonia*; (B) rocky intertidal, Fort Wetherill, Jamestonw RI—*Fucus vesiculosus* basiphyte, with *Ceramium virgatum* epiphyte and *Lacuna vincta*. (C) subtidal coastal lagoon, Charlestown, RI—*Fucus* sp. with numerous algal epiphytes and the colonial tunicate *Botryllus schlosseri* Pallas.

2006, Jones and Thornber 2010). We then address the cascading impacts of epibiont-basiphyte interactions on community structure of associated mobile animals, with an emphasis on how epiphyte-basiphyte interactions are modified by mobile mesograzers, and finish by investigating the impacts of human-induced stressors on some of these relationships. Our goal in creating this chapter was to provide a thorough (but not exhaustive) review of the existing literature on epibiont-basiphyte ecology, and include both classic and current studies that have illuminated our understanding of these important interactions. Our contribution thereby supplements recent, thorough reviews of epibionts on seagrasses by Borowitzka et al. (2006), of algal epiphytes, endophytes, and parasites on macroalgae by Potin (2012), and on Antarctic endophytes on macroalgae by Amsler et al. (2009). We also hope this chapter will stimulate thought and discussion on this topic by illustrating issues that are in need of further research and critical interpretation.

Epibiont diversity

Epiphytes (plants/algae living on plants)

Typically, most epiphytes are algae (not vascular plants), including, but not limited to, the Chlorophyta (green algae), Rhodophyta (red algae), Phaeophyceae (brown algae), Dinophyta (dinoflagellates), Bacillariophyceae (diatoms), and Cyanophyta (cyanobacteria). Macroscopic epiphytes can be categorized into different functional groups (*sensu* Littler 1980), including filamentous, foliose, corticated filamentous, corticated foliose, saccate, and coralline groups (Saunders et al. 2003). While macroscopic epiphytes are typically more thoroughly studied and quantified than their microscopic counterparts, microscopic epiphyte assemblages can also be quite diverse (Jernakoff and Nielsen 1997). On seagrass basiphytes, red algal epiphytes typically dominate in terms of biomass, while the presence of mainly cyanobacterial and/or green algal epiphytes frequently indicates eutrophic or other seasonal, high nutrient conditions (Lavery and Vanderklift 2002, Lapointe et al. 2004, Borowitzka et al. 2006).

   Epiphytes have a variety of mechanisms they employ to attach to a basiphyte, including via single cells or filamentous/rhizoidal basal structures (e.g., Leonardi et al. 2006), typically at a sporeling or juvenile stage (see **Adaptations** section). Other epiphytes can produce 'hooks' to 'grab' basophytes (e.g., *Hypnea musciformis* (Wulfen) J.V. Lamouroux, *Bonnemaisonia hamifera* Hariot), and some epibiotic species produce secondary rhizoids to re-attach, if needed, to new substrates/basiphytes (e.g., Perrone and Cecere 1997).

Epibiotic fauna (epizoans—animals living on plants)

The taxonomic diversity of sessile faunal epibiota includes groups such as colonial hydroids (Cnidaria: Hydrozoa), encrusting bryozoans (Bryozoa: Gymnolaemata), barnacles (Arthropoda: Cirripedia), sponges (Porifera), ascidians (Chordata: Ascidiacea), and polychaetes (Annelida: Polychaeta). We also include nonmotile sea anemones (Cnidaria: Anthozoa) in this group, while recognizing that some sea anemones occurring on basiphytes are motile. Sessile epizoans typically settle at the larval stage (Hadfield 1986), via the production of adhesive granules (Stricker 1989) and/or the growth of stolons (Cerrano et al. 2001). Similar to epiphytes, faunal epibiota have distinct zonation patterns on seagrass and macroalgal basiphytes; in some cases, faunal epibiota and epiphytes are inversely correlated in abundance, while on others they are positively correlated (e.g., Trautman and Borowitzka 1999). Individual basiphyte species can harbor a wide richness of sessile epibiotic fauna (e.g., > 20 species; Fredriksen et al. 2007).

One of the most well studied epizoans is the encrusting bryozoan *Membranipora membranacea* Linnaeus, which has been found on kelps such as *Macrocystis pyrifera* (Linnaeus) C. Agardh (Hepburn et al. 2006), *Saccharina longicruris* (Bachelot de la Pylaie) Kuntze (Saunders and Metaxas 2008), and *Laminaria digitata* (Hudson) J.V. Lamouroux and *Laminaria hyperborea* (Gunnerus) Foslie (Schultze et al. 1990), as well as on other large species such as the red alga *Dilsea carnosa* (Schmidel) Kuntze (Nylund and Pavia 2005), the fucoid *Fucus serratus* Linnaeus and the eelgrass *Zostera marina* Linnaeus (Fredriksen et al. 2007). Sessile epizoans such as *Membranipora* can benefit from their association with basiphytes by absorbing supplemental carbon from kelp exudates (De Burgh and Fankboner 1978), while mutualisms have been reported for other epizoan-basiphyte systems (Hepburn and Hurd 2005).

Co-existence mechanisms/interactions between basiphytes and epibionts

Marine macrophytes as secondary substrate

In marine communities, one of the most limiting resources for both plants and animals is space (Dayton 1971). Many marine macrophytes, including macroalgae, seagrasses, and mangroves, compete for primary substrate such as rock or soft-bottom habitat. However, these organisms also act as foundation species (*sensu* Dayton 1972), increasing substrate heterogeneity and the area available for settlement for smaller algae and invertebrates. These habitat-forming species play critical roles in facilitating epibiota (Stachowicz 2001, Bruno et al. 2003), as shown by the abundance and diversity of organisms that can grow attached to basiphyte hosts. For example, 15 to 30 species of macroalgae grow epiphytically on coralline turf algae in southern California (Stewart 1982), 60% of the macroalgal species in New England have the potential to grow epiphytically (Jones 2007), and over 500 algal species can grow attached to seagrasses (Harlin 1980). The ability for these epibiota and basiphytes to co-exist is based on a variety of mechanisms and interactions, which we will review in this section.

Facultative vs. *obligate associations*

Associations between epibiotic organisms can be either facultative, in which epibionts are able to settle and grow on a variety of biotic or abiotic substrates, or obligate, where epibionts depend on one or more specific hosts. The majority of marine epibiotic organisms are facultative on plant and/or animal basiphytes (Wahl 1989, 2009), with surveys showing that less than 5% of organisms are obligate on a single plant or animal (Wahl 2009). However, several species of obligate epiphytes (e.g., Abbott and Hollenberg 1976, Hallam et al. 1980, Harlin 1980, Gonzalez and Goff 1989, Pearson and Evans 1990) and invertebrates (e.g., Hughes et al. 1991) exist on marine macrophytes, and understanding these obligate interactions is important, as these epibiotic species may be unable to survive in the absence of a specific basiphyte.

Some species that are considered obligate epibionts may be able to settle and grow on other substrates; however, they may grow at very low abundances, be unable to reach full size, or not persist long-term. For example, the filamentous red alga *Smithora naidum* (C.L. Anderson) Hollenberg, an obligate epiphyte on the seagrass genera *Zostera* and *Phyllospadix*, can grow on artificial substrate at lower abundances (Harlin 1973) and on the alga *Plocamium cartilagineum* (Linnaeus) P.S. Dixon at a much smaller size (Hansen 1986). Thus, although *Smithora* may grow on other substrates, it appears to be most productive on its primary seagrass hosts. Because the occurrence of obligate epiphytes on other substrates may be rare, undocumented, or not easily seen (e.g., microscopic life stages) it is difficult to determine how many species are truly obligate on single host species. In addition, the underlying mechanisms causing many of these obligate interactions are not well understood. One of the most well-studied obligate macroalgal interactions is between the red alga *Vertebrata* (*Polysiphonia*) *lanosa* (Linnaeus) T.A. Christensen and the basiphyte *Ascophyllum nodosum* (Linnaeus) Le Jolis (Lobban and Baxter 1983). *Vertebrata* is primarily restricted to the mid-frond, lateral pits of the intertidal canopy-forming brown alga *Ascophyllum* (Lobban and Baxter 1983), as it can only grow on the distal ends of the thallus if wounds are present (Longtin and Scrosati 2009). This epiphyte-host interaction may also be mediated by an obligate fungus on *Ascophyllum*, *Mycophycias ascophylli* (Cotton) Kohlmeyer and Volkmann-Kohlmeyer, which minimizes tissue damage to the host and may aid in nutrient transfer between the two macroalgal species (Garbary et al. 2005). Despite this knowledge, however, we don't know why *Vertebrata* is not found on other macroalgal species or whether *Mycophycias* or other fungi or bacteria must be present for it to persist.

Although some epibiont species are primarily obligate, many species live on a variety of substrate types. The ability of some epiphytic algae to grow on different macrophyte species may depend on attachment mechanisms. In the Northeast Pacific, the red alga *Microcladia californica* Farlow is an obligate epiphyte on the main axis of the brown alga *Egregia menziesii* (Turner) Areschoug, while the closely related congener *Microcladia coulteri* Harvey is able to grow on over 25 genera of macroalgae (Gonzalez and Goff 1989). One reason for these differences is that *M. californica* attaches to substrates via a discoid holdfast, and is sloughed off or dislodged from other basiphytes. In contrast, in addition to discoid attachment, *M. coulteri* also uses rhizoidal attachment, which allows it to resist sloughing (Gonzalez and Goff 1989). Similarly, *Neosiphonia* (*Polysiphonia*) *harveyi* (J.W. Bailey) M.-S. Kim, H.-G. Choi, Guiry, and G.W. Saunders, one of the most abundant epiphytes on a variety of macroalgae in the Northwest Atlantic (Jones and Thornber 2010), shows plasticity for holdfast attachment (Wilson 1978). Wilson (1978) found that when *Neosiphonia* grows on smooth substrata such as *Chondrus crispus* Stackhouse, it uses discoid-like holdfast attachment, but when it grows on heterogenous substrate such as *Codium fragile* ssp. *tomentosoides* (van Goor) P.C. Silva, it uses rhizoidal attachment. Thus, epibionts may be substratum generalists depending on whether they attach to the surface of other organisms, penetrate into the tissues, or are capable of using both methods. However, despite these early studies, the ecological significance of this holdfast plasticity has largely been ignored; additional studies on what cues elicit these plastic responses and the frequency of this plasticity across species would aid in our understanding of epibiont-basiphyte interactions.

Negative vs. positive impacts

Interactions between epibionts and basiphytes can be negative (van Montfrans et al. 1984, D'Antonio 1985, Williams and Seed 1992), positive (Stewart 1982, Norton and Benson 1983, Karez et al. 2000), or neutral (Cattaneo 1983, Uku 2005), typically depending on environmental conditions (Bertness and Callaway 1994). The effects of basiphytes on epibionts are generally positive, as they provide substrate for these organisms to grow on when primary substrate is limiting (Wahl 1989). However, in addition to increasing settlement space, basiphytes raise organisms higher in the water column, increasing flow (Butman 1987), food and nutrient availability (Keough 1986, Laihonen and Furman 1986), and light levels for photosynthesis (Brouns and Heijs 1986). Some species can also act as an associational refuge for epibiotic organisms (Hay 1986), when consumers avoid unpalatable basiphytes.

Despite these positive impacts, basiphytes can also have negative impacts on their epibionts, especially when basiphyte tissue is removed due to physical stressors, seasonal changes, or consumption. For instance, the tissues of many perennial basiphyte species are shed during winter storms (e.g., Seed and O'Connor 1981), while many ephemeral macroalgal species that could potentially serve as hosts exist during only parts of the year. Epibiota may also face a "shared doom" (Wahl and Hay 1995) when epibionts and basiphytes are co-consumed (Karez et al. 2000). For example, large seagrass herbivores such as dugongs may consume epibiota while feeding on seagrass shoots. In addition to tissue losses of the basiphyte, the macrophyte species can also modify the environment and make it less suitable for epibiont species, relative to other substrate. Daleo et al. (2006) found that *Ulva lactuca* Linnaeus biomass was decreased when growing on the turf-forming alga *Corallina officinalis* Linnaeus compared to growing alone on hard substrate, due to an increase in desiccation stress.

Most research on the interactions between epibionts and basiphytes has focused on the negative impacts epibionts pose to basiphyte species such as decreased growth (Honkanen and Jormalainen 2005) and reproduction (Kraberg and Norton 2007) due to increased competition for light and nutrients (Sand-Jensen 1977, D'Antonio 1985, Cebrian et al. 1999). For instance, in seagrass communities, epibionts can reduce light availability by 10–90% due to shading of basiphyte surfaces (Sand-Jensen 1977, Borum et al. 1984), and epiphytes may preferentially absorb blue and red light before they reach the seagrass leaves (Drake et al. 2003). These negative effects may depend on where epibionts grow on the basiphyte. For example Cancino et al. (1987) found that the bryozoan *Jellyella* (*Membranipora*) *tuberculata* Taylor and Monks decreased light and photosynthesis rates for the alga *Gelidium rex* Santelices and I.A. Abbot, but did not affect net growth rates due to compensatory growth in other parts of the thallus. Epibiotic organisms can also increase physical stressors on basiphytes by increasing drag (D'Antonio 1985, Hemmi et al. 2005) and decreasing elasticity (Dixon et al. 1981), both of which can cause increased mortality. For example, encrusting species such as bryozoans can reduce blade motion and flexibility, increasing breakage in high flow environments (Dixon et al. 1981) and decreasing kelp abundances (Scheibling and Gagnon 2009). Basiphytes may also be damaged by consumers of both invertebrates (Bernstein and Jung 1979) and algae (Karez et al. 2000) if they remove host tissue while feeding on epibiotic organisms.

Although many epibiont effects are negative, epibiota can also benefit basiphyte species. In contrast to epiphytic algae that compete for nutrients, sessile invertebrates may increase nutrient exchange by excreting ammonium. Hepburn and Hurd (2005) found that during times of low nutrient concentrations, *Macrocystis pyrifera* colonized by hydroids actually increased biomass, relative to non-colonized individuals. Epibiota can also decrease physical stress, by retaining water and reducing desiccation during low tide (Stewart 1982). And similarly to unpalatable basiphytes acting as an associational defense for epibiota, epibiont growth can also protect basiphyte species from being consumed by herbivores (Wahl and Hay 1995, Karez et al. 2000). For example, Wahl and Hay (1995) found that urchins fed less on basiphyte species such as *Gracilaria tikvahiae* McLachlan when they were covered by low-palatability epiphytes such as *Ectocarpus* sp. and *Polysiphonia* sp.

Adaptations

Basiphyte adaptations

Marine macrophytes have developed a suite of avoidance, tolerance, and defensive traits to prevent colonization and/or overgrowth by epibiotic organisms (Wahl 1989). Some macrophyte species may be able to avoid epibiont settlement either temporally or spatially through rapid or ephemeral growth (den Hartog 1972), high tissue turnover (Bernstein and Jung 1979), or living in locations such as areas with turbulent conditions where epibionts are not able to survive (Seed and O'Connor 1981). Other macrophyte species use mechanical and chemical defenses to remove epibionts or deter them from settling. Sloughing or shedding cuticle and epidermal tissues is a common mechanism for removing epibiotic organisms (Filion-Myklebust and Norton 1981, Sieburth and Tootle 1981, Moss 1982, Russell and Veltkamp 1984, Craigie et al. 1992, Nylund and Pavia 2005). Some macrophyte species may produce mucus or slime that is sloughed off or prevents epibiont settlement (Sieburth and Tootle 1981, Dawes et al. 2000). The effectiveness of these mechanical defenses may be highly variable, however, depending on seasonality of epibiont recruitment and growth rates (Sieburth and Tootle 1981, Wahl 1989, Jones and Thornber 2010) or age of the host (Dawes et al. 2000). Finally, a variety of macrophyte species use chemical defenses to inhibit epibiont growth (e.g., Schmitt et al. 1995, Suzuki et al. 1998, Cho et al. 2001, Hellio et al. 2004, Kim et al. 2004, Wilkström and Pavia 2004, Paul et al. 2006, Nylund et al. 2007). Although many of these studies have used whole tissue extracts to test for chemical inhibition, Nylund et al. (2007) used more ecologically relevant tissue surface extracts to determine that non-polar metabolites from both *Delisea pulchra* (Greville) Montagne and *Caulerpa filiformis* (Suhr) Herring are capable of inhibiting both sessile invertebrates and epiphytic algae. Basiphytes may also use a combination of these life history, mechanical, or chemical strategies simultaneously, across different tissues of the macrophyte, or over time. For example, *Ulva* species are both ephemeral and are capable of cuticle peeling (Tootle 1974) while *Fucus vesiculosus* Linnaeus exhibits both tissue sloughing (Sieburth and Tootle 1981) and chemical inhibition (Wilkström and Pavia 2004).

Epibiont adaptations

The majority of epibiotic organisms are opportunistic, ephemeral species that have much shorter lifespans than their basiphyte hosts and are poor competitors for primary substrate. As a result of this high competition for space, many species have developed a variety of adaptations at both the settlement and attachment stages so that they can successfully colonize secondary substrata. Settlement of many epibiota is influenced by physical characteristics such as irradiance levels, substrate roughness, and hydrodynamics (reviewed by Fletcher and Callow 1992, Harder 2008). However, some species are also able to respond to chemical cues released from macroalgae (e.g., Kato et al. 1975, Bouarab et al. 2001) or surface bacteria (Joint et al. 2002). By these abiotic and biotic cues, epibiota are more likely to settle on favorable substrates.

Once epibiota have settled on a basiphyte, they must successfully attach in order to survive. As mentioned above, the ability of epibionts to attach to different hosts often depends on attachment mechanisms. Linskens (1963) defined holoepiphytes as algae that attach to the outer layer of a macrophyte, while amphiepiphytes are those that penetrate cell layers. Leonardi et al. (2006) described five different types of epiphyte attachment on the alga *Gracilaria chilensis* C.J. Bird, McLachlan and E.C. Oliveira: (1) weakly attached to surface, (2) strongly attached to surface, (3) penetrating the outer layer of the cell wall, (4) penetrating into the cortical tissue, and (5) penetrating deeply into the cortex. They found that the abundance of epiphytes was greatest for those strongly attached to the surface or penetrating the outer cell layer, while the other attachment types were more seasonal (Leonardi et al. 2006). This seasonality may be due to the ephemeral nature of the epiphytes themselves, or because there is a tradeoff associated with deeper tissue penetration. However, some of the species that penetrated deep into the cortex, but were only found in the summer (*Ceramium rubrum* C. Agardh, *Neosiphonia harveyi*) can persist throughout the year on other host species (Jones and Thornber 2010), suggesting that these interactions are variable on different basiphyte substrates or in different geographic locations. Epiphytic macroalgal genera including *Hypnea, Bonnemaisonia, Cystoclonium, Laurencia,* and *Chaetomorpha* can also attach to basiphytes using specialized hooks, tendrils, and/or secondary rhizoids. Not only do these mechanisms allow epiphyte species to attach without directly binding to basiphyte tissue, they also provide a means of reattachment if an epiphyte is dislodged via herbivory or wave action. For example, when severed, the alga *Solieria filiformis* (Kützing) P.W. Gabrielson produces new rhizoids from the damaged surfaces that allow it to reattach to secondary substratum (Perrone and Cecere 1997). In addition, if fragmented, *Hypnea musciformis* apical hooks can produce secondary rhizoids to reattach to basiphyte species (Thomsen 2004a). The ability of sessile invertebrates to attach to macrophyte substrates will also be important for epibiont survival, although in contrast to epiphytic algae, some invertebrates such as hydroids on seagrasses can use stolon transfer to "move" between substrates (Hughes et al. 1991).

Interaction strengths and community context

The strength of these basiphyte-epibiont interactions can be highly context dependent, depending on biotic and abiotic conditions. Epibiont communities on the same basiphytes can vary seasonally (Arrontes 1990, Rindi and Guiry 2004, Jones and Thornber 2010), geographically (Rindi and Guiry 2004), and across environmental gradients such as wave exposure (Kersen et al. 2011), salinity (Kendrick et al. 1988), and intertidal height (Longtin and Scrosati 2009, Longtin et al. 2009). This variation in time and space may affect the direction and magnitude of interactions between epibiont and basiphyte species. For example, when epibionts shade seagrass, they decrease the available light and the depth at which seagrasses are able to grow (Borum et al. 1984). However, this interaction may be positive if it allows seagrasses to grow in sunnier locations (Wiencke 1987) or expand into shallower habitats, by reducing photo-inhibition. Epibionts may have varying interactions within an individual basiphyte species as well. The brown alga *Soranthera ulvoidea* Postels and Ruprecht is commonly found attached to the red alga *Odonthalia floccosa* (Esper) Falkenberg in the mid to low intertidal zone from the Bering Sea to California (Abbott and Hollenberg 1976). During low tide, the presence of *Soranthera* can decrease desiccation rates of *Odonthalia* (Anderson 2012). However, when submerged, *Soranthera* increases drag on *Odonthalia* even at low flow, and dislodgment at high flow (Anderson 2012). Thus, the net effect of these interactions may vary with emergence time and flow conditions. However, the majority of research on basiphyte-epibiont interactions has focused on single stressors and response variables (e.g., growth) and additional studies are needed that focus on multiple stressors and response variables (Anderson 2012) to enhance our understanding of how basiphyte and epibiont species are able to co-exist.

The larger community-context is also important to consider when investigating the consequences of epibiota on basiphyte species. As mentioned previously, consumers may enhance negative effects of epibionts on basiphytes by co-consuming epibiont and basiphyte tissue (Karez et al. 2000). However, consumers can also mediate the interactions between epibionts, reducing epibiont effects. For instance, the removal of epibiota by predatory snails doubled the growth rate of *Fucus vesiculosus* (Honkanen and Jormalainen 2005). In another case, only the combination of two complementary consumers (the snails *Cotonopsis* (*Anachis*) *lafresnayi* P. Fisher and Bernardi and *Astyris* (*Mitrella*) *lunata*) Say successfully removed both solitary ascidians and encrusting bryozoans from the basiphyte *Chondrus crispus*, increasing basiphyte biomass (Stachowicz and Whitlatch 2005). Thus, community interactions among species and trophic levels can play important roles in altering the interactions between epibiota and basiphyte species.

Cascading effects of basiphyte-epibiont interactions

Implications for abundance and diversity of mobile animals

In the previous sections we discussed basiphyte–epibiota interactions, with an emphasis on how basiphytes increase the abundance and diversity of sessile epibiota by being a host/habitat for these species. Here, we note that the sessile epibiota species themselves also provide and modify habitats for more and/or different sessile and mobile epibiota (we here refer to this second group of epibiota as 'end-users'). This interaction chain is an example of cascading habitat formation—where indirect positive effects on focal organisms are mediated by successive facilitation in the form of biogenic formation or modification of habitat (Thomsen et al. 2010, Fig. 3). Or, in other words, in this chain reaction basiphytes have indirect positive effects on end-users mediated through positive effect on epibionts.

There are ample data to suggest that cascading habitat formation is common in marine systems dominated by epibiont-basiphyte interactions. For example, numerous studies show that herbivores have positive impacts on basiphytes by preferentially consuming epiphytes (e.g., Howard 1982, Shacklock and Doyle 1983, Orth and Montfrans 1984, Hootsmans and Vermaat 1985, Duffy 1990, Klumpp et al. 1992, Mukai and Iijima 1995, Alcoverro et al. 1997, Boström and Mattila 1999, Pavia et al. 1999, Worm and Sommer 2000, Hily et al. 2004, Tomas et al. 2005, Prado et al. 2007, Jones and Thornber 2010). These studies focus on how herbivores—through keystone consumption (Paine 1966, Thomsen et al. 2010)—control community structure. Thus, in keystone consumption, consumers (here grazers) control abundances and diversity of key species (here basiphytes) by consuming competitive dominants (here epiphytes). Importantly, all of above studies also provide support for cascading habitat formation; reversing focus to the basiphyte suggests that the basiphyte has indirect positive effects on herbivores (and other end-users) by providing high quality food, habitat, and protection from predators. Cascading habitat formation and keystone consumption, are therefore, at least for basiphyte-epiphyte-grazer interactions, 'mirror-processes'

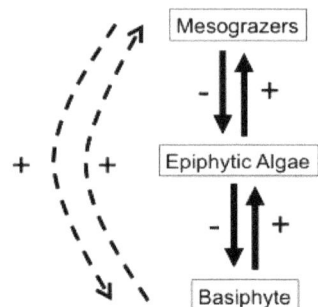

Figure 3. Diagram showing potential basiphyte-epibiont-mesograzer interactions. Arrows go toward the species experiencing the effect from the one causing it. Signs (+/–) indicate the type of effect. Solid lines represent direct efforts, while dashed lines represent indirect effects. The solid line pointing away from the mesograzer indicate keystone consumption, while the solid line pointing away from the basiphyte indicates cascading habitat formation.

(Fig. 3). Still, we address keystone consumption, where the main focus is on the pivotal role of herbivores, separately in the next section (as traditionally done). Here we review the relatively few studies that address basiphyte-epiphyte-enduser interactions explicitly as cascading habitat formation, i.e., where research focus is on how basiphytes—through control of epiphytes—control abundances and diversity of end-users. We include a few examples with drift macroalgae entangled around basiphytes, because the main interactors and the type of direct and indirect interactions are conceptually similar to classical basiphyte-epiphyte-end-user interactions. Also, some epiphytes can 'change' between attached and drifting states: for example, *Hypnea musciformis* can survive as drift algae, become entangled or re-attach to a basiphyte with hooks and/or secondary rhizoids, and break off and drift around again (Thomsen 2004a).

Cascading habitat formation is well documented from seagrass beds. For example, Hall and Bell (1988) documented positive effects on meiofauna associated with the seagrass *Thalassia testudinum* ex König and its epiphytes. More specifically, surveys and experiments were carried out to examine relationships between the biomass of epiphytic algae on *Thalassia testudinum* blades and density of end-users, in particular copepods, nematodes, amphipods, and crustacean nauplii. Colonization experiments documented a positive relationship between biomass of the epiphytic alga *Giffordia michelliae* (Harvey) G. Hamel and end-users, suggesting that results were density dependent (but length of colonization time did not vary among seagrass blades). Follow up experiments using artificial blades and several densities of artificial epiphytes produced similar results, i.e., with highest densities of end-users on blades in high epiphyte treatments. These results suggested that much of the relationship between these end-users and epiphytes could be attributed to the physical structure—not trophic subsidy—of the epiphyte. In another colonization experiment in a seagrass bed, Schneider and Mann (1991) also used plastic mimics to investigate the relative importance of basiphyte shape and epiphyte cover in determining the distribution of invertebrate end-users. Both epiphyte cover and shape were important, but end-user responses were highly species-specific where some species responded to basiphyte shape while others responded to epiphyte cover. Follow-up laboratory predation experiments suggested that, contrary to expectations, predation by fish and crabs was not affected by the presence of artificial macrophytes. Bologna and Heck (1999b) also used artificial basiphytes and manipulated both epiphytic structure and epiphytic food resources. However, somewhat in contrast to the previous studies, they found that end-user densities and diversity were higher on mimic basiphytes that were covered by live epiphytes, compared to artificial epiphytes. This response to live epiphytes was strong for herbivores and omnivores and weak for filter feeders and predator end-users. Epiphytic structure thereby appeared to play a limited role in determining the density of most mobile end-users, with one exception being the settlement of bivalves. Thus, this study suggested that epiphytes can have a dramatic positive impact on end-users via trophic subsidy (and supported by the extensive literature reviewed in the **Epibiont-mesograzer interactions** section).

Similar general positive effects of epiphytes on invertebrate end-users have also been found in subtropical *Amphibolis* seagrass beds (Edgar and Robertson 1992) where removal of epiphytes resulted in fewer species and lower abundances of focal organisms. Patches within a mixed bed of the seagrasses *Amphibolis antarctica* (Labillardière) Sonder and Ascherson ex Ascherson and *Amphibolis griffithii* (J.M. Black) den Hartog were manipulated by removing epiphytes, basiphyte leaves, and by reducing basiphyte density. Leaf and epiphyte removal decreased abundances of most end-users dramatically. Follow-up caging experiments showed a reduction in faunal densities on basiphytes both in open plots and enclosed in cages. Hence, predation by fish or decapod predators was unlikely to cause the faunal decline in the open seagrass plots. In this experiment, end-users associated with basiphyte leaves appeared to actively select dense basiphyte habitats, possibly because of evolutionary selection to minimize predation or to avoid high levels of solar radiation. In a somewhat unusual example, it was recently documented that the large invasive macroalga *Codium fragile* (Suringar) Hariot can be an abundant epiphyte on *Zostera marina*, by attaching to the seagrass' rhizomes (Drouin et al. 2011). More specifically, surveys and experiments documented higher abundance and diversity of invertebrates associated with the epiphyte than with the basiphyte alone, i.e., the abundance and taxonomic richness of end-users were positively correlated with the biomass of the epiphyte. Furthermore, experimental manipulation of the epiphyte morphology showed that end-users were not influenced by this factor, indicating that factors other than structural complexity may be important for end-user abundance and diversity.

In addition to seagrass beds, cascading habitat formation is also likely to be important on macroalgal-dominated rocky reefs. For example Martin-Smith (1993) removed epiphytes from two types of *Sargassum* macroalgal mimics in Queensland, Australia. Again, community composition differed between the epiphyte-covered and the clean mimics, leading to higher abundances of crustacean, polychaete and gastropod end-users in the presence of epiphytes. Finally, surveys by Leite and Turra (2003) found significant positive relationships between the combined biomass of the basiphyte *Sargassum cymosum* C. Agardh and epiphyte *Hypnea musciformis*, and the total density of all invertebrate end-users.

Similar to above basiphyte-epiphyte case studies, drift macroalgae that are entangled around seagrass stems and leaves (without being physically attached) provide analogue examples of cascading habitat formations. Thus, invasive (Thomsen 2010) and native (Thomsen et al. 2012a) coarsely branched *Gracilaria*, and a complex of various red algal species (Holmquist 1997) have, just like for typical epiphytic macroalgae, strong positive effects on many invertebrate end-users within seagrass beds. Similar results have also been highlighted in other studies for individual end-user species, showing positive effects of entangled macroalgae within seagrass beds, on the snails *Potamopyrgus antipodarum* J.E. Gray (Cummins et al. 2004) and *Peringia ulvae* Pennant (as *Hydrobia ulva*: Cardoso et al. 2004)—but these studies also report relatively strong negative effects on different invertebrate end-users. Drift seaweeds, like epiphytes, typically have negative impacts on the seagrass itself, competing for light and nutrients and increasing anoxia, sulphide and ammonia levels in the water column or sediment pore-water, and may, in some cases, even kill-off the host (Holmquist 1997, McGlathery 2001, Hauxwell et al. 2003, Thomsen et al. 2012b), thereby destabilizing the entire habitat cascade.

The above studies suggest that positive effects of epiphytes on mobile animal end-users can be caused by both quantitative and qualitative differences in habitat attributes and resource provisions between the epiphyte and basiphyte. It is likely that epiphytes simultaneously provide shelter from predators, food for grazers, attachment space for recruits and, potentially, also ameliorate abiotic stress, particularly in the intertidal zone (Norkko et al. 2000, Cardoso et al. 2004, Nyberg et al. 2009). It is important to note though that the reviewed studies used "addition/removal" (not "substitution") type experiments and surveys. The combined biomass of the basiphyte and epiphyte is therefore higher than for the basiphyte alone, making it difficult to separate quantitative and qualitative habitat effects. It may therefore be that the increase in total habitat space facilitates mobile animal end-users, irrespective of epiphyte traits. Facilitation could also be caused by qualitative trait differences between epiphytes and basiphytes. For example, the entangled macroalga *Gracilaria* contrasts its host *Zostera*, by being positioned horizontally within the seagrass bed, having cylindrical branching laterals that are not shed and by being more palatable. Thus, there are several co-occurring mechanisms whereby end-users can be facilitated by epiphytes, including stress-reduction from desiccation and extreme temperatures in the intertidal zone, consumer-avoidance and food subsidies. In summary, positive effects of epiphytes on mobile animal end-users are likely large, when the epiphyte is larger, more abundant, and ecologically different from the basiphyte, and when spatial heterogeneity or food web complexity are low and small—because the latter mechanisms represent larger scale alternative pathways for end-users to escape stress and enemies and find resources (Thomsen et al. 2010).

Epibiont-mesograzer interactions

Influence of epiphyte-basiphyte associations on mesograzers

As mentioned above, macrophyte species that support epibiota influence higher trophic levels. In this section, we focus on small herbivores, or mesograzers, which are typically crustaceans and gastropods 0.1 to 2.5 cm in size (Hay et al. 1987, Brawley 1992). These species serve as both a food sources for larger predators (Edgar and Shaw 1995, Taylor 1998, Heck et al. 2000), as well as important consumers of primary producers (Jernakoff et al. 1996, Valentine and Duffy 2006). Because many mesograzers use epiphytic algae for both food and habitat, basiphyte-epiphyte interactions can strongly influence these species and vice versa.

Mesograzer and epiphyte abundances are often strongly correlated, with higher abundances of grazers occurring on epiphytized basiphytes (Pavia et al. 1999, Orav-Kotta and Kotta 2004, Jones and Thornber 2010, Rohr et al. 2011). Large macroalgae and seagrasses are often less accessible and palatable to mesograzers than the small, filamentous epiphyte species (e.g., Steneck and Watling 1982, Chavanich and Harris 2002, Boström and Mattila 2005), so macrophytes with epiphytic algae may provide more suitable food resources compared to areas where epiphyte growth is minimal. In seagrass communities, a large number of organisms feed on epiphytic algae (Jernakoff et al. 1996), and epiphytes are primarily used as food, rather than habitat (Bologna and Heck 1999b; see also **Implications for abundance and diversity of mobile animals**). However, in macroalgal communities, these interactions often vary depending on mesograzer, basiphyte, and epiphyte identity. Pavia et al. (1999) found that abundances of the amphipod *Gammarus locusta* Linnaeus were five to eight times higher when epiphytes were present on the alga *Ascophyllum nodosum*, while isopod distributions were more variable (*Idotea granulosa* Rathke) or not correlated (*Jaera albifrons* Leach). Small individuals of both *Gammarus* and *Idotea* preferred the epiphytes *Ceramium nodulosum* Ducluzeau and *Pylaiella littoralis* (Linnaeus) Kjellman over *Ascophyllum*, while large individuals of *Idotea* also consumed *Ascophyllum*. In contrast to these results, Karez et al. (2000) found that *Gammarus* preferred the tips of the basiphyte *Fucus vesiculosus* over the epiphyte *Elachista fucicola* (Velley) Areschoug, while *Idotea* consumed both *Elachista* and older *Fucus* tissue. Similarly, Orav-Kotta and Kotta (2004) and Kotta et al. (2000) found that the isopod *Idotea baltica* preferentially consumed the epiphyte *Pylaiella littoralis* over the basiphyte *Fucus vesiculosus*, while Jormalainen et al. (2001a) found a preference for *Fucus* over these same epiphyte species. These contrasting feeding preferences for epiphytes and basiphytes may be due to structural or palatability differences among the macroalgal species (e.g., *Ascophyllum* may be better habitat and/or tougher than *Fucus*), or due to differences in mesograzer characteristics such as size (Pavia et al. 1999) and sex (Jormalainen et al. 2001b).

In addition to serving as a food source for mesograzers, epibiota can also increase habitat complexity (Hacker and Steneck 1990, Martin-Smith 1993; see also **Implications for abundance and diversity of mobile animals**), providing protection from predators and buffering abiotic stressors such as desiccation (Salemaa 1987, Boström and Mattila 1999, Williams et al. 2002). This increased habitat structure may be especially important for newly recruited individuals. For instance, recruitment of the snail *Lacuna vincta* Montagu is influenced by epiphyte abundance in the low intertidal zone of New England, with more individuals occurring on macroalgae or mimic substrates where epiphytes are present (Jones and Thornber 2010, Rohr et al. 2011). These snails may also use different epiphyte species for food and habitat, as abundances are positively correlated with *Neosiphonia harveyi* biomass, while they preferentially consume the epiphytic red alga *Ceramium virgatum* Roth (Jones and Thornber 2010). Similarly, using both feeding and habitat preference assays, Kotta et al. (2000) found that *Idotea balthica* Pallas prefers the epiphyte *Pilayella* as food, but the basiphyte *Furcellaria lumbricalis* (Hudson) J.V. Lamouroux as habitat.

Influence of mesograzers on basiphyte communities

A variety of studies have shown that mesograzers can influence primary production, trophic energy transfer, and biogeochemical cycling (Edgar and Aoki 1993, Edgar and Shaw 1995, Heck et al. 2000, Spivak et al. 2009). Mesograzer activity may be especially important in areas dominated by macrophyte foundation species such as macroalgae and seagrasses, as these grazers prevent competition with, or overgrowth by, epibiotic organisms (Duffy 1990, Stachowicz and Hay 1996, Miller 1998, Stachowicz and Hay 1999, Stachowicz and Whitlatch 2005). While mesograzer activity is generally non-lethal to large macroalgae and seagrasses, it can still affect basiphyte fitness and influence community structure, depending on the tissues consumed, grazer feeding rates, macrophyte growth rates, grazer interactions, and abiotic conditions (Arrontes 1990, Ilken 2012). By contrast, via the removal of epiphytes, mesograzers may increase light availability and nutrient acquisition for basiphytes (Duffy 1990), while providing nutrients via excretion (Fong et al. 1997, Bracken et al. 2007). However, mesocosm experiments clearly support the idea that grazers can control epiphytic algal growth (reviewed by Valentine and Duffy 2006). These effects can vary from strong (Duffy and Hay 2000, Bruno and O'Connor 2005) to weak or undetected (O'Connor and Bruno 2007, Douglass et al. 2008), and the ecological relevance of these studies in field conditions is not well understood.

Most studies investigating mesograzer impacts have manipulated grazer densities and diversity in laboratory mesocosms, or excluded grazers using mesh cages in the field (see review by Poore et al. 2009). Understanding the community-wide effects of these studies is difficult, however, as mesocosm studies fail to incorporate the high spatial and temporal variation of mesograzer densities (Ruesink 2000), species-specificity of feeding preferences for different epiphytes (Duffy and Harvilicz 2001), and consumption of specific basiphyte parts such as meristems (Poore 1994) or reproductive tissues (Nakaoka 2002), while cages present several experimental artifacts (Miller and Gaylord 2007). Two studies, however, have effectively decreased mesograzers abundances in the field using an insecticide, with contrasting results. Poore et al. (2009) found that in a temperate Australian algal bed, mesograzer reductions had no effect on the growth rate of the basiphyte *Sargassum linearfolium* (Turner) C. Agardh, epiphyte cover, or algal community composition. However, using similar techniques, Whalen et al. (2013) found that mesograzers can control seagrass epiphytes in natural seagrass communities, although the dominance of top-down and bottom-up effects shifted with season. During the summer, when mesograzer densities were typically high, removal of grazers led to a significant increase in epiphyte biomass and decrease in eelgrass shoot density. However, during the fall when natural mesograzer densities declined, epiphytes escaped from grazer control, hence removing grazers had no effect on epiphyte abundance. This escape from grazer control occurred at amphipod densities much higher than those found naturally on *Sargassum* by Poore et al. (2009), suggesting that one possible reason for these contrasting results between studies may be due differences in grazing pressure. Further field studies are needed to assess whether there is variation in top-down and bottom-up control between epiphyte-seagrass and epiphyte-macroalgal systems however. In addition, understanding the role of natural grazer variation may help us predict where or when nutrient enrichment may prevent grazer control of epiphytic algae (e.g., Kotta et al. 2000).

Influence of predators on mesograzer-epiphyte-basiphyte interactions

In addition to epiphyte abundance and habitat complexity, predation risk can be an important factor in structuring epiphyte communities, although studies on this topic are limited. Predators may influence mesograzer effects on epiphyte species both by direct consumption and by altering mesograzer behavior. For instance, if epiphytes provide a refuge for mesograzers from predators (Williams et al. 2002), predators may have greater effects on mesograzer abundances when epiphyte cover is low. Although empirical evidence for this idea is somewhat lacking, especially from field studies, Russo (1987) found that in the lab, Hawaiian amphipods survived better in more structurally complex mimic habitats in the presence of predators. Predation risk may also change the behavior of some, but not all mesograzer species. In the absence of predation, the isopods *Idotea balthica* and *Erichsonella attenuata* Harger both select less complex seagrass habitat with epiphytes, their preferred food source (Bologna and Heck 1999a). However, in the presence of a predator, *Idotea* still chose food over shelter, while *Erichsonella* chose the more complex seagrass habitat that provided better protection from predators. These behaviors have the potential to influence both epiphyte abundances and mesograzer mortality, although these effects were not directly measured in this study. One area of research that is lacking in these interactions is how spatial and temporal changes in epiphyte abundances influence mesograzer behavior, indirectly affecting their susceptibility to predation. For example, if mesograzers prefer to feed on epiphyte species, do they increase foraging movement as epiphyte abundance declines, increasing predation risk? More research is needed to understand how these predator-prey interactions change with variation in resources over space and time.

Human threats to stable basiphyte-epibiont co-existence

Human activities modify basiphyte-epibiont interactions

Human activities have a fundamental influence on almost all aspects of marine ecology. The main human stressors in coastal ecosystems are eutrophication, habitat-alterations, fisheries, invasions, and climate changes (Worm et al. 2006) and the impact of most of these stressors on marine plants (as basiphytes) are described in separate chapters of this book. In this section we briefly outline how these stressors may also modify basiphytes-epibiota interactions. Unfortunately, few studies have specifically tested how human stressors modify basophyte-epibiont interactions, and much of our discussion below is therefore circumstantial, being derived from the more common studies that focus on human impacts on either the basophyte, the epibiont or on organisms with epibiont-like traits.

Eutrophication

Eutrophication is the ecosystem response to the addition of excessive amounts of nutrients, such as nitrates and phosphates, through fertilizers or sewage, to aquatic systems. Of the different human stressors, eutrophication has the most straightforward-to-predict effects on basiphyte-epiphyte interactions. Large amounts of inorganic nutrients favor plants with high Surface area-to-Volume (SV)-ratios and rapid nutrient uptake and growth more than plants with low SV ratios and slow nutrient uptake and growth (i.e., species with classic r-strategy are favored over species with K-strategy; Duarte 1995, Pedersen and Borum 1996, 1997). Given that many epiphytes (and turf and drift macroalgae) have these r-strategy traits, eutrophication typically favors epiphytes over basiphytes (Lotze et al. 2000, Thomsen et al. 2012b). Indeed, eutrophication can cause population booms of epiphytes (and drift macroalgae) that shade basiphytes and cause low oxygen levels at night or when epiphyte populations decay (Krause-Jensen et al. 1999, Holmer and Nielsen 2007). On longer and larger spatio-temporal scales, epiphytic blooms have caused strong reductions in basiphyte populations in coastal zones around the world (Kangas et al. 1982, Cambridge et al. 1986, Kautsky et al. 1986, Valiela et al. 1997, McGlathery 2001, Hauxwell et al. 2003). These nutrient-driven changes in the relative dominance between basiphytes and epiphytes may ultimately cause strong impacts on the entire ecosystem, which are sometimes referred to as regime shifts (Troell et al. 2005, Andersen et al. 2009). Typically, most nutrients are derived from diffusive and point sources of the local watershed. This implies that it may be possible to locally manage and rectify problems associated with eutrophication by shifting competitiveness back to the basiphyte, although factors such as 'hysteresis' (Andersen et al. 2009), local extinctions of basiphytes (Smale and Wernberg 2013), non-local atmospheric pollution (Paerl 1995), and accumulated nutrient banks in sediments (Schelske et al. 1986, Reddy et al. 1993) can make this a complicated and slow recovery process.

Habitat alterations

Habitat alterations is probably the most important human stressor in terrestrial ecosystems, but is of less direct importance in marine systems. Direct habitat alterations can be associated with boating activities (e.g., anchoring) and local construction projects, such as building harbours, jetties, groynes, and oil rigs (Bulleri and Chapman 2010). However, of more general importance for basiphyte-epiphyte interactions are those habitat alterations that occur in adjacent terrestrial drainage basins. Importantly, run-off and sedimentation levels have increased many-fold around the world as coastal forests have been converted to urban centers and agricultural land (Airoldi 2003). These enhanced sediment levels in coastal systems cause decreased light levels and can smother plants and sessile animals. However, slow growing basiphytes are typically more negatively affected by high sediment loads compared to ephemeral epiphytes that may 'outgrow' enhanced sedimentation. Epiphytes typically also have broader habitat requirements than basiphytes, for example typically being able to survive both drifting and attached to man-made structures, and as epiphytes. Indeed, it appears to be a global trend that where sedimentation levels are high, canopy-forming basiphytes are being rapidly replaced with species that have turf and epiphytic algal traits (Airoldi and Cinelli 1997, Airoldi and Virgilio 1998, Irving and Connell 2002, Eriksson and Johansson 2003, 2005, Balata et al. 2007). Problems associated with enhanced sedimentation can be managed locally by altering those catchment practices that cause increased sediments and by constructing sediment deposition buffer zones to further reduce sedimentation on basiphytes (e.g., by restoring salt-marshes and mangrove forests).

Fishing

Fishing does not appear to affect basiphyte-epiphyte interactions directly, but could cause complex indirect effects. Fishing alters the local communities and abundances of different fish species, typically at first reducing populations of large apex predators, and later, if fishing pressure continues to be intense, also smaller fish lower on the food chain (Jackson 1997, Caddy et al. 1998). Alteration to the local fish community can be expected to modify basiphyte-epiphyte relations in complex ways, but in particular, through long and short consumption cascades (Post 2002). For example, removal of big predatory fish may cause population increases of smaller fish which typically consume mesograzers like small crustaceans and gastropods. Given that mesograzers prefer to consume epiphytes over basiphytes (see **Epibiont-mesograzer interactions**), the removal of large predators could indirectly increase epiphytes and thereby increase competitive stress on basiphytes. However, if continued fishing effort subsequently also removes small fish, basiphytes could be indirectly facilitated because predation on mesograzers would thereby be reduced, potentially resulting in less epiphytism. These long and short consumption cascade scenarios are highly simplified, as many fish species are omnivores, can shift diets, and may be replaced by functionally similar species, thereby blurring our ability to predict indirect cascading impacts of fishing on basiphytes and epiphytes (Duffy and Hay 2001, Hay et al. 2004, Burkepile and Hay 2006). Coastal fishing is also a local stressor that, like sedimentation and eutrophication, at least theoretically, can be managed locally.

Invasive species

Invasions by non-native species also modify basiphyte-epiphyte interactions. Invasive species can be both basiphytes and epiphytes, as well as grazers and predators that can affect basiphyte-epiphyte interactions directly (e.g., by consumption) or indirectly (e.g., by altering abiotic environments or consuming grazers). This complexity implies that both basiphytes and epiphytes can be both facilitated and inhibited, and that the net effect on local basiphyte-epiphyte interactions is highly context-dependent (Thomsen et al. 2011). For example, many invasive algae, like *Gracilaria vermiculophylla* (Ohmi) Papenfuss (Nyberg et al. 2009), *Sargassum muticum* (Yendo) Fensholt (Viejo 1999, Wernberg et al. 2004, Thomsen et al. 2006), *Fucus evanescens* C. Agardh (Wikström and Kautsky 2004), *Grateloupia turuturu* Yamada (Jones and Thornber 2010, Janiak and Whitlatch 2012), and *Codium fragile* (Mathieson et al. 2003, Schmidt and Scheibling 2006, Jones and Thornber 2010) are basiphytes that provide habitat for numerous sessile or mobile organisms. However, although these case studies suggest invaders are common basiphytes—with positive effects on epiphytes—they probably represent rather atypical invasion impacts. Notable, these invasive macrophytes are all relatively large, relatively easy to see and identify, and have conspicuous impacts, and are therefore obvious to study. However, examinations of typical non-native species (and their traits) found on boat hulls, in ballast water, and on aquacultural molluscs, more often points to non-native species being small, inconspicuous, and difficult to detect and identify (small turf and filamentous macroalgae and sessile invertebrates; Carlton 1996, Hewitt et al. 2004). It is therefore possible that many epiphytes are non-native species that have been translocated prior to scientific data collections or remain unrecognized as non-native species, implying that the importance of invasive epiphytic species may be significantly underestimated. Despite a likely underestimation, there are still many case studies on invasive epiphytes, in particular represented by bryozoa (Saunders and Metaxas 2008), tunicates (Dijkstra et al. 2007, McCarthy et al. 2007) and macroalgae (Piazzi et al. 2002, Deudero et al. 2010). Invaders can also be herbivores that may preferentially graze on habitat forming basiphytes or on epiphytes and other small macroalgae that compete with the basiphytes (Lubchenco 1978, Eastwood et al. 2007). Finally, many invaders are predators or omnivores, that through their consumption of herbivores further modify basiphyte-epiphyte interactions (Trussell et al. 2004, Eastwood et al. 2007, Albins and Hixon 2008). Overall, we can therefore, at least not yet, predict simple effects of invasions on basiphyte-epiphyte interactions. Furthermore, the control options for local managers are poor, in part because even extreme border control cannot eliminate invasions (many invasive species arrive by natural dispersal from adjacent less controlled invaded regions), and in part because following a successful establishment, invaders, are extremely difficult to eradicate or even just control.

Climate change

Finally, climate changes also modify basiphyte-epiphyte interactions. Climate change is a global phenomena with complex and difficult to predict local effects. In marine systems, the main climate changes relate to increased storminess, ocean acidification and warming. Effects associated with increased storminess are difficult to predict in part because effects are expected to be highly localized. However, if increased storminess occurs, both basiphytes and epiphytes could be favored, depending on the specific changes to hydrodynamic regimes. For example, increased storminess could cause increased dislodgement of weakly attached epiphytes (Schanz et al. 2000, Schanz et al. 2002, Thomsen 2004b) but could alternatively also dislodge or prune the entire basiphyte (Gaylord et al. 1994, Thomsen and Wernberg 2005, de Bettignies et al. 2012). More complex indirect interactions can also be envisioned because increased storminess may change consumer-basiphyte-epiphyte interactions, as grazers and predators are likely to be dislodged more frequently and their feeding patterns altered (Schanz et al. 2000, Schanz et al. 2002). Finally, species with high colonization abilities are likely to be favored as storms will increasingly create gaps and open spaces within beds of canopy-forming basiphytes. Importantly, early gap colonizers can reproduce year round and/or can re-attach to substratum from fragments (e.g., Perrone and Cecere 1997). Many of these species can also exist as epiphytes. As oceans and seas become more acidic, acidification will inhibit calcifying organisms, including calcifying basiphytes (Martin and Gattuso 2009, Guerra-García et al. 2012), calcifying epiphytes (Saunders and Metaxas 2008), and calcifying consumers (Bibby et al. 2007). It is therefore difficult to predict broad acidification effects on basiphyte-epiphyte interactions, being further complicated by recent findings that non-calcified organisms can also be dramatically influenced by acidification (Russel et al. 2009). Finally, oceans and seas are becoming warmer. Warming is probably the most important driver of climatic changes and will generally favor warm-water species with wide temperature tolerances, and rapid growth. Importantly, these traits are more often associated with epiphytic species (and turf forming and drift macroalgae) than large canopy forming basiphytes. For example, kelp basiphytes are generally adapted to relatively cold waters and can be negatively impacted by warming, particularly near their poleward ranges (Wernberg et al. 2012, Smale and Wernberg 2013). Similarly, slow growing seagrass basiphytes may also be more susceptible to heat stress than many epiphytic species (Short and Neckles 1999) and heating can reduce their resistance to other stressors, such as competition with macroalgae (Hoeffle et al. 2011, Holmer et al. 2011) or sulphide stress (Koch et al. 2007). In contrast to the previous stressors, local managers have virtually no options for controlling or rectifying global climate change impacts on basiphyte-epiphyte interactions, but will instead have to focus on adaptations and mitigations to expected changes.

Co-occurring stressors

Most importantly, human stressors do not occur in isolation, and to understand how they modify basiphyte-epiphyte interactions, their combined effects need to be considered. Unfortunately, relatively few studies have tested for combined effects of multiple human stressors, and we therefore typically have to make simple predictions from their individual effects (but see Piazzi et al. 2005, Russel et al. 2009, Gennaro and Piazzi 2011, Hoeffle et al. 2011, Holmer et al. 2011, Hoeffle et al. 2012 for examples, suggesting that combined stress effects typically are additive or synergistic). As reviewed above, fishing, invasions, acidification and increased storminess all have complex effects that can both favor and inhibit both basiphytes and epiphytes. However, eutrophication, enhanced sedimentation and warming generally favor epiphytes over basiphytes, suggesting that the net effect from all human stressors combined will be decreased abundances of basiphytes and increased abundances of epiphytes, or species with similar traits and strategies (Steneck et al. 2002, Orth et al. 2006, Waycott et al. 2009, Wernberg et al. 2012). However, if basiphytes decrease, this may negatively impact species that have adapted to being primarily (or exclusively) epibionts. Furthermore, a simple comparison of attributes associated with epiphytes vs. basiphytes, under increased stress and disturbances, support this notion; typically large basiphytes have, in contrast to epiphytes, relatively slow growth, narrow temperature requirements, are cold-water adapted, perennial, have seasonal recruitment, complex life-history, depend on sexual reproduction, cannot reattach to hard substratum, cannot survive long-term as unattached populations, and have relatively narrow habitat requirements. These traits generally point to basiphytes being less resistant and resilient, compared to epiphytes, to storminess, warming, sedimentation, eutrophication and high water turbidity. Furthermore, epiphytes (including other ephemeral and opportunistic macroalgae) are more speciose compared to large canopy-forming basiphytes being represented by relatively few seagrasses, kelps, fucalean and large red and green algae. This implies that if a basiphytic species is becoming locally extinct, there is relatively little chance it can be replaced by a functionally similar species. By contrast, the more speciose epiphytic species are typically not only better adapted and more resilient to human stressors, but, simply because there are more of them, also have a higher chance that a functionally similar species can replace a lost epiphyte. In conclusion, we suggest that human stressors will continue to favor ephemeral opportunistic epiphytes over perennial slow growing basiphytes, and that an integrated management approach is needed to tackle the many complex direct and indirect effects whereby human stressors can modify basiphyte-epiphyte interactions.

Conclusions

Epibionts and their basiphyte hosts encompass a wide range of taxonomic diversity and occur in most benthic marine habitats within the photic zone. Epibiont-basiphyte associations are commonplace in marine habitats and impact numerous ecological interactions and processes, from physiology to ecosystem ecology. While their interactions may range from negative to neutral to positive, the consequences of these interactions on the basiphyte are necessarily limited, as extreme epiphyte overgrowth that kills the host can result in host dislodgement, thereby likely causing the death of the epiphytes. In addition, host-epiphyte interactions can impact higher trophic levels.

While the distributions and abundances of epiphytes and basiphytes have been fairly well documented, less is known about the reciprocal impacts of basiphytes and epiphytes on each other—many studies have focused on the impact of epiphytes on a basiphyte, or vice versa. More laboratory and field studies that incorporate this ecological relevance, and/or test for specific interactions (e.g., chemical modifications between epiphytes and basiphytes) are needed. In addition, it is unknown how the species or genetic diversity of epiphytes interacts with the genetic diversity of the host species (Reusch and Hughes 2006). Lastly, epiphytes and basiphytes are not immune to anthropogenic impacts, which may affect them unequally. Understanding how these relationships may be affected as climate change occurs will be of critical importance to preserving these important ecosystem engineers.

Acknowledgements

This chapter is dedicated to the memory of Dr. Scott Nixon (1943–2012). Research on macroalgae and algal epiphyte-host interactions in the Thornber lab has been funded by agencies including Rhode Island Sea Grant, Woods Hole Sea Grant, RI Science and Technology Advisory Council, and National Science Foundation-EPSCOR. MST was supported by the Marsden Fund Council from Government funding, administered by the Royal Society of New Zealand.

References

Abbott, I.A. and G. Hollenberg. 1976. Marine Algae of California. Stanford University Press, Stanford, California.

Airoldi, L. 2003. The effects of sedimentation on rocky coast assemblages. Oceanography and Marine Biology Annual Review 41: 161–236.

Airoldi, L. and F. Cinelli. 1997. Effects of sedimentation on subtidal macroalgal assemblages: an experimental study from mediterranean rocky shore. J. Exp. Mar. Biol. Ecol. 215: 269–288.

Airoldi, L. and M. Virgilio. 1998. Responses of turf-forming algae to spatial variations in the deposition of sediments. Mar. Ecol. Prog. Ser. 165: 271–282.

Albins, M.A. and M.A. Hixon. 2008. Invasive Indo-Pacific lionfish *Pterois volitans* reduce recruitment of Atlantic coral-reef fishes. Mar. Ecol. Prog. Ser. 367: 233–238.

Alcoverro, T., C.M. Duarte and J. Romero. 1997. The influence of herbivores on *Posidonia oceanica* epiphytes. Aquat. Bot. 56: 93–104.

Amsler, C.D., M.O. Amsler, J.B. McClintock and B.J. Baker. 2009. Filamentous algal endophytes in macrophytic Antarctic algae: prevalence in hosts and palatability to mesoherbivores. Phycologia 48: 324–334.

Andersen, T., J. Carstensen, E. Hernandez-Garzia and C.M. Duarte. 2009. Ecological thresholds and regime shifts: approaches to identification. Trends in Ecology and Evolution 24: 49–57.

Anderson, L. 2012. Cost and benefits of intertidal algal epiphytism. M.S. Thesis. University of British Columbia, Vancouver.

Arrontes, J. 1990. Composition, distribution on host, and seasonality of epiphtyes on three intertidal algae. Bot. Mar. 33: 205–211.

Ávila, E., M. del Carmen Méndez-Trejo, R. Riosmena-Rodríguez, J.M. López-Vivas and A. Sentíes. 2012. Epibiotic traits of the invasive red seaweed *Acanthophora spicifera* in La Paz Bay, South Baja California (Eastern Pacific). Mar. Ecol. 33: 470–480.

Balata, D., L. Piazzi and F. Cinelli. 2007. Increase of sedimentation in a subtidal system: Effects on the structure and diversity of macroalgal assemblages. J. Exp. Mar. Biol. Ecol. 351: 73–82.

Bernstein, B.B. and N. Jung. 1979. Selective pressures and coevolution in a kelp canopy community in southern California. Ecol. Monogr. 49: 335–355.

Bertness, M.D. and R. Callaway. 1994. Positive interactions in communities. Trends in Ecology and Evolution 9: 191–193.

Bibby, R., P. Cleall-Harding, S. Rundle, S. Widdlcombe and J. Spicer. 2007. Ocean acidification disrupts induced defences in the intertidal gastropod *Littorina littorea*. Biol. Lett. 3: 699–701.

Bologna, P.A. and K.L. Heck. 1999a. Macrofaunal associations with seagrass epiphytes—Relative importance of trophic and structural characteristics. Journal of Experimental Marine Biology and Ecology 242: 21–39.

Bologna, P.A.X. and K.L. Heck. 1999b. Macrofaunal associations with seagrass epiphytes: relative importance of trophic and structural characteristics. J. Exp. Mar. Biol. Ecol. 242: 21–39.

Borowitzka, M.A., P.S. Lavery and M. van Keulen. 2006. Epiphytes of seagrasses. pp. 441–461. *In*: Larkum, A.W.D., R.J. Orth and C.M. Duarte (eds.). Seagrasses: Biology, Ecology, and Conservation. Springer, Dordrecht.

Borum, J., H. Kaas and S. Wium-Anderson. 1984. Biomass variation and autotrophic production of an epiphyte-macrophyte community in a coastal Danish area: II. Epiphyte species composition, biomass and production. Ophelia 23: 165–179.

Boström, C. and J. Mattila. 1999. The relative importance of food and shelter for seagrass-associated invertebrates: a latitudinal comparison of habitat choise by isopod grazers. Oecologia 120: 162–170.

Boström, C. and J. Mattila. 2005. Effects of isopod grazing: an experimental comparison in temperate (*Idotea balthica*, Baltic Sea, Finland) and subtropical (*Erichsonella attenuata*, Gulf of Mexico, USA) ecosystems. Crustaceana 78: 185–200.

Bouarab, K., P. Potin, F. Weinberger, J. Correa and B. Kloareg. 2001. The *Chondrus crispus-Acrochaete operculata* host-pathogen association, a novel model in glycobiology and applied phycopathology. J. Appl. Phycol. 13: 185–193.

Bracken, M.E., C.A. Gonzalez-Dorantes and J.J. Stachowicz. 2007. Whole-community mutualism: associated invertebrates facilitate a dominant habitat-forming seaweed. Ecology 88: 2211–2219.

Brawley, S.H. 1992. Mesoherbivores. Plant-animal interactions in the marine benthos. Clarendon Press, Oxford: 235–263.

Brouns, J.J.W.M. and F.M.L. Heijs. 1986. Production and biomass of the seagrass *Enhalus acaroides* (Lf) Royle and its epiphytes. Aquat. Bot. 25: 21–45.

Bruno, J.F., J.J. Stachowicz and M.D. Bertness. 2003. Inclusion of facilitation into ecological theory. Trends Ecol. Evol. 18: 119–125.

Bruno, J.F. and M.I. O'Connor. 2005. Cascading effects of predator diversity and omnivory in a marine food web. Ecol. Lett. 8: 1048–1056.

Bulleri, F. and M.G. Chapman. 2010. The introduction of coastal infrastructure as a driver of change in marine environments. J. Appl. Ecol. 47: 26–35.

Burke, C., T. Thomas, M. Lewis, P. Steinberg and S. Kjelleberg. 2011. Composition, uniqueness and variability of the epiphytic bacterial community of the green alga *Ulva australis*. Isme Journal 5: 590–600.

Burkepile, D.E. and M.E. Hay. 2006. Herbivore vs. nutrient control of marine primary producers: context-dependent effects. Ecology 87: 3128–3139.

Butman, C.A. 1987. Larval settlement of soft-sediment invertebrates—the spatial scales of pattern explained by active habitat selection and the emerging role of hydrodynamical processes Oceanogr. Mar. Biol. Annu. Rev. 25: 113–165.

Caddy, J., J. Csirke, S. Garcia and R. Grainger. 1998. How pervasive is "fishing down marine food webs"? Science 282: 1383–1383.

Cambridge, M.L., A.W. Chiffings, C. Brittan, L. Moore and A.J. McComb. 1986. The loss of seagrass in Cockburn Sound, Western Australia. II. Possible causes of seagrass decline. Aquat. Bot. 24: 269–285.

Cancino, J.M., M. Muñoz and M.C. Orellana. 1987. Effects of the bryozoan *Membranipora tuberculata* (Bose) on the photosynthesis and growth of *Gelidium rex* Santelices et Abbott. J. Exp. Mar. Biol. Ecol. 113: 105–112.

Cardoso, P.G., M.A. Pardal, D. Raffaelli, A. Baeta and J.C. Marques. 2004. Macroinvertebrate response to different species of macroalgal mats and the role of disturbance history. J. Exp. Mar. Biol. Ecol. 308: 207–220.

Carlton, J.T. 1996. Biological invasions and cryptogenic species. Ecology 77: 1653–1655.

Cebrian, J., S. Enriquez, M. Fortes, N. Agawin, J.E. Vermaat and C.M. Duarte. 1999. Epiphyte accrual on *Posidonia oceanica* (L.) Delile leaves: implications for light absorption. Bot. Mar. 42: 123–128.

Cebrian, J., J. Stutes and B. Christiaen. 2013. Effects of grazing and fertilization on epiphyte growth dynamics under moderately eutrophic conditions: implications for grazing rate estimates. Mar. Ecol. Prog. Ser. 474: 121–133.

Cerrano, C., S. Puce, M. Chiantore, G. Bavestrello and R. Cattaneo-Vietti. 2001. The influence of the epizoic hydroid *Hydractinia angusta* on the recruitment of the Antarctic scallop *Adamussium colbecki*. Polar Biol. 24: 557–581.

Chavanich, S. and L.G. Harris. 2002. The influence of macroalgae on seasonal abundance and feeding preference of a subtidal snail, *Lacuna vincta* (Montagu) (Littorinidae) in the Gulf of Maine. J. Molluscan Stud. 68: 73–78.

Cho, J.Y., E. Kwon, J. Choi, S. Hong and H. Shin. 2001. Antifouling activity of seaweed extracts on the green alga *Enteromorpha prolifera* and the mussel *Mytilus edulis*. J. Appl. Phycol. 13: 117–125.

Cook, K., M.A. Vanderklift and A.G.B. Poore. 2011. Strong effects of herbivorous amphipods on epiphyte biomass in a temperate seagrass meadow. Mar. Ecol. Prog. Ser. 442: 263–269.

Craigie, J.S., J. Correa and M.E. Gordon. 1992. Cuticles from *Chondrus crispus* (Rhodophyta). J. Phycol. 28: 777–786.

Cummins, S.P., D.E. Roberts and K.D. Zimmerman. 2004. Effects of the green macroalgae *Enteromorpha intestinalis* on macrobenthic and seagrass assemblages in a shallow coastal estuary. Mar. Ecol. Prog. Ser. 266: 77–87.

D'Antonio, C. 1985. Epiphytes on the rocky intertidal red alga *Rhodomela larix* (Turner) C. Agardh: Negative effects on the host and food for herbivores? J. Exp. Mar. Biol. Ecol. 86: 197–218.

Daleo, P., M. Escapa, J. Alberti and O. Iribarne. 2006. Negative effects of an autogenic ecosystem engineer: interactions between coralline turf and an ephemeral green alga. Mar. Ecol. Prog. Ser. 315: 67–73.

Dawes, C.J., B.W. Teasdale and M. Friedlander. 2000. Cell wall structure of the agarophytes *Gracilaria tikvahiae* and *G. cornea* (Rhodophyta) and penetration by the epiphyte *Ulva lactuca* (Chlorophyta). J. Appl. Phycol. 12: 567–575.

Dayton, P. 1971. Competition, disturbance, and community organization: the provision and subsequent utilization of space in a rocky intertidal community. Ecological Monographs 41: 351–389.

Dayton, P.K. 1972. Toward an understanding of community resilience and the potential effects of enrichment to the benthos of McMurdo Sound, Antarctica. pp. 81–95. *In*: Proceedings of the Colloquium on Conservation Problems in Antarctica. Allen Press, Lawrence, KS.

de Bettignies, T., M.S. Thomsen and T. Wernberg. 2012. Wounded kelps: patterns and susceptibility to breakage. Aquatic Biology 17: 223–233.

De Burgh, M.E. and P.V. Fankboner. 1978. A nutritional association between the bull kelp *Nereocystis luetkeana* and its epizoic bryozoan. Oikos 31: 69–72.

den Hartog, C. 1972. Substratum: plants-multicellular plants. Mar. Ecol. 1: 1277–1289.

Deudero, S., A. Blanco, A. Box, G. Mateu-Vicens, M. Cabanellas-Reboredo and A. Sureda. 2010. Interaction between the invasive macroalga *Lophocladia lallemandii* and the bryozoan *Reteporella grimaldii* at seagrass meadows: density and physiological responses. Biol. Invasions 12: 41–52.

Dijkstra, J., H. Sherman and L.G. Harris. 2007. The role of colonial ascidians in altering biodiversity in marine fouling communities. J. Exp. Mar. Biol. Ecol. 342: 169–171.

Dixon, J., S.C. Schroeter and J. Kastendiek. 1981. Effects of the encrusting bryozoan, *Membranipora membranacea*, on the loss of blades and fronds by the giant kelp, *Macrocystis pyrifera* (Laminariales). J. Phycol. 17: 341–345.

Douglass, J.G., J.E. Duffy and J.F. Bruno. 2008. Herbivore and predator diversity interactively affect ecosystem properties in an experimental marine community. Ecol. Lett. 11: 598–608.

Drake, L.A., F.C. Dobbs and R.C. Zimmerman. 2003. Effects of epiphyte load on optical properties and photosynthetic potential of the seagrasses *Thalassia testudinum* Banks ex Konig and *Zostera marina* L. Limnol. Oceanogr. 48: 456–463.

Drouin, A., C.W. McKindsey and L.E. Johnson. 2011. Higher abundance and diversity in faunal assemblages with the invasion of *Codium fragile* ssp. *fragile* in eelgrass meadows. Mar. Ecol. Prog. Ser. 424: 105–117.

Duarte, C.M. 1995. Submerged aquatic vegetation in relation to different nutrient regimes. Ophelia 41: 87–112.

Duffy, J.E. 1990. Amphipods on seaweeds: partners or pests? Oecologia 83: 267–276.

Duffy, J.E. and M.E. Hay. 2000. Strong impacts of grazing amphipods on the organization of a benthic community. Ecol. Monogr. 70: 237–263.

Duffy, J.E. and A.M. Harvilicz. 2001. Species-specific impacts of grazing amphipods in an eelgrass-bed community. Mar. Ecol. Prog. Ser. 223: 201–211.

Duffy, J.E. and M.E. Hay. 2001. The ecology and evolution of marine consumer-prey interactions. pp. 131–157. *In*: Bertness, M.D., S.D. Gaines and M.E. Hay (eds.). Marine Community Ecology. Sinauer Associates, Inc., Sunderland, Massachussetts.

Eastwood, M.M., M.J. Donahue and A.E. Fowler. 2007. Reconstructing past biological invasions: niche shifts in response to invasive predators and competitors. Biol. Invasions 9: 397–407.

Edgar, G. and M. Aoki. 1993. Resource limitation and fish predation: their importance to mobile epifauna associated with Japanese *Sargassum*. Oecologia 95: 122–133.

Edgar, G.J. 1990. The influence of plant structure on the species richness, biomass, and secondary production of macrofaunal assemblages associated with Western Australia seagrass beds. J. Exp. Mar. Biol. Ecol. 137: 215–240.

Edgar, G.J. and A.I. Robertson. 1992. The influence of seagrass structure on the distribution and abundance of mobile epifauna: pattern and processes in a Western Australian *Amphibolis* bed. J. Exp. Mar. Biol. Ecol. 160: 13–31.

Edgar, G.J. and C. Shaw. 1995. The production and trophic ecology of shallow-water fish assemblages in southern Australia III. General relationships between sediments, seagrasses, invertebrates and fishes. J. Exp. Mar. Biol. Ecol. 194: 107–131.

Engelen, A.H., A.L. Primo, T. Cruz and R. Santos. 2013. Faunal differences between the invasive brown macroalga *Sargassum muticum* and competing native macroalgae. Biol. Invasions 15: 171–183.

Eriksson, B.K. and G. Johansson. 2003. Sedimentation reduces recruitment success of *Fucus vesiculosus* (Phaeophyceae) in the Baltic Sea. Eur. J. Phycol. 38: 217–222.

Eriksson, B.K. and G. Johansson. 2005. Effects of sedimentation on macroalgae: species-specific responses are related to reproductive traits. Oecologia 143: 438–448.

Filion-Myklebust, C. and T.A. Norton. 1981. Epidermis shedding in the brown seaweed *Ascophyllum nodosum* (L.) Le Jolis and its ecological significance. Marine Biology Letters 1981: 45–51.

Fletcher, R. and M. Callow. 1992. The settlement, attachment, and establishment of marine algal spores. Eur. J. Phycol. 27: 303–329.

Fong, P., J.S. Desmond and J.B. Zedler. 1997. The effect of a horn snail on *Ulva expansa* (Chlorophyta): Consumer or facilitator of growth? J. Phycol. 33: 353–359.

Fredriksen, S., H. Christie and B. Sæthre. 2007. Species richness in macroalgae and macrofauna assemblages on *Fucus serratus* L. (Phaeophyceae) and *Zostera marina* L. (Angiospermae) in Skagerrak, Norway. Mar. Biol. Res. 1: 2–19.

Fricke, A., T.V. Titlyanova, M.M. Nugues and K. Bischof. 2011. Depth-related variation in epiphytic communities growing on the brown alga *Lobophora variegata* in a Caribbean coral reef. Coral Reefs 30: 967–973.

Garbary, D.J., R.J. Deckert and C.B. Hubbard. 2005. *Ascophyllum* and its symbionts. VII. Three-way interactions among *Ascophyllum nodosum* (Phaeophyceae), *Mycophycias ascophylli* (Ascomycetes) and *Vertebrata lanosa* (Rhodophyta). Algae 20: 353–361.

Gaylord, B., C.A. Blanchette and M.W. Denny. 1994. Mechanical consequences of size in wave swept algae. Ecol. Monogr. 64: 287–313.

Gennaro, P. and L. Piazzi. 2011. Synergism between two anthropic impacts: *Caulerpa racemosa* var. *cylindracea* invasion and seawater nutrient enrichment Mar. Ecol. Prog. Ser. 427: 59–70.

Gonzalez, M.A. and L.J. Goff. 1989. The red algal epiphytes *Microcladia coulterii* and *M. californica* (Rhodophyceae: Ceramiaceae). II. Basiphyte specificity. J. Phycol. 25: 558–567.

Guerra-García, J.M., M. Ros, D. Izquierdo and M.M. Soler-Hurtado. 2012. The invasive *Asparagopsis armata* versus the native *Corallina elongata*: Differences in associated peracarid assemblages. J. Exp. Mar. Biol. Ecol.: 121–128.

Hacker, S.D. and R.S. Steneck. 1990. Habitat architecture and the abundance and body-size-dependent habitat selection of a phytal amphipod. Ecology 71: 2269–2285.

Hadfield, M.G. 1986. Settlement and recruitment of marine invertebrates: A perspective and some proposals. Bull. Mar. Sci. 39: 418–425.

Hall, M. and S. Bell. 1988. Response of small motile epifauna to complexity of epiphytic algae on seagrass blades. J. Mar. Res. 46: 613–630.

Hallam, N.D., M.N. Clayton and D. Parish. 1980. Studies on the association between *Notheia anomala* and *Hormosira banksii* (Phaeophyta). Aust. J. Bot. 28: 239–248.

Hansen, G.I. 1986. A newly discovered host of the seagrass epiphyte *Smithora naiadum* (Bangiophyceae, Rhodophyta). Canadian Journal of Botany 64: 900–901.

Harder, T. 2008. Marine epibioisis: concepts, ecological consequences, and host defence. Springer Series on Biofilms 4: 219–231.

Harlin, M. 1973. 'Obligate' algal epiphyte: *Smithora naiadum* grows on synthetic substrate. J. Phycol. 9: 230–232.

Harlin, M. 1980. Seagrass epiphytes. pp. 117–152. *In*: Phillips R.C. and G.P. McRoy (eds.). Handbook of Seagrass Biology: An Ecosystem Perspective. Taylor & Frances, New York.

Hauxwell, J., J. Cebrian and I. Valiela. 2003. Eelgrass *Zostera marina* loss in temperate estuaries: relationship to land-derived nitrogen loads and effect of light limitation imposed by algae. Mar. Ecol. Prog. Ser. 247: 59–73.

Hay, M.E. 1986. Associational plant defenses and the maintenance of species diversity: turning competitors into accomplices. Am. Nat. 128: 617–641.

Hay, M.E., J.E. Duffy, C.A. Pfister and W. Fenical. 1987. Chemical defense against different marine herbivores: Are amphipods insect equivalents? Ecology 68: 1567–1580.

Hay, M.E., J.D. Parker, D.E. Burkepile, C.C. Caudill, A.E. Wilson, Z.P. Hallinan and A.D. Chequer. 2004. Mutualisms and aquatic community structure: the enemy of my enemy is my friend. Annual Review of Ecology, Evolution, and Systematics 35: 175–197.

Heck, K.L., J.R. Pennock and J.F. Valentine. 2000. Effects of nutrient enrichment and small predator density on seagrass ecosystems: An experimental assessment. Limnol. Oceanogr. 45: 1041–1057.

Hellio, C., J.-P. Marechal, B. Véron, G. Bremer, A.S. Clare and Y. Le Gal. 2004. Seasonal variation of antifouling activities of marine algae from the Brittany coast (France). Mar. Biotechnol. 6: 67–82.

Hemmi, A., A. Makinen, V. Jormalainen and T. Honkanen. 2005. Responses of growth and phlorotannins in *Fucus vesiculosus* to nutrient enrichment and herbivory. Aquat. Ecol. 39: 201–211.

Hepburn, C. and C. Hurd. 2005. Conditional mutualism between the giant kelp *Macrocystis pyrifera* and colonial epifauna. Mar. Ecol. Prog. Ser. 302: 37–48.

Hepburn, C., C. Hurd and R.D. Frew. 2006. Colony structure and seasonal differences in light and nitrogen modify the impact of sessile epifauna on the giant kelp *Macrocystis pyrifera* (L.) C. Agardh. Hydrobiologia 560: 373–384.

Hewitt, C.L., M.L. Campbell, R.E. Thresher, R.B. Martin, S. Boyd, B.R. Cohen, D.R. Currie, M.F. Gomon, M.J. Keough, J.A. Lewis, M.M. Lockett, N. Mays, M.A. McArthur, T. O'Hara, G.C. Poore, J. Ross, M.J. Storey, J.E. Watson and R.S. Wilson. 2004. Introduced and cryptogenic species in Port Phillip Bay, Victoria, Australia. Mar. Biol. 144: 183–202.

Hily, C., S. Connan, C. Raffin and S. Wyllie-Echeverria. 2004. *In vitro* experimental assessment of the grazing pressure of two gastropods on *Zostera marina* L. epiphytic algae. Aquat. Bot. 78: 183–195.

Hoeffle, H., M.S. Thomsen and M. Holmer. 2011. High mortality of *Zostera marina* under high temperature regimes but minor effects of the invasive macroalgae *Gracilaria vermiculophylla*. Estuar. Coast. Shelf Sci. 92: 35–46.

Hoeffle, H., T. Wernberg, M.S. Thomsen and M. Holmer. 2012. Drift algae, an invasive snail and elevated temperature reduces the ecological performance of a warm-temperate seagrass via additive effects. Mar. Ecol. Prog. Ser. 450: 67–80.

Holmer, M. and R.M. Nielsen. 2007. Effects of filamentous algal mats on sulfide invasion in eelgrass (*Zostera marina*). J. Exp. Mar. Biol. Ecol. 353: 245–252.

Holmer, M., P. Wirachwong and M.S. Thomsen. 2011. Negative effects of stress-resistant drift algae and high temperature on a small ephemeral seagrass species. Mar. Biol. 158: 297–309.

Holmquist, J.G. 1997. Disturbance and gap formation in a marine benthic mosaic—influence of shifting macroalgal patches on seagrass structure and mobile invertebrates. Mar. Ecol. Prog. Ser. 158: 121–130.

Honkanen, T. and V. Jormalainen. 2005. Genotypic variation in tolerance and resistance to fouling in the brown alga *Fucus vesiculosus*. Oecologia 144: 196–205.

Hootsmans, M.J.M. and J.E. Vermaat. 1985. The effect of periphyton-grazing by three epifaunal species on the growth of *Zostera marina* L. under experimental conditions. Aquat. Bot. 22: 83–88.

Howard, R.K. 1982. Impact of feeding activities of epibenthic amphipods on surface-fouling of eelgrass leaves. Aquat. Bot. 14: 91–97.

Hughes, R.G., S. Johnson and I.D. Smith. 1991. The growth patterns of some hydroids that are obligate epiphytes of seagrass leaves. Hydrobiologia 216-217: 205–210.

Ilken, K. 2012. Grazers on benthic seaweeds. pp. 157–175. *In*: Wiencke, C. and K. Bischof (eds.). Seaweed Biology. Springer, Berlin Heidelberg.

Irving, A.D. and S.D. Connell. 2002. Interactive effects of sedimentation and microtopography on the abundance of subtidal turf-forming algae. Phycologia 41: 517–522.

Jackson, J.B.C. 1997. Reefs since Columbus. Coral Reefs 16: S23–S32.

Janiak, D.S. and R.B. Whitlatch. 2012. Epifaunal and algal assemblages associated with the native *Chondrus crispus* (Stackhouse) and the non-native *Grateloupia turuturu* (Yamada) in eastern Long Island Sound. J. Exp. Mar. Biol. Ecol. 413: 38–44.

Jernakoff, P. and J. Nielsen. 1997. The relative importance of amphipod and gastropod grazers in *Posidonia sinuosa* meadows. Aquat. Bot. 56: 183–202.

Jernakoff, P., A. Brearley and J. Nielsen. 1996. Factors affecting grazer-epiphyte interactions in temperate seagrass meadows. Oceanogr. Mar. Biol. Annu. Rev. 34.

Joint, I., K. Tait, M. Callow, J.A. Callow, D. Milton, P. Williams and M. Camara. 2002. Cell-to-cell communication across the prokaryote-eukaryote boundary. Science 298: 1207.

Jones, E. 2007. Impacts of habitat-modifying invasive macroalgae on epiphytic algal communities. M.S. Thesis. University of Rhode Island, Kingston, Rhode Island.

Jones, E. and C.S. Thornber. 2010. Effects of habitat-modifying invasive macroalgae on epiphytic algal communities. Mar. Ecol. Prog. Ser. 400: 87–100.

Jormalainen, V., T. Honkanen and N. Heikkilä. 2001a. Feeding preferences and performance of a marine isopod on seaweed hosts: cost of habitat specialization. Mar. Ecol. Prog. Ser. 220: 219–230.

Jormalainen, V., T. Honkanen, A. Mäkinen, A. Hemmi and O. Vesakoski. 2001b. Why does herbivore sex matter? Sexual differences in utilization of *Fucus vesiculosus* by the isopod *Idotea baltica*. Oikos 93: 77–86.

Kangas, P., H. Autio, G. Hälfors, H. Luther, Å. Niemi and H. Salemaa. 1982. A general model of the decline of *Fucus vesiculosus* at Tvärminne, south coast of Finland in 1977–81. Acta Bot. Fenn. 118: 1–27.

Karez, R., S. Engelbert and U. Sommer. 2000. 'Co-consumption' and 'protective coating': two new proposed effects of epiphytes on their macroalgal hosts in mesograzer-epiphyte-host interactions. Mar. Ecol. Prog. Ser. 205: 85–93.

Kato, T., A.S. Kumanireng, I. Ichinose, Y. Kitahara, Y. Kakinuma, M. Nishihira and M. Kato. 1975. Active components of *Sargassum tortile* effecting the settlement of *Coryne Uchidai*. Experientia 31: 433–434.

Kautsky, N., H. Kautsky, U. Kautsky and M. Waern. 1986. Decreased depth penetration of *Fucus vesiculosus* (L.) since the 1940's indicates eutrophication of the Baltic Sea. Mar. Ecol. Prog. Ser. 28: 1–8.

Kendrick, G., D. Walker and J. Arthur. 1988. Changes in distribution of macro-algal epiphytes on stems of the seagrass *Amphibolis antarctica* along a salinity gradient in Shark Bay, Western Australia. Phycologia 27: 201–208.

Keough, M.J. 1986. The distribution of the bryozoan *Bugula neritina* on seagrass blades: settlement, growth, and mortality. Ecology 67: 846–657.

Kersen, P., J. Kotta, M. Bučas, N. Kolesova and Z. Dekere. 2011. Epiphytes and associated fauna on the brown alga *Fucus vesiculosus* in the Baltic and the North Seas in relation to different abiotic and biotic variables. Mar. Ecol. 32: 87–95.

Kim, M.J., J.S. Choi, S.E. Kang, J.Y. Cho, H. Jin, B.S. Chun and Y.K. Hong. 2004. Multiple allelopathic activity of the crustose coralline alga *Lithophyllum yessoense* against settlement and germination of seaweed spores. J. Appl. Phycol. 16: 175–179.

Klumpp, D.W., J.S. Salita-Espinosa and M.D. Fortes. 1992. The role of epiphytic periphyton and macroinvertebrate grazers in the trophic flux of a tropical seagrass community. Aquat. Bot. 43: 327–349.

Koch, M.S., S. Schopmeyer, C. Kyhn-Hansen and C.J. Madden. 2007. Synergistic effects of high temperature and sulfide on tropical seagrass. J. Exp. Mar. Biol. Ecol. 341: 91–101.

Kotta, J., T. Paalme, G. Martin and A. Mäkinen. 2000. Major changes in macroalgae community composition affect the food and habitat preference of *Idotea baltica*. Int. Rev. Hydrobiol. 85: 697–705.

Kraberg, A.C. and T.A. Norton. 2007. Effect of epiphytism on reproductive and vegetative lateral formation in the brown, intertidal seaweed *Ascophyllum nodosum* (Phaeophyceae). Phycol. Res. 55: 17–24.

Krause-Jensen, D., P.B. Christensen and S. Rysgaard. 1999. Oxygen and nutrient dynamics within mats of the filamentous macroalgae *Chaetomorpha linum*. Estuaries 22: 31–38.

Lachnit, T., D. Meske, M. Wahl, T. Harder and R. Schmitz. 2011. Epibacterial community patterns on marine macroalgae are host-specific but temporally variable. Environ. Microbiol. 13: 655–665.

Laihonen, P. and E.R. Furman. 1986. The site of settlement indicates commensalism between blue mussel and its epibiont. Oecologia 71: 38–40.

Lapointe, B.E., P.J. Barile and W.R. Matzie. 2004. Anthropogenic nutrient enrichment of seagrass and coral reef communities in the Lower Florida Keys: discrimination of local versus regional nitrogen sources. J. Exp. Mar. Biol. Ecol. 308: 23–58.

Lavery, P.S. and M.A. Vanderklift. 2002. A comparison of spatial and temporal patterns in epiphytic macroalgal assemblages of the seagrasses *Amphibolis griffithii* and *Posidonia coriacea*. Mar. Ecol. Prog. Ser. 236: 99–112.

Leite, F.P.P. and A. Turra. 2003. Temporal variation in *Sargassum* biomass, *Hypnea* epiphytism and associated fauna. Brazilian Archives of Biology and Technology 46: 665–671.

Leonardi, P.I., A.B. Miravalles, S. Faugeron, V. Flores, J. Beltrán and J. Correa. 2006. Diversity, phenomenology and epidemiology of epiphytism in farmed *Gracilaria chilensis* (Rhodophyta) in northern Chile. Eur. J. Phycol. 41: 247–257.

Linskens, H.F. 1963. Beitrag zur Frage der Beziehungen zwischen Epiphyte und Basiphyt bei marinen Algen. Publ. Staz. Zool. Napoli 33: 274–293.

Littler, M.M. 1980. Morphological form and photosynthetic performances of marine macroalgae: Tests of a functional form hypothesis. Bot. Mar. 23: 161–166.

Lobban, C.S. and D.M. Baxter. 1983. Distribution of the red algal epiphyte *Polysiphonia lanosa* and its brown algal host *Ascophyllum nodosum* in the Bay of Fundy. Bot. Mar. 26: 533–538.

Lobelle, D., E.J. Kenyon, K. Cook and J.C. Bull. 2013. Local competition and metapopulation processes drive long-term seagrass-epiphyte population dynamics. PLoS ONE 8: e57072.

Longtin, C.M. and R.A. Scrosati. 2009. Role of surface wounds and brown algal epiphytes in the colonization of *Ascophyllum nodosum* (Phaeophyceae) fronds by *Vertebrata lanosa* (Rhodophyta). J. Phycol. 45: 535–539.

Longtin, C.M., R.A. Scrosati, G.B. Whalen and D.J. Garbary. 2009. Distribution of algal epiphytes across environmental gradients at different scales: intertidal elevation, host canopies, and host fronds. J. Phycol. 45: 820–827.

Lotze, H.K., B. Worm and U. Sommer. 2000. Propagule banks, herbivory and nutrient supply control population development and dominance patterns in macroalgal blooms. Oikos 89: 46–58.

Lubchenco, J. 1978. Plant species diversity in a marine intertidal community: Importance of herbivore food preference and algal competitive abilities. The American Naturalist 112: 23–39.

Martin, S. and J.P. Gattuso. 2009. Response of Mediterranean coralline algae to ocean acidification and elevated temperature. Global Change Biol. 15: 2089–2100.

Martin-Smith, K.M. 1993. Abundance of mobile epifauna: the role of habitat complexity and predation by fishes. Journal of Experimental Marine Biology and Ecology 174: 243–260.

Mathieson, A.C., C.J. Dawes, L.G. Harris and E.J. Hehre. 2003. Expansion of the Asiatic green algal *Codium fragile* ssp. *tomentosoides* in the Gulf of Maine. Rhodora 105: 1–53.

McCarthy, A., R.W. Osman and R.B. Whitlatch. 2007. Effects of temperature on growth rates of colonial ascidians: A comparison of *Didemnum* sp. to *Botryllus schlosseri* and *Botrylloides violaceus*. J. Exp. Mar. Biol. Ecol. 342: 172–174.

McGlathery, K. 2001. Macroalgal blooms contribute to the decline in seagrasses in nutrient-enriched coastal waters. J. Phycol. 37: 453–456.

Miller, L.P. and B. Gaylord. 2007. Barriers to flow: the effects of experimental cage structures on water velocities in high-energy subtidal and intertidal environments. J. Exp. Mar. Biol. Ecol. 344: 215–228.

Miller, M.W. 1998. Coral/seaweed competition and the control of community structure within and between latitudes. Oceanography and Marine Biology: An Annual Review 36: 65–96.

Moss, B.L. 1982. The control of epiphtyes by *Halidrys siliquosa* (L.) Lyngb. (Phaeophyta, Cystoseiraceae). Phycologia 21: 185–191.

Mukai, H. and A. Iijima. 1995. Grazing effects of a gammaridean amphipod, *Ampithoe* sp., on the seagrass, *Syringodium isoetifolium*, and epiphytes in a tropical seagrass bed of Fiji. Ecol. Res. 10: 243–257.

Nakaoka, M. 2002. Predation on seeds of seagrasses *Zostera marina* and *Zostera caulescens* by a tanaid crustacean *Zeuxo* sp. Aquat. Bot. 72: 99–106.

Norkko, J., E. Bonsdorff and A. Norkko. 2000. Drifting algal mats as an alternative habitat for benthic invertebrates: Species specific responses to a transient resource. J. Exp. Mar. Biol. Ecol. 248: 79–104.

Norton, T.A. and M.R. Benson. 1983. Ecological interactions between the brown seaweed *Sargassum muticum* and its associated fauna. Mar. Biol. 75: 169–177.

Nyberg, C.D., M.S. Thomsen and I. Wallentinus. 2009. Flora and fauna associated with the introduced red alga *Gracilaria vermiculophylla*. Eur. J. Phycol. 44: 395–403.

Nylund, G. and H. Pavia. 2005. Chemical versus mechanical inhibition of fouling in the red alga *Dilsea carnosa*. Mar. Ecol. Prog. Ser. 299: 111–121.

Nylund, G., P. Gribben, R. de Nys, P.D. Steinberg and H. Pavia. 2007. Surface chemistry versus whole-cell extracts: antifouling tests with seaweed metabolites. Mar. Ecol. Prog. Ser. 329: 73–84.

O'Connor, N.E. and J.F. Bruno. 2007. Predatory fish loss affects the structure and functioning of a model marine food web. Oikos 116: 2027–2038.

Orav-Kotta, H. and J. Kotta. 2004. Food and habitat choice of the isopod *Idotea baltica* in the northeastern Baltic Sea. Hydrobiologia 514: 79–85.

Orth, R., T.J.B. Carruthers, W.C. Dennison, C.M. Duarte, J.W. Fourqurean, J.K.L. Heck, A.R. Hughes, G. Kendrick, W.J. Kenworthy, S. Olyarnik, F.T. Short, M. Waycott and S.L. Williams. 2006. A global crisis for seagrass ecosystems. Bioscience 56: 987–996.

Orth, R.J. and J.V. Montfrans. 1984. Epiphyte-seagrass relationships with an emphasis on the role of micrograzing: a review. Aquat. Bot. 18: 43–69.

Paerl, H.W. 1995. Coastal eutrophication in relation to atmospheric nitrogen deposition: current perspectives. Ophelia 41: 237–259.

Paine, R.T. 1966. Food web complexity and species diversity. The American Naturalist 100: 65–75.

Paul, V.J., M.P. Puglisi and R. Ritson-Williams. 2006. Marine chemical ecology. Natural Product Reports 23: 153–180.

Pavia, H., H. Carr and P. Åberg. 1999. Habitat and feeding preferences of crustacean mesoherbivores inhabiting the brown seaweed *Ascophyllum nodosum* (L.) Le Jol. and its epiphytic macroalgae. J. Exp. Mar. Biol. Ecol. 236: 15–32.

Pearson, G.A. and L.F. Evans. 1990. Settlement and survival of *Polysiphonia lanosa* (Ceramiales) spores on *Ascophyllum nodosum* and *Fucus vesiculosus* (Fucales). J. Phycol. 26: 597–603.

Pedersen, M.F. and J. Borum. 1996. Nutrient control of algal growth in estuarine waters. Nutrient limitation and the importance of nitrogen requirements and nitrogen storage among phytoplanton and species of macroalgae. Mar. Ecol. Prog. Ser. 142: 261–272.

Pedersen, M.F. and J. Borum. 1997. Nutrient control of estuarine macroalgae: growth strategy and the balance between nitrogen requirements and uptake. Mar. Ecol. Prog. Ser. 161: 155–163.

Pereg-Gerk, L., N. Sar and Y. Lipkin. 2002. *In situ* nitrogen fixation associated with seagrasses in the Gulf of Elat (Red Sea). Aquat. Ecol. 36: 387–394.

Perrone, C. and E. Cecere. 1997. Regeneration and mechanisms of secondary attachment in *Solieria filiformis* (Gigartinales, Rhodophyta). Phycologia 36: 120–127.

Piazzi, L., D. Balata and F. Cinelli. 2002. Epiphytic macroalgal assemblages of *Posidonia oceanica* rhizomes in the western Mediterranean. Eur. J. Phycol. 37: 69–76.

Piazzi, L., D. Balata, G. Ceccherelli and F. Cinellia. 2005. Interactive effect of sedimentation and *Caulerpa racemosa* var. *cylindracea* invasion on macroalgal assemblages in the Mediterranean Sea. Estuar. Coast. Shelf Sci. 64: 467–474.

Poore, A.G. 1994. Selective herbivory by amphipods inhabiting the brown alga *Zonaria angustata*. Mar. Ecol. Prog. Ser. 107: 113–113.

Poore, A.G.B., A.H. Campbell and P.D. Steinberg. 2009. Natural densities of mesograzers fail to limit growth of macroalgae or their epiphytes in a temperate algal bed. J. Ecol. 97: 164–175.

Post, D.M. 2002. The long and short of food-chain length. Trends Ecol. Evol. 17: 269–277.

Potin, P. 2012. Intimate associations between epiphytes, endophytes, and parasites of seaweeds. pp. 203–234. *In*: Wiencke, C. and K. Bischof (eds.). Seaweed Biology. Springer-Verlag, Berlin.

Prado, P., T. Alcoverro, B. Martínez-Crego, A. Vergés, M. Pérez and J. Romero. 2007. Macrograzers strongly influence patterns of epiphytic assemblages in seagrass meadows. J. Exp. Mar. Biol. Ecol. 350: 130–143.

Reddy, K., W. DeBusk, R. DeLaune and M. Koch. 1993. Long-term nutrient accumulation rates in the Everglades. Soil Sci. Soc. Am. J. 57: 1147–1155.

Reusch, T.B.H. and A.R. Hughes. 2006. The emerging role of genetic diversity for ecosystem functioning: Estuarine macrophytes as models. Estuaries and Coasts 29: 159–164.

Rindi, F. and M.D. Guiry. 2004. Composition and spatio-temporal variability of the epiphytic macroalgal assemblage of *Fucus vesiculosus* Linnaeus at Clare Island, Mayo, western Ireland. J. Exp. Mar. Biol. Ecol. 311: 233–252.

Rohr, N.E., C.S. Thornber and E. Jones. 2011. Epiphyte and herbivore interactions impact recruitment in a marine subtidal system. Aquat. Ecol. 45: 213–219.

Ruesink, J.L. 2000. Intertidal mesograzers in field microcosms: linking laboratory feeding rates to community dynamics. J. Exp. Mar. Biol. Ecol. 248: 163–176.

Russel, B.D., J.A. Thompson, L.J. Falkenberg and S.D. Connell. 2009. Synergistic effects of climate change and local stressors: CO_2 and nutrient driven change in subtidal rocky habitats. Global Change Biol.

Russell, G. and C. Veltkamp. 1984. Epiphyte survival on skin-shedding macrophytes. Mar. Ecol. Prog. Ser. 18: 149–153.

Salemaa, H. 1987. Herbivory and microhabitat preferences of *Idotea* spp. (Isopoda) in the northern Baltic Sea. Ophelia 27: 1–15.

Sand-Jensen, K.A. 1977. Effect of epiphytes on eelgrass photosynthesis. Aquat. Bot. 3: 55–63.

Saunders, J.E., M.J. Attrill, S.M. Shaw and A.A. Rowden. 2003. Spatial variability in the epiphytic algal assemblages of *Zostera marina* seagrass beds. Mar. Ecol. Prog. Ser. 249: 107–115.

Saunders, M. and A. Metaxas. 2008. High recruitment of the introduced bryozoan *Membranipora membranacea* is associated with kelp bed defoliation in Nova Scotia, Canada. Mar. Ecol. Prog. Ser. 369: 139–151.

Schanz, A., P. Polte, H. Asmus and R. Asmus. 2000. Currents and turbulence as a top-down regulator in intertidal seagrass communities. Biol. Mar. Medit. 7: 278–281.

Schanz, A., P. Polte and H. Asmus. 2002. Cascading effects of hydrodynamics on an epiphyte-grazer system in intertidal seagrass beds of the Wadden Sea. Mar. Biol. 141: 287–297.

Scheibling, R.E. and P. Gagnon. 2009. Temperature-mediated outbreak dynamics of the invasive bryozoan *Membranipora membranacea* in Nova Scotian kelp beds. Mar. Ecol. Prog. Ser. 390: 1–13.

Schelske, C.L., D.J. Conley, E.F. Stoermer, T.L. Newberry and C. Campbell. 1986. Biogenic silica and phosphorus accumulation in sediments as indices of eutrophication in the Laurentian Great Lakes. Hydrobiologia 143: 79–86.

Schmidt, A.L. and R.E. Scheibling. 2006. A comparison of epifauna and epiphytes on native kelps (*Laminaria* species) and an invasive alga (*Codium fragile* ssp. *tomentosoides*) in Nova Scotia, Canada. Bot. Mar. 49: 315–330.

Schmitt, T.M., M. Hay and N. Lindquist. 1995. Constraints on chemically mediated coevolution: multiple functions for seaweed ecology metabolites. Ecology 76: 107–123.

Schneider, F.I. and K.H. Mann. 1991. Species specific relationships of invertebrates to vegetation in a seagrass bed. II, Experiments on the importance of macrophyte shape, epiphyte cover and predation. J. Exp. Mar. Biol. Ecol. 145: 119–139.

Schultze, K., K. Janke, A. Krüß and W. Weidemann. 1990. The macrofauna and macroflora assoiated with *Laminaria digitata* and *Laminaria hyperborea* at the island of Helgoland (German Bight, North Sea). Helgolander Meeresuntersuchungen 44: 39–51.

Seed, R. and R.J. O'Connor. 1981. Community organization in marine algal epifaunas. Annu. Rev. Ecol. Syst. 12: 49–74.

Shacklock, P.F. and R.W. Doyle. 1983. Control of epiphytes in seaweed cultures using grazers. Aquaculture 31: 141–151.

Short, F.T. and H.A. Neckles. 1999. The effects of global climate change on seagrasses. Aquat. Bot. 63: 169–196.

Sieburth, J.M. and J.L. Tootle. 1981. Seasonality of microbial fouling on *Asophyllum nodosum* (L.) Lejol., *Fucus vesiculosus* L., *Polysiphonia lanosa* (L.) Tandy and *Chondrus crispus* Stackh. J. Phycol. 17: 57–64.

Smale, D.A. and T. Wernberg. 2013. Extreme climatic event drives range contraction of a habitat-forming species. Proceedings of the Royal Society B: Biological Sciences 280: 2012–2829.

Spivak, A.C., E.A. Canuel, J.E. Duffy and J.P. Richardson. 2009. Nutrient enrichment and food web composition affect ecosystem metabolism in an experimental seagrass habitat. PloS ONE 4: e7473.

Stachowicz, J.J. 2001. Mutualism, facilitation, and the structure of ecological communities. BioSciences 51: 235–246.

Stachowicz, J.J. and M.E. Hay. 1996. Facultative mutualism between an herbivorous crab and a coralline alga: advantages of eating noxious seaweeds. Oecologia 105: 377–387.

Stachowicz, J.J. and M.E. Hay. 1999. Mutualism and coral persistence: the role of herbivore resistance to algal chemical defense. Ecology 80: 2085–2101.

Stachowicz, J.J. and R.B. Whitlatch. 2005. Multiple mutualists provide complementary benefits to their seaweed host. Ecology 86: 2418–2427.

Steneck, R.S. and L. Watling. 1982. Feeding capabilities and limitation of herbivorous molluscs: a functional group approach. Mar. Biol. 68: 299–319.

Steneck, R.S., M.H. Graham, B.J. Bourque, D. Corbett, J.M. Erlandson, J.A. Estes and M.J. Tegner. 2002. Kelp forest ecosystems: biodiversity, stability, resilience and future. Environ. Conserv. 29: 436–459.

Stewart, J.G. 1982. Anchor species and epiphytes in intertidal algal turf. Pac. Sci. 36: 45–60.

Stricker, S.A. 1989. Settlement and metamorphosis of the marine bryozoan *Membranipora membranacea*. Bull. Mar. Sci. 45: 387–405.

Suzuki, Y., T. Takabayashi, T. Kawaguchi and K. Matsunaga. 1998. Isolation of an allelopathic substance from the crustose coralline algae, *Lithophyllum* spp., and its effect on the brown alga *Laminaria religiosa* Miyabe (Phaeophyta). J. Exp. Mar. Biol. Ecol. 225: 69–77.

Taylor, P.D. and M.A. Wilson. 2003. Palaeoecology and evolution of marine hard substrate communities. Earth-Sci. Rev. 62: 1–103.

Taylor, R.B. 1998. Density, biomass and productivity of animals in four subtidal rocky reef habitats: the importance of small mobile invertebrates. Mar. Ecol. Prog. Ser. 172: 37–51.

Thomsen, M. 2004a. Macroalgal distribution patterns and ecological performances in a tidal coastal lagoon, with emphasis on the non-indigenous *Codium fragile* ssp. *tomentosoides*. Ph.D. Thesis. University of Virginia, Charlottesville, Virginia.

Thomsen, M.S. 2004b. Species, thallus size and substrate determine macroalgal break forces and break places in a low-energy soft-bottom lagoon. Aquat. Bot. 80: 153–161.

Thomsen, M.S. 2010. Experimental evidence for positive effects of invasive seaweed on native invertebrates via habitat-formation in a seagrass bed. Aquatic Invasions 5: 341–346.

Thomsen, M.S. and T. Wernberg. 2005. What affects the forces required to break or dislodge macroalgae? A minireview. Eur. J. Phycol. 40: 1–10.

Thomsen, M.S., T. Wernberg, P.A. Stæhr and M.F. Pedersen. 2006. Spatio-temporal distribution patterns of the invasive macroalga *Sargassum muticum* within a Danish *Sargassum*-bed. Helgol. Mar. Res. 60: 50–58.

Thomsen, M.S., T. Wernberg, A.H. Altieri, F. Tuya, D. Gulbransen, K.J. McGlathery, M. Holmer and B.R. Silliman. 2010. Habitat cascades: The conceptual context and global relevance of facilitation cascades via habitat formation and modification. Integr. Comp. Biol. 50: 158–175.

Thomsen, M.S., J.D. Olden, T. Wernberg, J.N. Griffin and B.R. Silliman. 2011. A broad framework to organize and compare ecological invasion impacts. Environ. Res. 111: 899–908.

Thomsen, M.S., T. de Bettignies, T. Wernberg, M. Holmer and B. Debeuf. 2012a. Harmful algae are not harmful to everyone. Harmful Algae 16: 74–80.

Thomsen, M.S., T. Wernberg, A.H. Engelen, F. Tuya, M.A. Vanderklift, M. Holmer, K.J. McGlathery, F. Arenas, J. Kotta and B.R. Silliman. 2012b. A meta-analysis of seaweed impacts on seagrasses: generalities and knowledge gaps. PLoS ONE 7: e28595.

Tomas, F., X. Turon and J. Romero. 2005. Effects of herbivores on a *Posidonia oceanica* seagrass meadow: importance of epiphytes. Mar. Ecol. Prog. Ser. 287: 115–125.

Tootle, J.L. 1974. The fouling of intertidal seaweeds: Seasonal species and cuticular regulation. M.S. Thesis. University of Rhode Island, Kingston.

Trautman, D.A. and M.A. Borowitzka. 1999. The distribution of the epiphytic organisms on *Posidonia australis* and *P. sinuosa*, two seagrasses with differing leaf morphology. Mar. Ecol. Prog. Ser. 179: 215–229.

Troell, M., L. Pihl, P. Roenbaek, H. Wennhage, T. Soederquist and N. Kautsky. 2005. Regime shifts and ecosystem services in Swedish coastal soft bottom habitats: when resilience is undesirable. Ecol. Soc. 10: 13 pp., online journal.

Trussell, G., P. Ewanchuk, M.D. Bertness and B.R. Silliman. 2004. Trophic cascades in rocky shore tide pools: distinguishing lethal and non-lethal effects. Oecologia 139: 427–432.

Valentine, J.F. and J.E. Duffy. 2006. The central role of grazing in seagrass ecology. pp. 463–501. *In*: Larkum, A.W.D., R.J. Orth and C.M. Duarte (eds.). Seagrasses: Biology, Ecology, and Conservation. Springer, Dordrecht, The Netherlands.

Valiela, I., J. McClelland, J. Hauxwell, P.J. Behr, D. Hersh and K. Foreman. 1997. Macroalgal blooms in shallow estuaries: controls and ecophysiological and ecosystem consequences. Limnol. Oceanogr. 42: 1105–1118.

van Montfrans, J., R.L. Wetzel and R.J. Orth. 1984. Epiphyte-grazer relationships in seagrass meadows: consequences for seagrass growth and production. Estuaries 7: 289–309.

Viejo, R.M. 1999. Mobile epifauna inhabiting the invasive *Sargassum muticum* and two local seaweeds in northern Spain. Aquat. Bot. 64: 131–149.

Wahl, M. 1989. Marine epibiosis. I. Fouling and antifouling: some basic aspects. Mar. Ecol. Prog. Ser. 58: 175–189.

Wahl, M. 2009. Epibiosis: Ecology, effects, and defences. pp. 61–72. *In*: Wahl, M. (ed.). Marine Hard Bottom Communities. Springer-Verlag, Berlin.

Wahl, M. and M. Hay. 1995. Associational resistance and shared doom: effects of epibiosis on herbivory. Oecologia 102: 329–240.

Waycott, M., C.M. Duarte, T.J.B. Carruthers, R.J. Orth, W.C. Dennison, S. Olyarnik, A. Calladine, J.W. Fourqurean, J.K.L. Heck, A.R. Hughes, G.A. Kendrick, W.J. Kenworthy, F.T. Short and S.L. Williams. 2009. Accelerating loss of seagrasses across the globe threatens coastal ecosystems. Proceedings of the National Academy of Sciences 106: 12377–12381.

Wernberg, T., M.S. Thomsen, P.A. Staerh and M.F. Pedersen. 2004. Epibiota communities of the introduced and indigenous macroalgal relatives *Sargassum muticum* and *Halidrys siliquosa* in Limfjorden (Denmark). Helgol. Mar. Res. 58: 154–161.

Wernberg, T., D.A. Smale, F. Tuya, M.S. Thomsen, T.J. Langlois, T. de Bettignies, S. Bennett and C.S. Rousseaux. 2012. An extreme climatic event alters marine ecosystem structure in a global biodiversity hotspot. Nature Climate Change 3: 78–82.

Whalen, M.A., J.E. Duffy and J.B. Grace. 2013. Temporal shifts in top-down versus bottom-up control of epiphytic algae in a seagrass ecosystem. Ecology 94: 510–520.

Wiencke, C. 1987. Respiration and photosynthesis in the intertidal alga *Cladophora rupestris* (L.) Kistz under fluctuating salinity regimes. J. Exp. Mar. Biol. Ecol. 114: 183–197.

Wikström, S.A. and L. Kautsky. 2004. Invasion of a habitat-forming seaweed: effects on associated biota. Biol. Invasions 6: 141–150.

Wilkström, S.A. and H. Pavia. 2004. Chemical settlement inhibition versus post-settlement mortality as an explanation for differential fouling of two congeneric seaweeds. Oecologia 138: 223–230.

Williams, B.S., J.E. Hughes and K. Hunter-Thomson. 2002. Influence of epiphytic algal coverage on fish predation rates in simulated eelgrass habitats. Biological Bulletin 203: 248–249.

Williams, G.A. and R. Seed. 1992. Interactions between macrofaunal epiphytes and their host algae. pp. 189–211. *In*: John, D.M., S.J. Hawkins and J.H. Price (eds.). Plant-animal Interactions in the Marine Benthos. Clarendon Press, Oxford.

Wilson, L. 1978. Epiphytes on *Codium fragile* (Chlorophyceae). M.S. Thesis. University of Rhode Island, Kingston, RI.

Worm, B. and U. Sommer. 2000. Rapid direct and indirect effects of a single nutrient pulse in a seaweed-epiphyte-grazer system. Mar. Ecol. Prog. Ser. 202: 283–288.

Worm, B., E.B. Barbier, N. Beaumont, J.E. Duffy, C. Folke, B.S. Halpern, J.B.C. Jackson, H.K. Lotze, F. Micheli, S.R. Palumbi, E. Sala, K.A. Selkoe, J.J. Stachowicz and R. Watson. 2006. Impacts of biodiversity loss on ocean ecosystem services. Science 314: 787–790.

York, P.H., B.P. Kelaher, D.J. Booth and M.J. Bishop. 2012. Trophic responses to nutrient enrichment in a temperate seagrass food chain. Mar. Ecol. Prog. Ser. 449: 291–296.

Zhang, Y., J. Mu, Y. Feng, J. Zhang, P.-J. Gu, Y. Wang, L.-F. Ma and Y.-H. Zhu. 2009. Broad-spectrum antimicrobial epiphytic and endophytic fungi from marine organisms: Isolation, bioassay, and taxonomy. Mar. Drugs 7: 97–112.

The Role of Floating Plants in Dispersal of Biota Across Habitats and Ecosystems

Martin Thiel[1,2,3,]* and *Ceridwen Fraser*[4]

When descends on the Atlantic
The gigantic
Storm-wind of the equinox,
Landward in his wrath he scourges
The toiling surges,
Laden with seaweed from the rocks...

Ever drifting, drifting, drifting
On the shifting
Currents of the restless main;
Till in sheltered coves, and reaches
Of sandy beaches,
All have found repose again.

From the poem 'Seaweed' by Henry Wadsworth Longfellow

[1] Universidad Católica del Norte, Larrondo 1281, 1781421 Coquimbo, Chile.
[2] MN Ecology and Sustainable Management of Oceanic Island (ESMOI), Coquimbo, Chile.
[3] Centro de Estudios Avanzados en Zonas Áridas CEAZA, Coquimbo, Chile.
[4] Fenner School of Environment and Society, Australian National University, Building 141, Linnaeus way, Canberra, ACT 2601, Australia.
* Corresponding author: thiel@ucn.cl

Introduction

At about the time of the publication of the poem 'Seaweed' by Henry Wadsworth Longfellow (1850) his contemporanean Charles R. Darwin pondered over the dispersal of floating seaweeds and plant seeds, and he concluded that "considering the above means of transport, and several other means, which without doubt remain to be discovered, have been in action year after year, for centuries and tens of thousands of years, it would I think be a marvellous fact if many plants had not thus become widely transported" (Darwin 1859).

During the years leading up to the publication of the Origin of Species, Darwin also conducted extensive floating experiments on seeds of many different terrestrial plants (Darwin 1857). He combined his data on floating times and germination rates with velocity estimates of well-known ocean currents to estimate the distances of open ocean that these seeds could potentially cross. Although he focused on plant dispersal in his floating experiments, Darwin (1859) was fully aware that accumulations of floating plants could serve as dispersal vehicles for a wide range of animals: "though we do not know how sea-shells are dispersed, yet we can see that their eggs or larvae, perhaps attached to seaweed or floating timber, ... might be transported ... across three or four hundred miles of open sea".

Recent studies have confirmed that dispersal distances of macrophytes and associated organisms may even surpass the cautious estimates of Darwin (1859), extending in some cases over thousands of miles of open sea (Fraser et al. 2009, Macaya and Zuccarello 2010, Nikula et al. 2010). For these extensive voyages to be successful for a range of species, the buoyant macrophytes must naturally be able to survive at the sea surface for the duration of the journey, and their passengers must also be able to cope with conditions in the open ocean. What are the specific traits of these macrophytes that can survive extended trips, floating at the mercy of wind and waves? Who are the castaways that are able to persist on these scrubby rafts and potentially colonize new territory at the end of these voyages? What are the conditions needed for successful dispersal via floating macrophytes?

Floating macrophytes

Floating plants and algae can be found in all major oceans, sometimes in great abundances. Indeed, the Sargasso Sea gained its name from the seaweed *Sargassum* that floats as vast mats on its surface; this seaweed also forms habitat for a wide range of other species (Fine 1970, Butler et al. 1983). Many plants or parts thereof have positive buoyancy and float at the sea surface when detached from benthic habitats or flushed into the sea by rivers or flood events. Terrestrial plant material floats better and longer after having dried up on river banks or seashores before reaching the open ocean, as experimentally confirmed by Darwin (1857). Most marine macrophytes (algae and seagrasses), though, quickly dry up and are consumed by numerous beachdwellers when stranding on the shore, and their potential to refloat is very limited (Thiel and Gutow 2005a).

Sizes of plants floating in the oceans range from minuscule seeds of coastal plants to giant kelps and large trees. Enormous rafts of entangled trees and branches have been reported (Rouch 1954, King 1962, Hardy 1982). The majority of the rafts, though, are of small and intermediate sizes, rarely exceeding a few meters in diameter (Kingsford 1992, 1993, Komatsu et al. 2008, Hinojosa et al. 2011). Terrestrial plants and parts thereof (e.g., branches and seeds) are common in coastal areas with large rivers and extended estuaries (Fig. 1). Trees are most commonly observed as driftwood in tropical regions and along Arctic shore lines (Johansen 1999, Alix 2005, Doong et al. 2011). Mangroves line the shorelines of many tropical coasts, and large amounts of floating mangrove debris are frequently observed after storm events (Heatwole and Levins 1972).

Macrophytes originating from inland sources that are transported into the sea by large rivers or flood events can be inhabited by terrestrial organisms (King 1962). Supply of these terrestrial or freshwater plants also is often linked to flood or storm events. For example, rivers usually transport floating wood to the sea after strong rainfall or after the annual snowmelt (Giddings 1952, Hinojosa et al. 2011). During flood events, large rivers transport freshwater macrophytes into the sea, such as for example extensive floating mats of water hyacinths (King 1962). In tropical regions vast accumulations of driftwood are found after hurricanes and typhoons (Doong et al. 2011).

Seeds of mangroves and of many other plants (including terrestrial and freshwater plants) are distributed along nearby seashores by tidal currents (e.g., Clarke 1993, Cadée 2005, Minchinton 2006), and transported to distant coasts by oceanic currents (e.g., Mason 1961, Smith et al. 1990, Brochard and Cadée 2003). Epibenthic rhizomes

Figure 1. Woody debris with seeds and marsh vegetation along an estuarine shore.

(viviparous propagules) that occur in some seagrasses may remain positively buoyant for more than one month (Thomson et al. 2014). Depending on the sizes of these coastal macrophytes, diverse organisms may be transported on them (e.g., Worcester 1994). Seeds range in size from a few millimeters to a few centimeters in diameter, and as they usually travel individually, they can only carry relatively few and small organisms. Rivers, extensive saltmarshes and seagrass beds along temperate coasts provide considerable amounts of plant debris, especially at the end of the annual growth seasons. Autumn storms redistribute large amounts of this debris along the coast or pull it out to sea where it may eventually sink to the seafloor, subsidizing benthic communities.

Large marine macroalgae (e.g., kelps) grow most extensively along temperate coasts, where floating species are especially common among the brown seaweeds, many of which have gas-filled structures that provide positive buoyancy. One of the best-known floating algae is the giant kelp, *Macrocystis pyrifera*, which can form extensive floating rafts, composed of large numbers of entangled individuals. These large kelp rafts are especially common in the eastern boundary currents of the Pacific Ocean and in the Antarctic Circumpolar Current (Hobday 2000a, Smith 2002, Rothäusler et al. 2012). In the warmer waters of the western boundary currents, large patches of floating seaweeds from the genus *Sargassum* are commonly observed (Fine 1970, Komatsu et al. 2008, 2014).

Marine algae can in general survive reasonably well afloat at sea; unlike terrestrial plants, they do not have a root system that requires attachment to a substratum in order to obtain nutrients, but instead absorb nutrients through their frond tissue (Norton 1991). As long as they remain buoyant at or near the surface—either through their own inherent buoyancy or via attachment to other floating objects—they can continue to photosynthesize and can in some cases survive for several months afloat at sea (Fraser et al. 2011). Several studies indicate that macroalgae remain reproductively viable while afloat (e.g., *Hormosira banksii*: McKenzie and Bellgrove 2008; *Turbinaria ornata*: Stewart 2008; *Macrocystis pyrifera*: Macaya et al. 2005, Hernández-Carmona et al. 2006; *Durvillaea antarctica*: Tala et al. 2013), although viability can be compromised at high temperatures (Rothäusler et al. 2009).

Passively floating macrophytes driven by ocean currents and wind typically accumulate in frontal systems (Hinojosa et al. 2010, Garden et al. 2014, Komatsu et al. 2014), which can have important implications for the dynamics of the rafting communities. Rafts have been estimated to travel at average speeds of between 0.1 and 0.3 meters per second (roughly 0.4 to 1.0 kilometers per hour) (e.g., Segawa et al. 1962, Thiel and Gutow 2005a), or faster (Minchinton 2006, Fraser et al. 2011). Although the speed of travel for floating macrophytes may seem slow when the life expectancy of some potential epifaunal passengers is considered, some organisms are particularly well suited to survive long periods afloat on or in macrophyte rafts (see next sections).

Organisms being dispersed via floating plants

"The number of living creatures of all orders, whose existence intimately depends on the kelp, is wonderful... I can only compare these great aquatic forests of the southern hemisphere with the terrestrial ones in the intertropical regions. Yet if in any country a forest was destroyed, I do not believe nearly so many species of animals would perish as would here, from the destruction of the kelp. Amidst the leaves of this plant numerous species of fish live, which nowhere else could find food or shelter; with their destruction the many cormorants and other fishing birds, the otters, seals, and porpoises, would soon perish also"...

Charles Darwin describing the voyage of the Beagle, June 1834
(Darwin 1839)

A diverse range of organisms has been reported from floating macrophytes (Thiel and Gutow 2005b). Some rafting organisms are only opportunistic travellers, whereas others are obligate rafters, i.e., they are only found on floating substrata, including a wide range of different plants. For example, facultative rafting on floating macrophytes has been reported for crustaceans, molluscs, polychaete worms, echinoderms, bryozoans, non-pelagic fish (e.g., Helmuth et al. 1994, Ingólfsson 1995, Hobday 2000b, Ólafsson et al. 2001, Fraser et al. 2011), and—less commonly—terrestrial taxa such as insects (e.g., Wheeler 1916, Heatwole and Levins 1972) and even larger vertebrates (e.g., Prescott 1959, King 1962, Censky et al. 1998).

Crustaceans are commonly reported from macrophytic rafts; peracarid crustaceans, in particular, often dominate the rafting communities (Fig. 2), even though harpacticoid copepods can also be very abundant and species-rich (Ólafsson et al. 2001). Some of the best-known obligate rafters are also crustaceans, such as goose barnacles (e.g., *Lepas* species) which colonize any objects floating at the sea surface including diverse macrophytes (Macaya et al. 2005, Hinojosa et al. 2006, Fraser et al. 2011), and the isopod *Idotea metallica* which is often found on floating macrophytes as well as other floating items (Gutow and Franke 2003).

The numbers and types of rafters also depend on the characteristics of macrophyte rafts. The larger the rafts, the higher the number and diversity of associated organisms. This relationship has been documented for floating algae from many different regions

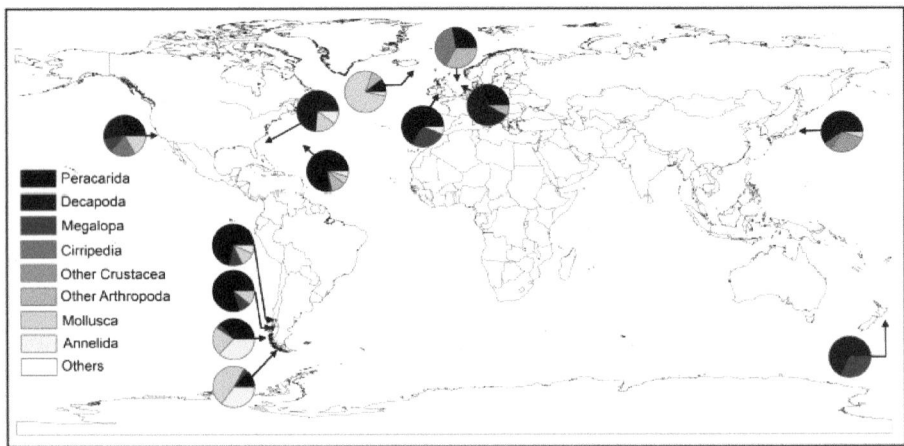

Figure 2. Proportions of the principal associated animals from macroalgal rafts. Data from Fine 1970, Kingsford and Choat 1985, Ingólfsson 1995, Hobday 2000b, Sano et al. 2003, Vandendriessche et al. 2006a, 2006b, Clarkin et al. 2012b, Wichmann et al. 2012.

in the world (for overview see Thiel and Gutow 2005b), including, most recently, from Irish coastal waters (Fig. 3). The size of the rafting organisms is also likely to depend on the raft size: small organisms can travel on rafts of all sizes, but large terrestrial vertebrates such as reptiles or mammals have only been reported from very large rafts. For example, King (1962) mentioned several reports of large vertebrates on extensive macrophyte rafts, and Censky et al. (1998) documented observations of iguanas travelling to an island on a mat of buoyant logs. Future studies should further examine the hypothesis that there is a positive relationship between raft size and the maximum size of associated organisms.

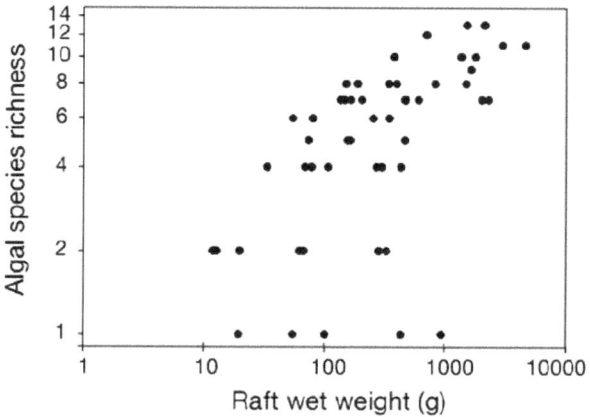

Figure 3. Relationship between raft size and species richness of associated algae. From Clarkin et al. 2012b.

Colonization of macrophyte rafts

The pattern of colonization of macrophytic rafts differs substantially between positively buoyant macrophytes from marine habitats and those from freshwater and terrestrial environments. Marine macrophytes are usually colonized by a wide range of organisms while growing attached to benthic habitats. In particular, marine algae and plants (seagrasses and mangroves) are commonly inhabited by diverse epibiont communities (e.g., Worcester 1994, Farnsworth and Ellison 1996, Buschbaum et al. 2006, Gutow et al. 2014). These organisms themselves may contribute to the breakage or detachment of their hosts during storm events (Black 1976, Svavarsson et al. 2002, Viejo and Åberg 2003), but failure of the rock substratum can also lead to detachment of macroalgae (Black 1976, Garden et al. 2011). Following detachment, many macrophytes growing in nearshore habitats may immediately be thrown onto nearby shores. Depending on coastal morphology and oceanographic conditions, however, a fraction of these detached macrophytes may be pulled out to sea, where they are moved by the currents and can become colonized by other organisms.

When macrophytes first begin to float, important changes happen to the rafting community. Many of the mobile organisms rapidly depart after detachment of the macrophytes and return to the seafloor (e.g., Miranda and Thiel 2008, Gutow et al.

2009). Other organisms may stay initially, but disappear during the first days of the trip either due to fish predation or because they are not adequately adapted to survive life on a raft. Some organisms, though, persist on the rafts, where they might even benefit from the absence of competitors common to benthic habitats. Additionally, rafts can be colonized by new species (Clarkin et al. 2012a, Gutow et al. 2014), and thus communities on older rafts may be substantially different from those initially starting the journey at the sea surface (Fig. 4).

Organisms on freshwater vegetation or on terrestrial plants may suffer when first exposed to saltwater. Many will immediately die upon exposure to seawater (Barbour 1916), but some are relatively resistant to the conditions at the sea surface. For example, insects living in self-excavated galleries in tree trunks appear capable of surviving voyages at sea (Wheeler 1916, Heatwole and Levins 1972). Similarly, reptiles appear to cope particularly well with the living conditions on macrophyte rafts, as is evidenced by numerous studies showing or suggesting dispersal of reptiles via rafting (Censky et al. 1998, Raxworthy et al. 2002, Chapple et al. 2009).

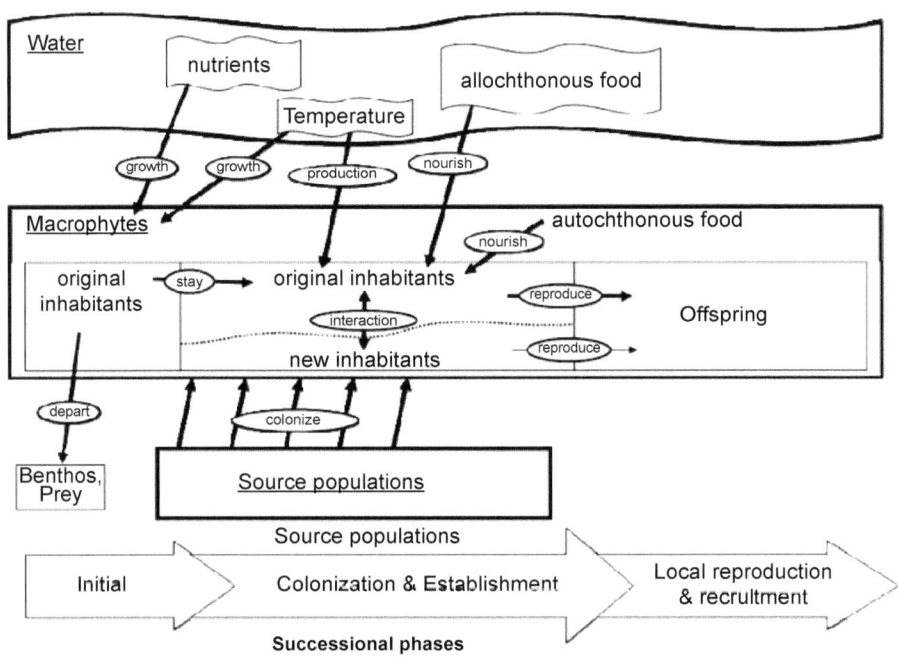

Figure 4. Succession of the rafting community on floating macrophytes after introduction to coastal waters. Modified after Thiel (2003).

Mobility

Both mobile and sessile species can be found on floating plants. Sessile organisms are, naturally, attached to their rafts, and cannot easily leave; they usually colonize rafts during their mobile, planktonic life stages, and—when these rafts are stranding in new environments—colonization of the new territory must also be via planktonic (not adult) stages. An exception might be some sessile, clonal organisms that can grow onto new substrata after floating plants make contact with attached vegetation, as for example shown for compound ascidians rafting on seagrass shoots (Worcester 1994). In contrast, mobile organisms can rapidly leave rafts or hop onto other rafts or benthic macrophytes.

Raft hopping may be a particularly important strategy for travellers on macrophyte rafts, because most floating plants and algae have only limited life spans. When a raft starts to disintegrate due to abiotic factors or intense grazing, mobile rafters may abandon the sinking raft and hop onto nearby, intact rafts; this process has been suggested by Hobday (2000b) for mobile fauna capable of swimming. Macrophyte rafts are not homogeneously distributed in the ocean, but rather are usually concentrated in frontal systems and drift rows (Hinojosa et al. 2010, Marmorino et al. 2011, Witherington et al. 2012). In such situations, large numbers of floating macrophytes of all ages and successional stages come together, offering ample opportunity for mobile organisms to move between rafts. Indeed, results from community studies suggest that mobile organisms frequently do switch between rafts at sea (Ingólfsson 1998, Clarkin et al. 2012b, Gutow et al. 2014).

Trophic ecology

Floating plants can sustain many different organisms with a wide range of feeding types. As floating brown algae continue to grow while afloat (Vandendriessche et al. 2007a, Rothäusler et al. 2009), they provide a renewable food supply for herbivorous raft associates. Other buoyant plant materials (seeds, wood, seagrasses, saltmarshes) often have a much more limited growth potential and food supply for rafters must thus be sustained by organic material that accumulates in the frontal systems where floating items are concentrated by converging currents (Fig. 5). The large number of suspension feeders on rafts (e.g., the ever-present goose barnacles, hydrozoans, or bryozoans) give testimony to the fact that suspended particles are abundantly available; this suspended organic matter (dead or alive) is recycled by the rafting community. Floating macrophytes (and other floating objects) can thus be considered catalysts for trophic processes in oceanic fronts where they are the only available attachment substrata for many consumers (Thiel and Gutow 2005b). The high productivity on rafts (compared to surrounding oceanic waters) also attracts diverse vertebrate consumers. Fishes, sea turtles and seabirds commonly concentrate along drift lines where they feed on the associated organisms (Mitchell and Hunter 1970, Kingsford 1995, Vandendriessche et al. 2007b, 2007c, Moser and Lee 2012, Witherington et al. 2012).

Consumer interactions may be more intense during the summer months when the density of associated travellers is significantly higher than during the winter months (Ingólfsson 2000). In particular, grazers can have a strong impact on floating algae (Rothäusler et al. 2009), which is further enhanced at high temperatures (Vandendriessche et al. 2007a). In addition, during summer floating macrophytes are exposed to environmental stress caused by abiotic factors (Graiff et al. 2013, Tala

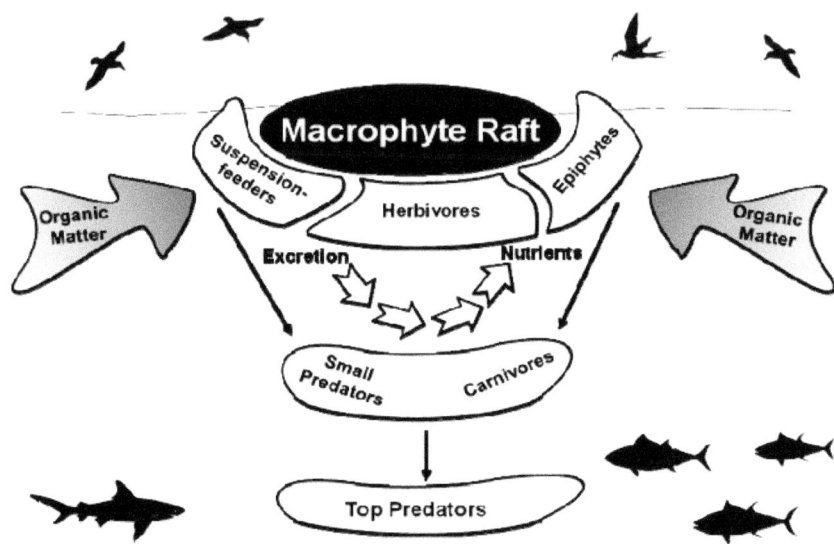

Figure 5. Trophic relationships around macrophyte rafts. Modified after Thiel and Gutow 2005b.

et al. 2013). Thus, the combination of high grazer densities and abiotic stress probably accelerate the decay of floating algae during summer months. Indeed, during winter months, when the floating potential of the algae is significantly higher than during summer, many rafters only occur in very low numbers or are absent altogether from the floating algae (Ingólfsson 2000).

The rafting organisms best adapted to survive long trips at sea are those that are able to utilize the food resources available on rafts and in the surrounding waters without destroying the raft. Small grazers (such as harpacticoid copepods, amphipods, and isopods) that live on actively growing macrophytes, making small burrows where they can also effectively hide from common vertebrate predators, can thus be considered ideal rafters.

Reproductive biology

The success of dispersal of organisms with short life spans depends on their capacity to maintain or produce viable offspring during prolonged voyages or after returning to coastal habitats. Long trips across the open ocean can last several months (and even over a year), exceeding the life span of many of the travellers. Species with internal fertilization and direct development are best suited to achieve successful dispersal under these conditions. Indeed, many species travelling on floating macrophytes have short planktonic stages or direct development; species with asexual reproduction are also commonly reported as rafters (Thiel and Gutow 2005b), probably because they can easily proliferate and persist for long periods on rafts. There is a high likelihood that, in these species, the descendants of the original travellers are the only potential colonizers of new habitats when their rafts reach shore.

Scales of dispersal

The distances that macrophytes can disperse via floating depends on environmental conditions and on the ability of the floating plants to survive at the sea surface. In general, population connectivity should be negatively correlated with dispersal distances (Fig. 6). For the vast majority of travellers, dispersal distances via floating macrophytes are short, as suggested by dispersal studies with mangrove seeds, marsh vegetation, kelp rafts or propagule mimics (Steinke and Ward 2003, Minchinton 2006, Hernández-Carmona et al. 2006, Muhlin et al. 2008). Thus, populations could be expected to be well connected over small spatial scales. However, extensive habitat discontinuities, e.g., long, muddy estuaries and beaches between rocky shores, appear to disrupt population connectivity for floating macrophytes and their passengers, as confirmed by a number of recent studies (e.g., Coleman and Kelaher 2009, Alberto et al. 2010, Fraser et al. 2010).

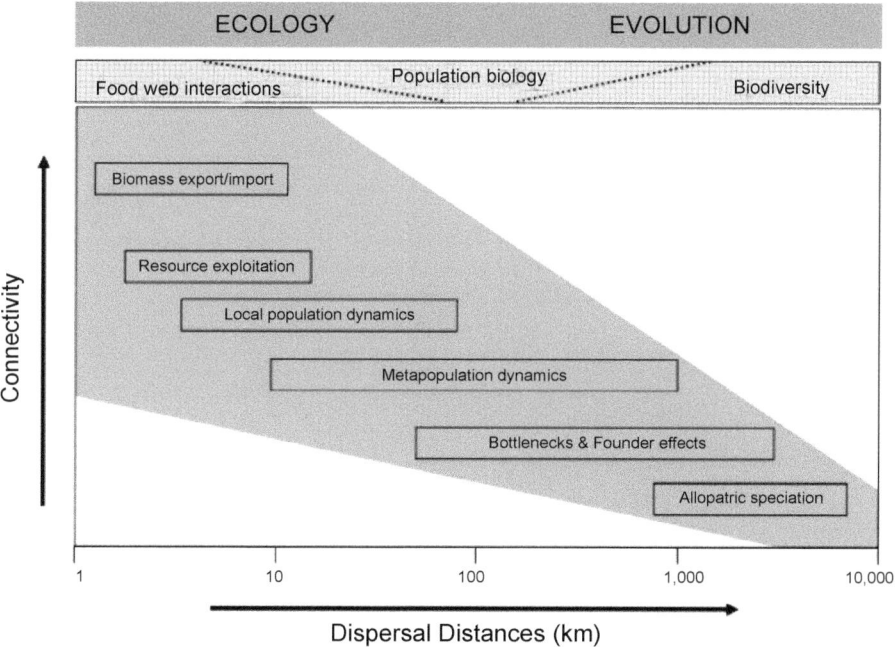

Figure 6. Hypothetical relationship between rafting distances and population connectivity. Modified after Thiel and Haye (2006).

Although most studies that have released artificial rafts or examined dispersing macrophytes show that transport is most common over only short distances, in some of these studies individual rafts were observed to travel over considerably larger distances, into adjacent estuaries and rocky points (Segawa et al. 1962, Steinke

1986, Minchinton 2006, Muhlin et al. 2008), but also to much more distant shores (for additional reports see Thiel and Gutow 2005a). Several macroalgal species have been inferred to have floated across vast ocean basins; for example, the buoyant kelps *Durvillaea antarctica* (Fraser et al. 2010) and *Macrocystis pyrifera* (Macaya and Zuccarello 2010), and the non-buoyant turfing algae *Wittrockiella lyallii* (Boedeker et al. 2010), *Adenocystis utricularis* and *Bostrychia intricata* (Fraser et al. 2013), have all been found to show close genetic affinities and even shared DNA sequences between the south-west Pacific region (New Zealand or sub-Antarctic Macquarie Island) and South America, suggesting recent trans-Pacific dispersal events involving more than 7000 km of open ocean. Molecular work on southern bull-kelp (*D. antarctica*) has also demonstrated successful oceanic rafting of both the living macroalga and diverse epifaunal communities in the kelp's holdfasts across some 600 km of ocean near New Zealand; a journey inferred, based on the size of attached pelagic goose barnacles, to have taken at least nine weeks (Fraser et al. 2011) (Fig. 7).

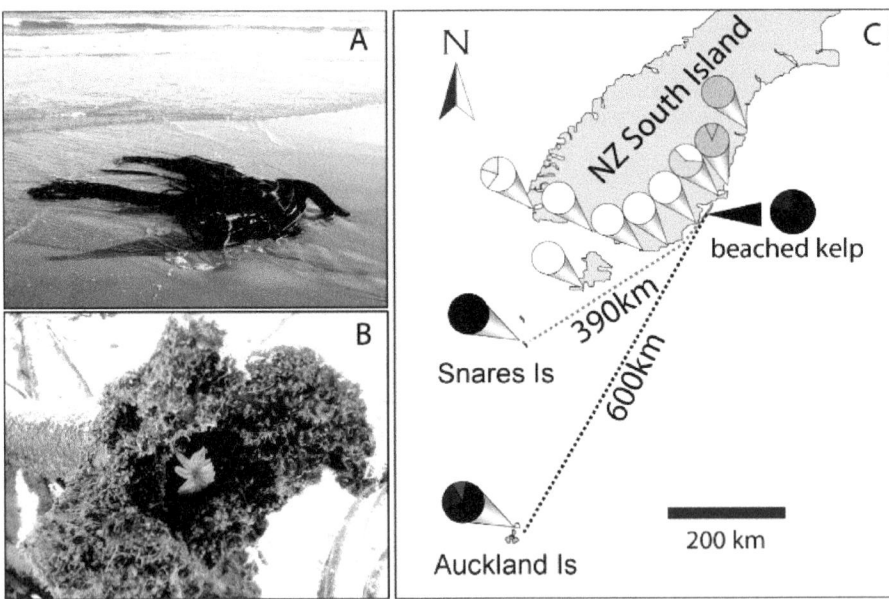

Figure 7. Seaweed can drift across oceans, sometimes carrying animals with it. In February 2009, in New Zealand, several bull kelp (*Durvillaea antarctica*) rafts, encrusted with goose barnacles indicating a long duration drifting at sea, were found by DNA analyses to have travelled several hundred kilometers from the sub-Antarctic. The kelp holdfasts housed many species of marine invertebrates. (A) *D. antarctica* washed up on a beach in southern New Zealand. (B) a *Stichaster* sea-star was among several species recovered from the sub-Antarctic rafts found in February 2009. (C) map indicating genetic differentiation between New Zealand and sub-Antarctic populations of the kelp, with shades in the pie charts indicating different haplotypes and their proportions; see Fraser et al. (2011).

Biogeography

Colonization

The ability of some marine macrophytes to survive long periods afloat at sea makes them ideal vectors for the transport of a range of flora and fauna across vast distances of open ocean. Although some such rafts will inevitably sink, degrade or wash up on hostile shores such as sandy beaches, some will achieve landfall at sites capable of being colonized by themselves or their progeny. Disturbed patches or new islands can therefore quickly be colonized by a wide range of organisms carried by dispersing macrophyte rafts.

Successful colonization of new habitats depends strongly on the biological traits of the travellers. Similar traits that facilitate persistence of a species on a raft (asexual reproduction, internal fertilization, sperm storage, direct development) also favor colonization once rafts have made land contact again (Johannesson 1988, Thiel and Haye 2006). A few individuals of a successfully travelling species can be sufficient to found small populations, often resulting in founder effects (e.g., Johannesson and Warmoes 1990, Islam et al. 2012). The genetic diversity of the new population thus depends largely on the genetic diversity of its founders. Where the new territory is not already occupied by conspecifics or direct competitors, the founders can rapidly multiply to fill available niche space, and the resulting dense populations (e.g., as shown for *Sargassum muticum*: Deysher and Norton 1982) can block establishment of late-arriving dispersers (the 'founder takes all' principle: Waters et al. 2013). Such density-dependent colonization processes have been invoked to explain why, despite demonstrated frequent long-distance dispersal events by buoyant macroalgae, these rarely result in colonization by the dispersers unless a disturbance has created new or newly available habitat (Fraser et al. 2009, Fraser et al. 2010, Olsen et al. 2010, Waters et al. 2013). For example, rafts of *Durvillaea antarctica* from the sub-Antarctic islands frequently arrive at the New Zealand mainland, but travellers leave no genetic signal in resident populations (Collins et al. 2010, Fraser et al. 2011).

The first colonization of newly available areas of shore by dispersing macroalgae and rafting fauna can lead to patchy distribution of species and lineages (e.g., Deysher and Norton 1982, Johnson et al. 2001, Muhlin et al. 2008). Local disturbances, such as storms or denudement of rock substratum by mechanical means, e.g., scouring by grazers or boulders, will lead to small-scale patchiness as dispersing organisms recolonize. In contrast, large-scale homogeneity can be an indication of past large-scale disturbances; for example, the genetic homogeneity observed throughout most of the sub-Antarctic region for the large kelps *Durvillaea antarctica* (Fraser et al. 2009) and *Macrocystis pyrifera* (Macaya and Zuccarello 2010) has been linked to large-scale extirpation of these macroalgae from Southern Hemisphere high latitudes by ice scouring at the peak of the last Ice Age. Similarly, large areas of relatively low genetic diversity in Northern Hemisphere high latitudes have been linked to Ice Age impacts followed by recolonization by the macroalgae *Fucus vesiculosus* (Hoarau et al. 2007, Coyer et al. 2011) and *Palmaria palmata* (Provan et al. 2005) (see also Maggs et al. 2008).

Colonization success of macroalgae thus depends not only on dispersal, but on subsequent establishment (Thiel and Haye 2006). Establishment relies largely on the availability of unoccupied habitat, which usually results from disturbance. For faunal passengers on dispersing, floating algae, colonization success may similarly be influenced by the existence and densities of populations of local counterpart species, and/or by environmental conditions. For example, despite evidence that the brooding isopod species *Limnoria stephenseni* frequently disperses on rafting bull-kelp from the sub-Antarctic islands to New Zealand, no local, established populations of this species have been detected, yet the congeneric *L. segnis* is common (Fraser et al. 2011). Whether it is the presence of dense populations of *L. segnis* in New Zealand that inhibits establishment of newly-arrived *L. stephenseni*, or whether establishment fails due to unfavourable environmental conditions, has yet to be determined.

Population connectivity

Gene flow and isolation

Dispersal by rafting with buoyant macroalgae has been observed or inferred for a wide variety of taxa (Ingólfsson 1995, Hobday 2000b, Thiel and Gutow 2005b, Fraser et al. 2011), and can in some cases promote long-distance population connectivity and gene flow. For example, the limpet *Cellana strigilis*, which has been observed to raft on southern bull-kelp (Fraser et al. 2011), shows apparent genetic connectivity among several isolated island groups in the New Zealand region (e.g., Campbell, Auckland, Snares and Stewart islands), suggesting that establishment of dispersing individuals plays a role in maintaining genetic connectivity among distant populations for this species complex (Reisser et al. 2011). Comparable patterns of long-distance connectivity have recently been inferred for a range of other kelp-associated invertebrates that lack pelagic life history stages, such as the snail *Cantharidus roseus*, the chiton *Onithochiton neglectus* and the amphipod *Parawaldeckia karaka* (Nikula et al. 2013). Indeed, population connectivity via rafting dispersal has been inferred for a number of herbivorous kelp associates with direct development (Wares 2001, Nikula et al. 2010, Haye et al. 2012), which feed on their macroalgal hosts. As the kelps continue to grow under suitable environmental conditions at sea (moderate temperatures and solar radiation—Rothäusler et al. 2011, 2012), travellers can rely on abundant food supply that even could allow population growth during prolonged journeys (see, e.g., Edgar 1987, Vásquez 1993). Consequently, upon arrival, hundreds of potential colonizers might disembark on new coastlines.

Nonetheless, frequent rafting does not always result in maintenance of genetic connectivity of distant populations. Genetic affinities among distant populations, apparently as a result of long-distance dispersal via rafting, can indicate past or recent colonization events following disturbances (see above) rather than ongoing genetic connectivity. Indeed, considering the vast amounts of floating macrophytes at sea, and the frequency with which they and their passengers must make landfall, genetic variation on small spatial scales is remarkably high for many taxa. For example, relatively high genetic variation within and among southern bull-kelp (*D. antarctica*) populations along the coasts of New Zealand and central Chile suggest these populations experience little genetic connectivity, yet long-distance rafting events in these regions are common (Collins et al. 2010, Fraser et al. 2010, Fraser et al. 2011). Similarly, high genetic diversity on small spatial scales has been maintained in the brown macroalga *Carpophyllum maschalocarpum* in New Zealand, despite evidence that long-distance dispersal by floating adult thalli occurs for this species (Buchanan and Zuccarello 2012). In the north-west Atlantic, patchy genetic variation among *Fucus vesiculosus* populations along the Maine coast was not found to follow an isolation by distance pattern, but rather reflected long-distance dispersal events followed by establishment of migrants (Muhlin et al. 2008). Although Muhlin et al. (2008) inferred some interbreeding of migrant and local *F. vesiculosus*, the genetic heterogeneity noted among populations nonetheless suggested that connectivity had been rather limited. Studies that have assessed genetic connectivity and rafting success for the kelp-associated, brooding isopods of the genus *Limnoria* have indicated that genetic connectivity may be maintained by rafting in some of these species/lineages (Nikula et al. 2010, Haye et al. 2012). Shared mitochondrial haplotypes were detected among *Limnoria* populations across broad spatial scales (thousands of kilometers) in both sub-Antarctic regions thought to have been scoured by ice during the last Ice Age (Nikula et al. 2010), and in presumably more persistently populated regions such as northern-central Chile (Haye et al. 2012).

Floating plants and climate change

Past climate change

Variations in temperature regimes and sea level have caused important shifts in the distribution of coastal macrophytes. The high colonization potential of many species via floating have allowed them to track shifting coastlines (Graham et al. 2010), generally moving poleward in warming, interglacial periods, and towards the equator as the Earth approaches glacial maxima. Glacial and interglacial periods result not only in changed climate, but in changed geography; during past glacial periods, sea levels were much lower than today, as much of the world's water was drawn up onto the land as glaciers. At the Last Glacial Maximum (LGM), sea levels are estimated to have been approximately 120 m lower than they currently are, changing the shape of many continents, and in some cases connecting landmasses via 'land bridges'. For example, Australia's large, southern island, Tasmania, was connected to the mainland at the LGM, and all three of the large islands of New Zealand (North Island, South Island and Stewart Island) were joined. Such geographical changes had major impacts on marine species, sundering and uniting populations, and altering the courses of oceanic currents. The Pleistocene witnessed repeated glacial-interglacial cycles of varying intensity. For species able to disperse, and especially for those able to disperse across oceans, these cycles will have been echoed in cyclical changes to ecosystem structure. For example, in the Southern Hemisphere, the major ecosystem-structuring kelps *Macrocystis pyrifera* and *Durvillaea antarctica*, along with their associated invertebrate communities, are thought to have been extirpated from most sub-Antarctic islands at the LGM. Consequently, at Pleistocene glacial maxima, nearshore ecosystems of the sub-Antarctic islands would have been drastically different, populated only by mobile, seasonal or otherwise ice-scour resistant species such as those found within the range of Antarctic sea ice today (Fraser 2012). During interglacials, the highly dispersive kelps would have returned from lower-latitude glacial refugia (e.g., New Zealand) by rafting across oceans, carrying invertebrate passengers and repopulating depauperate sub-Antarctic shores. The impacts of such massive ecosystem changes would likely have had significant carry-over effects for many terrestrial species dependent (directly or indirectly) on nutrition from nearshore marine systems (Fraser 2012).

In the Northern Hemisphere, the Bering land bridge connecting North America (Alaska) to Eurasia (Siberia) divided the North Atlantic from the North Pacific during Pleistocene glacial maxima, and opened as the Bering Strait during interglacial periods (Fig. 8). Marine species that occurred in both oceans were repeatedly segregated into Atlantic and Pacific populations for considerable periods, and some dispersive species—such as macroalgae capable of floating as adults at sea—were cyclically reunited at glacial minima. Indeed, vicariant speciation between Atlantic and Pacific macroalgae that predates the formation of the Bering Strait (~ 5.3 Ma: Gladenkov et al. 2002) has been inferred for relatively few seaweed groups (Lindstrom 2001),

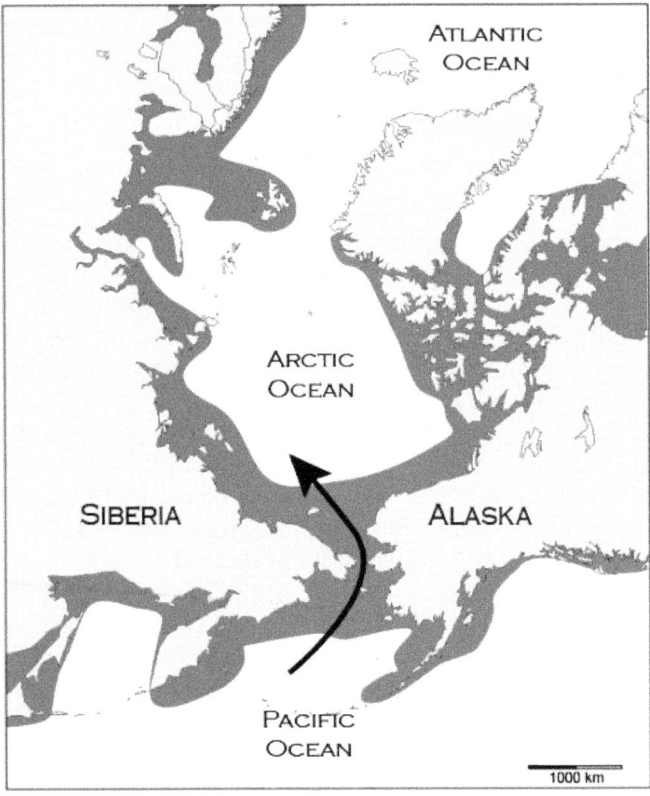

Figure 8. The Bering Land Bridge connected Siberia to Alaska during Pleistocene ice ages, and effectively blocked dispersal of marine species between the North Atlantic and North Pacific oceans. In interglacial periods, with higher sea levels, dispersal across the Bering Strait was possible. Light grey: current land. Dark grey: approximate coastlines during the Last Glacial Maximum.

suggesting that the Strait allowed many species to establish or maintain Atlantic-Pacific population connections during interglacial periods. As an example, molecular research on the brown seaweed *Fucus distichus* indicates that the species dispersed from the North Pacific into the North Atlantic through the Bering Strait during one of the Pleistocene interglacial periods preceding the LGM (Coyer et al. 2011). Furthermore, Coyer et al. (2011) predict that, with reduced ice cover in the intervening Arctic Ocean brought about by global warming, contact between Pacific and Atlantic populations of this species will be increased via floating adult fronds.

These changes also affected the supply of terrestrial macrophytes to the sea. Long-term variations (centuries to millenia) in driftwood supply on Arctic shores are generally attributed to changes in ocean currents, sea level and possibly river run-off (Dyke et al. 1997, Johansen 1999, Alix 2005, Hellmann et al. 2013). Such changes are likely to affect colonization of Arctic islands and shores (Johansen and Hytteborn 2001, Alsos et al. 2007) by a range of rafting species.

Predicted impacts of future climate change

Global warming will not only continue to change the shape of continental coastlines and the oceanographic connectivity of distant marine populations, but is also expected to result in altered environmental factors such as temperature, solar radiation and salinity. These changes are likely to affect the distributional ranges of benthic populations, as well as the dispersal potential of floating macrophytes. Macreadie et al. (2011) recently reviewed macrophyte rafting literature and developed a set of predictions on how dispersal via rafting may change with future climate change. Firstly, they suggest that we can expect an increase in the amount of floating macrophytes at sea, with detachment of seaweed and seagrasses facilitated by an increase in storms and grazing herbivore activity, which in turn could improve dispersal opportunities for rafts and their passengers. However, they further predict that raft quality and persistence will decrease with higher water temperatures reducing buoyancy, reproductive capacity and the size of rafting plants. Finally, changes to oceanographic conditions (e.g., major currents changing course and strength) are expected. A recent study (Huffard et al. 2014) reports important changes (decline in species numbers and coverage of sessile bryozoans) in the rafting community of the Sargasso Sea; they discussed the observed increase in water temperature and decline in pH as possible causes for these shifts.

Across almost all taxonomic groups, both marine and terrestrial, broad-scale genetic and ecological research indicates that species move towards the Equator as the Earth cools, and towards the poles as it warms. Global warming has been underway since the peak of the last Ice Age about 20,000 years ago, but is now accelerating at hitherto-unknown rates as humans pump carbon dioxide into the atmosphere (IPCC 2007, Anderegg et al. 2010). Species have to track changing coastlines and environmental conditions faster than before, and naturally, some slow-dispersers will be left behind in the race to move poleward, or be blocked by oceanic barriers (Fraser et al. 2012). Large kelps, which form the most robust macrophytic rafts for trans-oceanic transport (Macreadie et al. 2011), are most common in temperate and cool-temperate regions. As a large number of seaweed species' ranges are indeed moving poleward

(Müller et al. 2009, Wernberg et al. 2011), we can expect a shift in temperate regions to smaller seaweeds as the kelps depart, and in polar regions to larger seaweeds as the kelps arrive. Naturally, these distributional changes will impact broad-scale dispersal patterns not only of the kelps, but of associated invertebrates. In some cases poleward dispersal could be greatly assisted by floating buoyant kelps; indeed, recolonization of the southern sub-Antarctic islands by a wide range of marine species after the LGM is thought to have been facilitated by kelp rafting (Fraser 2012). However, trans-oceanic dispersal of seaweeds and rafting passengers relies on their buoyancy and robustness, and not all float well. Many seaweed species in Australia, for example, are expected to 'drop off' the southern edge of the continent with climate-driven range shifts, and those that are endemic will therefore face extinction, along with their obligate floral and faunal associates (Wernberg et al. 2010, Fraser et al. 2012).

At low latitudes the increasing frequency and intensity of tropical cyclones (Bender et al. 2010) will enhance the supply of floating macrophytes to coastal waters. In particular, the expected increase in rainfall associated with tropical cyclones is expected to lead to higher runoff, flushing large amounts of terrestrial and freshwater macrophytes out to the sea. In tropical regions dispersal of terrestrial biota is often associated with large rafts of macrophytes after storm events (King 1962, Heatwole and Levins 1972, Censky et al. 1998, Elsey and Aldrich 2009). Increased frequency and intensity of dispersal is expected to reduce spatial separation of species, possibly limiting the potential for allopatric speciation in tropical regions.

Outlook

Marine macrophytes have strongly influenced the evolutionary history of the global biota. A wide range of different species have been or are presently being dispersed by floating macrophytes. Variations in ice cover, river runoff, sea surface temperature and solar radiation across the globe will result in unprecedented changes in dispersal opportunities via macrophyte rafts. In simplified form these changes can be generalized as follows: although the supply of floating macrophytes will be enhanced in many areas (due to larger storm frequencies and higher river runoff), their survival at the sea surface is expected to decrease (higher degradation rates due to warmer waters and stronger solar radiation). Consequently, higher dispersal opportunities (and population connectivity) can be expected for many organisms in coastal waters. However, the possibility of long-distance dispersal across the open ocean via floating macrophytes will be reduced because their survival at the sea surface will be shortened in the future. Whether this can be compensated by increasing loads of persistent anthropogenic litter, prone to float for long times and over long distances across the open ocean, is questionable. Many of the small herbivorous organisms rafting on large macrophytes simply could not survive on food-poor inorganic litter, and the rest of the rafting community entirely depends on allochthonous food supply.

Rafting transport is arguably the most efficient natural means of long-distance dispersal in the world's oceans (Fig. 9). While larval dispersal may facilitate mass dispersal over short and medium distances, the larvae of few species are capable of travelling the extensive distances covered by buoyant rafts. Not surprisingly,

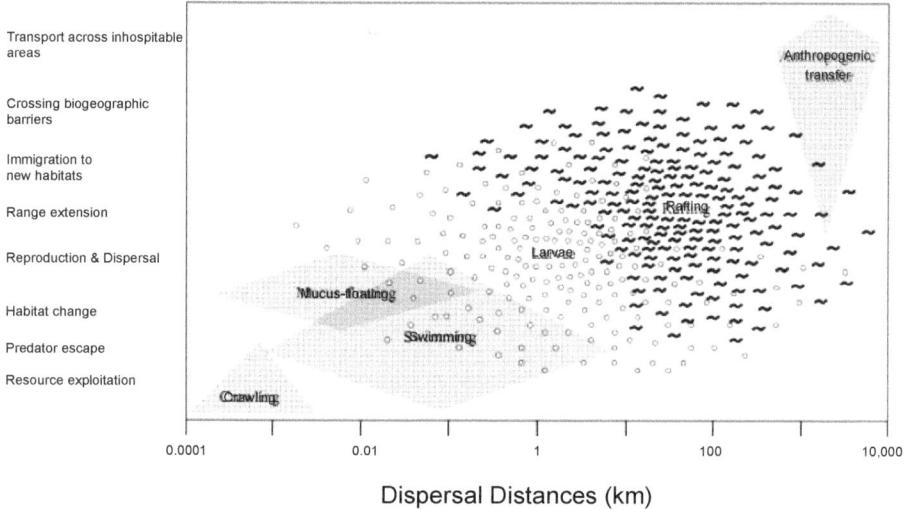

Figure 9. Schematic overview dispersal distances and processes in marine organisms. Modified after Thiel and Haye (2006).

colonization of remote islands is dominated by rafting organisms, whereas larval colonists drive the colonization patterns on nearshore islands (van den Hoek 1987, Davenport and Stevenson 1998). The expected changes in the dispersal potential of direct developers via rafting can also influence the structure of intertidal communities (Johnson et al. 2001).

Conclusions

Floating plants transport diverse organisms over short and long distances, and thus perform a critically important ecosystem service, facilitating long-distance dispersal of numerous organisms that would otherwise be unable to achieve long voyages across the sea. There are many types of floating plants that can act as rafts to take passengers among landmasses, including terrestrial and marine species as whole (e.g., trees and kelp) or partial (e.g., seeds, branches and fronds) organisms from direct-developing marine species to terrestrial vertebrates. Macrophyte rafts carry many different organisms, but small herbivores with direct development have the highest potential for long-distance dispersal. The degree of population connectivity achieved via macrophyte rafting does not only depend on the traits of the travellers, but also on the abundance and survival of macrophyte rafts, and on the local densities of benthic populations and communities. Climate change effects are thought to lead to higher abundances of floating macrophytes in coastal waters, but increasing water temperatures and higher solar radiation is expected to suppress their survival at the sea surface. This should lead to higher population connectivity via rafting in coastal waters, but lower potential of long-distance dispersal via floating macrophytes.

Acknowledgements

We are grateful to Jose Pantoja for help with Fig. 2. MT was supported through FONDECYT grants 1100749 and 1131082 during the preparation of this contribution.

References

Alberto, F., P.T. Raimondi, D.C. Reed, N.C. Coelho, R. Leblois, A. Whitmer and E.A. Serrao. 2010. Habitat continuity and geographic distance predict population genetic differentiation in giant kelp. Ecology 91: 49–56.

Alix, C. 2005. Deciphering the impact of change on the driftwood cycle: contribution to the study of human use of wood in the Arctic. Glob. Planet. Change 47: 83–98.

Alsos, I.G., P.B. Eidesen, D. Ehrich, I. Skrede, K. Westergaard, G.H. Jacobsen, J.Y. Landvik, P. Taberlet and C. Brochmann. 2007. Frequent long-distance plant colonization in the changing Arctic. Science 316: 1606–1609.

Anderegg, W.R.L., J.W. Prall, J. Harold and S.H. Schneider. 2010. Expert credibility in climate change. Proc. Natl. Acad. Sci. USA 107: 12107–12109.

Barbour, T. 1916. Some remarks upon Matthew's "Climate and Evolution". Ann. N.Y. Acad. Sci. 27: 1–10.

Bender, M.A., T.R. Knutson, R.E. Tuleya, J.J. Sirutis, G.A. Vecchi, S.T. Garner and I.M. Held. 2010. Modeled impact of anthropogenic warming on the frequency of intense Atlantic hurricanes. Science 327: 454–458.

Black, R. 1976. The effects of grazing by the limpet, *Acmaea insessa*, on the kelp, *Egregia laevigata*, in the intertidal zone. Ecology 57: 265–277.

Boedeker, C., M.E. Ramirez and W.A. Nelson. 2010. *Cladophoropsis brachyartra* from southern South America is a synonym of *Wittrockiella lyallii* (Cladophorophyceae, Chlorophyta), previously regarded as endemic to New Zealand. Phycologia 49: 525–536.

Brochard, C.J.E. and G.C. Cadée. 2003. Tropical drift-seeds from the French coast: an overview. Cah. Biol. Mar. 44: 61–65.

Buchanan, J. and G.C. Zuccarello. 2012. Decoupling of short and long distance dispersal pathways in the endemic New Zealand seaweed *Carpophyllum maschalocarpum* (Phaeophyceae, Fucales). J. Phycol. 48: 518–529.

Buschbaum, C., A.S. Chapman and B. Saier. 2006. How an introduced seaweed can affect epibiota diversity in different coastal systems. Mar. Biol. 148: 743–754.

Butler, J.N., B.F. Morris, J. Cadwallader and A.W. Stoner. 1983. Studies of *Sargassum* and the *Sargassum* community. Special Publication 22, Bermuda: Bermuda Biological Research Station.

Cadée, G.C. 2005. Drifting branches of *Crambe maritima* L. with fruits. Cah. Biol. Mar. 46: 217–219.

Censky, E.J., K. Hodge and J. Dudley. 1998. Over-water dispersal of lizards due to hurricanes. Nature 395: 556.

Chapple, D.G., P.A. Ritchie and C.H. Daugherty. 2009. Origin, diversification, and systematics of the New Zealand skink fauna (Reptilia: Scincidae). Mol. Phyl. Evol. 52: 470–487.

Clarke, P.J. 1993. Dispersal of grey mangrove (*Avicennia marina*) propagules in southeastern Australia. Aquat. Bot. 45: 195–204.

Clarkin, E., C.A. Maggs, G. Arnott, S. Briggs and J.D. Houghton. 2012a. The colonization of macroalgal rafts by the genus *Idotea* (sub-phylum Crustacea; order Isopoda): an active or passive process? J. Mar. Biol. Assoc. UK 92: 1273–1282.

Clarkin, E., C.A. Maggs, A.L. Allcock and M.P. Johnson. 2012b. Environment, not characteristics of individual algal rafts, affects composition of rafting invertebrate assemblages in Irish coastal waters. Mar. Ecol. Prog. Ser. 470: 31–40.

Coleman, M.A. and B.P. Kelaher. 2009. Connectivity among fragmented populations of a habitat-forming alga, *Phyllospora comosa* (Phaeophyceae, Fucales) on an urbanised coast. Mar. Ecol. Prog. Ser. 381: 63–70.

Collins, C.J., C.I. Fraser, A. Ashcroft and J.M. Waters. 2010. Asymmetric dispersal of southern bull-kelp (*Durvillaea antarctica*) adults in coastal New Zealand: testing an oceanographic hypothesis. Mol. Ecol. 19: 4572–4580.

Coyer, J.A., G. Hoarau, J.F. Costa, B. Hogerdijk, E.A. Serrao, E. Billard, M. Valero, G.A. Pearson and J.L. Olsen. 2011. Evolution and diversification within the intertidal brown macroalgae *Fucus spiralis/F. vesiculosus* species complex in the North Atlantic. Mol. Phylogenet. Evol. 58: 283–296.

Darwin, C.R. 1839. Narrative of the surveying voyages of his Majesty's ships Adventure and Beagle, between the years 1826 and 1836, describing their examination of the southern shores of South America, and the Beagle's circumnavigation of the globe. Vol. III. Journal and Remarks. Henry Colburn, London, UK.

Darwin, C.R. 1857. On the action of sea-water on the germination of seeds. J. Proc. Linn. Soc. London (Botany) 1: 130–140.

Darwin, C.R. 1859. On the origin of species by natural selection. John Murray, London, UK.

Davenport, J. and T.D.I. Stevenson. 1998. Intertidal colonization rates. A matched latitude, north v. south, remote v. near shore island experiment. Divers. Distrib. 4: 87–92.

Deysher, L. and T.A. Norton. 1982. Dispersal and colonization in *Sargassum muticum* (Yendo) Fensholt. J. Exp. Mar. Biol. Ecol. 56: 179–195.

Doong, D.J., H.C. Chuang, C.L. Shieh and J.H. Hu. 2011. Quantity, distribution, and impacts of coastal driftwood triggered by a typhoon. Mar. Pollut. Bull. 62: 1446–1454.

Dyke, A.S., J. England, E. Reimnitz and H. Jette. 1997. Changes in driftwood delivery to the Canadian Arctic Archipelago: the hypothesis of postglacial oscillations of the Transpolar Drift. Arctic 50: 1–16.

Edgar, G.J. 1987. Dispersal of fauna and floral propagules associated with drifting *Macrocystis pyrifera* plants. Mar. Biol. 95: 599–610.

Elsey, R.M. and C. Aldrich. 2009. Long-distance displacement of a juvenile alligator by hurricane Ike. Southeast. Nat. 8: 746–749.

Farnsworth, E.J. and A.M. Ellison. 1996. Scale-dependent spatial and temporal variability in biogeography of mangrove root epibiont communities. Ecol. Monogr. 66: 45–66.

Fine, M.L. 1970. Faunal variation on pelagic *Sargassum*. Mar. Biol. 7: 112–122.

Fraser, C.I. 2012. The impacts of past climate change on sub-Antarctic nearshore ecosystems. Pap. Proc. Royal Soc. Tas. 146: 89–93.

Fraser, C.I., R. Nikula, H.G. Spencer and J.M. Waters. 2009. Kelp genes reveal effects of subantarctic sea ice during the Last Glacial Maximum. Proc. Natl. Acad. Sci. USA 106: 3249–3253.

Fraser, C.I., M. Thiel, H. Spencer and J.M. Waters. 2010. Contemporary habitat discontinuity and historic glacial ice drive genetic divergence in Chilean kelp. BMC Evol. Biol. 10: 203.

Fraser, C.I., R. Nikula and J.M. Waters. 2011. Oceanic rafting by a coastal community. Proc. R. Soc. B 278: 649–655.

Fraser, C.I., R. Nikula, D.E. Ruzzante and J.M. Waters. 2012. Poleward bound: biological impacts of Southern Hemisphere glaciation. Trends Ecol. Evol. 27: 462–471.

Fraser, C.I., G.C. Zuccarello, H.G. Spencer, L.C. Salvatore, G.R. Garcia and J.M. Waters. 2013. Genetic affinities between trans-oceanic populations of non-buoyant macroalgae in the high latitudes of the southern hemisphere. PloS ONE 8(7): e69138.

Garden, C.J., D. Craw, J.M. Waters and A. Smith. 2011. Rafting rocks reveal marine biological dispersal: a case study using clasts from beach-cast macroalgal holdfasts. Est. Coast. Shelf Sci. 95: 388–394.

Garden, C.J., K. Currie, C.I. Fraser and J.M. Waters. 2014. Rafting dispersal constrained by an oceanographic boundary. Mar. Ecol. Prog. Ser. 501: 297–302.

Giddings, J.L., Jr. 1952. Driftwood and problems of arctic sea currents. Am. Phil. Soc. Proc. Philadelphia 96: 129–142.

Gladenkov, A.Y., A.E. Oleinik, L. Marincovich and K.B. Barinov. 2002. A refined age for the earliest opening of Bering Strait. Palaeogeogr. Palaeocl. 183: 321–328.

Graham, M.H., B.P. Kinlan and R.K. Grosberg. 2010. Post-glacial redistribution and shifts in productivity of giant kelp forests. Proc. Roy. Soc. B 277: 399–406.

Graiff, A., U. Karsten, S. Meyer, D. Pfender, F. Tala and M. Thiel. 2013. Seasonal variation in the floating persistence of detached *Durvillaea antarctica* (Chamisso) Hariot. Bot. Mar. 56: 3–14.

Gutow, L. and H.D. Franke. 2003. Metapopulation structure of the marine isopod *Idotea metallica*, a species associated with drifting habitat patches. Helgoland Mar. Res. 56: 259–264.

Gutow, L., L. Gimenez, K. Boos and R. Saborowski. 2009. Rapid changes in the epifaunal community after detachment of buoyant benthic macroalgae. J. Mar. Biol. Assoc. UK 89: 323–328.

Gutow, L., J. Beermann, C. Buschbaum, M.M. Rivadeneira and M. Thiel. 2014. Castaways can't be choosers—Homogenization of rafting assemblages on floating seaweeds. J. Sea Res. http://dx.doi.org/10.1016/j.seares.2014.07.005.

Hardy, J.D. 1982. Biogeography of Tobago, West Indies, with special reference to amphibians and reptiles: a review. Bull. Maryland Herp. Soc. 18: 37–142.

Haye, P.A., A.I. Varela and M. Thiel. 2012. Genetic signatures of rafting dispersal in algal-dwelling brooders *Limnoria* spp. (Isopoda) along the SE Pacific (Chile). Mar. Ecol. Prog. Ser. 455: 111–122.

Heatwole, H. and R. Levins. 1972. Biogeography of the Puerto Rican Bank: flotsam transport of terrestrial animals. Ecology 53: 112–117.

Hellmann, L., W. Tegel, Ó. Eggertsson, F.H. Schweingruber, R. Blanchette, A. Kirdyanov, H. Gärtner and U. Büntgen. 2013. Tracing the origin of Arctic driftwood. J. Geophys. Res. Biogeosci. 118: doi:10.1002/jgrg.20022, 2013.

Helmuth, B., R.R. Veit and R. Holberton. 1994. Long-distance dispersal of a subantarctic brooding bivalve (*Gaimardia trapesina*) by kelp-rafting. Mar. Biol. 120: 421–426.

Hernández-Carmona, G., B. Hughes and M.H. Graham. 2006. Reproductive longevity of drifting kelp *Macrocystis pyrifera* (Phaeophyceae) in Monterey Bay, USA. J. Phycol. 42: 1199–1207.

Hinojosa, I., S. Boltana, D. Lancellotti, E. Macaya, P. Ugalde, N. Valdivia, N. Vásquez, W.A. Newman and M. Thiel. 2006. Zoogeography of four pelagic barnacles along the SE-Pacific coast of Chile. Rev. Chil. Hist. Nat. 79: 13–27.

Hinojosa, I.A., M. Pizarro, M. Ramos and M. Thiel. 2010. Spatial and temporal distribution of floating kelp in the channels and fjords of southern Chile. Est. Coast. Shelf Sci. 87: 367–377.

Hinojosa, I.A., M.M. Rivadeneira and M. Thiel. 2011. Temporal and spatial distribution of floating objects in coastal waters of central-southern Chile and Patagonian fjords. Cont. Shelf Res. 31: 172–186.

Hoarau, G., J.A. Coyer, J.H. Veldsink, W.T. Stam and J.L. Olsen. 2007. Glacial refugia and recolonization pathways in the brown seaweed *Fucus serratus*. Mol. Ecol. 16: 3606–3616.

Hobday, A.J. 2000a. Abundance and dispersal of drifting kelp *Macrocystis pyrifera* rafts in the Southern California Bight. Mar. Ecol. Prog. Ser. 195: 101–116.

Hobday, A.J. 2000b. Persistence and transport of fauna on drifting kelp (*Macrocystis pyrifera* (L.) C. Agardh) rafts in the Southern California Bight. J. Exp. Mar. Biol. Ecol. 253: 75–96.

Huffard, C.L., S. von Thun, A.D. Sherman, K. Sealey and K.L. Smith, Jr. 2014. Pelagic *Sargassum* community change over a 40-year period: temporal and spatial variability. Mar. Biol. DOI 10.1007/s00227-014-2539-y.

Ingólfsson, A. 1995. Floating clumps of seaweed around Iceland: natural microcosms and a means of dispersal for shore fauna. Mar. Biol. 122: 13–21.

Ingólfsson, A. 1998. Dynamics of macrofaunal communities of floating seaweed clumps off western Iceland: a study patches on the surface of the sea. J. Exp. Mar. Biol. Ecol. 231: 119–137.

Ingólfsson, A. 2000. Colonization of floating seaweed by pelagic and subtidal benthic animals in southwestern Iceland. Hydrobiologia 440: 181–189.

IPCC. 2007. Climate change 2007: the physical science basis. *In*: Solomon et al. (eds.). Contribution of Working Group I to the 4th Assessment Report of the Intergovernmental Panel on Climate Change. Intergovernmental Panel on Climate Change Cambridge, UK.

Islam, M.S., C. Lian, N. Kameyama and T. Hogetsu. 2012. Analyses of genetic population structure of two ecologically important mangrove tree species, *Bruguiera gymnorrhiza* and *Kandelia obovata* from different river basins of Iriomote Island of the Ryukyu Archipelago, Japan. Tree Genet. Genomes 8: 1247–1260.

Johansen, S. 1999. Origin of driftwood in north Norway and its relevance for transport routes of drift ice and pollution to the Barents Sea. Sci. Total Environ. 231: 201–225.

Johansen, S. and H. Hytteborn. 2001. A contribution to the discussion of biota dispersal with drift ice and driftwood in the North Atlantic. J. Biogeogr. 28: 105–115.

Johannesson, K. 1988. The paradox of Rockall: why is a brooding gastropod (*Littorina saxatilis*) more widespread than one having a planktonic larval dispersal stage (*L. littorea*)? Mar. Biol. 99: 507–513.

Johannesson, K. and T. Warmoes. 1990. Rapid colonization of Belgian breakwaters by the direct developer, *Littorina saxatilis* (Olivi) (Prosobranchia, Mollusca). Hydrobiologia 193: 99–108.

Johnson, M.P., A.L. Allcock, S.E. Pye, S.J. Chambers and D.M. Fitton. 2001. The effects of dispersal mode on the spatial distribution patterns of intertidal molluscs. J. Anim. Ecol. 70: 641–649.

King, W. 1962. The occurrence of rafts for dispersal of land animals into the West Indies. Q. J. Florida Acad. Sci. 25: 45–52.

Kingsford, M.J. 1992. Drift algae and small fish in coastal waters of northeastern New Zealand. Mar. Ecol. Prog. Ser. 80: 41–55.

Kingsford, M.J. 1993. Biotic and abiotic structure in the pelagic environment: importance to small fishes. Bull. Mar. Sci. 53: 393–415.

Kingsford, M.J. 1995. Drift algae—A contribution to near-shore habitat complexity in the pelagic environment and an attractant for fish. Mar. Ecol. Prog. Ser. 116: 297–301.

Kingsford, M.J. and J.H. Choat. 1985. The fauna associated with drift algae captured with a plankton-mesh purse seine net. Limnol. Oceanogr. 30: 618–630.

Komatsu, T., D. Matsunaga, A. Mikami, T. Sagawa, E. Boisnier, K. Tatsukawa, M. Aoki, T. Ajisaka, S. Uwai, K. Tanaka, K. Ishida, H. Tanoue and T. Sugimoto. 2008. Abundance of drifting seaweeds in eastern East China Sea. J. Appl. Phycol. 20: 801–809.

Komatsu, T., S. Mizuno, A. Natheer, A. Kantachumpoo, K. Tanaka, A. Morimoto, S.T. Hsiao, E.A. Rothäusler, H. Shishidou, M. Aoki and T. Ajisaka. 2014. Unusual distribution of floating seaweeds in the East China Sea in the early spring of 2012. J. Appl. Phycol. 26: 1169–1179.

Lindstrom, S.C. 2001. The Bering Strait connection: dispersal and speciation in boreal macroalgae. J. Biogeogr. 28: 243–251.

Macaya, E.C. and G.C. Zuccarello. 2010. Genetic structure of the giant kelp *Macrocystis pyrifera* along the southeastern Pacific. Mar. Ecol. Prog. Ser. 420: 103–112.

Macaya, E.C., S. Boltaña, I.A. Hinojosa, J.E. Macchiavello, N.A. Valdivia, N. Vásquez, A.H. Buschmann, J. Vásquez, J.M.A. Vega and M. Thiel. 2005. Presence of sporophylls in floating kelp rafts of *Macrocystis* spp. Phaeophyceae) along the Chilean Pacific coast. J. Phycol. 41: 913–922.

Macreadie, P.I., M.J. Bishop and D.J. Booth. 2011. Implications of climate change for macrophytic rafts and their hitchhikers. Mar. Ecol. Prog. Ser. 443: 285–292.

Maggs, C.A., R. Castilho, D. Foltz, C. Henzler, M.T. Jolly, J. Kelly, J. Olsen, K.E. Perez, W. Stam, R. Vainola, F. Viard and J. Wares. 2008. Evaluating signatures of glacial refugia for North Atlantic benthic marine taxa. Ecology 89: S108–S122.

Marmorino, G.O., W.D. Miller, G.B. Smith and J.H. Bowles. 2011. Airborne imagery of a disintegrating *Sargassum* drift line. Deep Sea Res. 58: 316–321.

Mason, R. 1961. Dispersal of tropical seeds by ocean currents. Nature 191: 408–409.

McKenzie, P.F. and A. Bellgrove. 2008. Dispersal of *Hormosira banksii* (Phaeophyceae) via detached fragments: Reproductive viability and longevity. J. Phycol. 44: 1108–1115.

Minchinton, T.E. 2006. Rafting on wrack as a mode of dispersal for plants in coastal marshes. Aquat. Bot. 84: 372–376.

Miranda, L. and M. Thiel. 2008. Active and passive migration in boring isopods *Limnoria* spp. (Crustacea, Peracarida) from kelp holdfasts. J. Sea Res. 60: 176–183.

Mitchell, C.T. and J.R. Hunter. 1970. Fishes associated with drifting kelp, *Macrocystis pyrifera*, off the coast of southern California and northern Baja California. Calif. Fish Game 56: 288–297.

Moser, M.L. and D.S. Lee. 2012. Foraging over *Sargassum* by Western North Atlantic Seabirds. Wilson J. Ornithol. 124: 66–72.

Muhlin, J.F., C.R. Engel, R. Stessel, R.A. Weatherbee and S.H. Brawley. 2008. The influence of coastal topography, circulation patterns, and rafting in structuring populations of an intertidal alga. Mol. Ecol. 17: 1198–1210.

Müller, R., T. Laepple, I. Bartsch and C. Wiencke. 2009. Impact of oceanic warming on the distribution of seaweeds in polar and cold-temperate waters. Bot. Mar. 52: 617–638.

Nikula, R., C.I. Fraser, H.G. Spencer and J.M. Waters. 2010. Circumpolar dispersal by rafting in two subantarctic kelpdwelling crustaceans. Mar. Ecol. Prog. Ser. 405: 221–230.

Nikula, R., H.G. Spencer and J.M. Waters. 2013. Passive rafting is a powerful driver of transoceanic gene flow. Biol. Lett. http://dx.doi.org/10.1098/rsbl.2012.0821.

Norton, T.A. 1991. Conflicting constraints on the form of intertidal algae. Brit. Phycol. J. 26: 203–218.

Ólafsson, E., A. Ingólfsson and M.B. Steinarsdóttir. 2001. Harpacticoid copepod communities of floating seaweed: controlling factors and implications for dispersal. Hydrobiologia 453: 189–200.

Olsen, J.L., F.W. Zechman, G. Hoarau, J.A. Coyer, W.T. Stam, M. Valero and P. Åberg. 2010. The phylogeographic architecture of the fucoid seaweed *Ascophyllum nodosum*: an intertidal 'marine tree' and survivor of more than one glacial-interglacial cycle. J. Biogeogr. 37: 842–856.

Prescott, J.H. 1959. Rafting of jack rabbit on kelp. J. Mammal. 40: 443.

Provan, J., R.A. Wattier and C.A. Maggs. 2005. Phylogeographic analysis of the red seaweed *Palmaria palmata* reveals a Pleistocene marine glacial refugium in the English Channel. Mol. Ecol. 14: 793–803.

Raxworthy, C.J., M.R. Forstner and R.A. Nussbaum. 2002. Chameleon radiation by oceanic dispersal. Nature 415: 784–787.

Reisser, C.M.O., A.R. Wood, J.J. Bell and J.P.A. Gardner. 2011. Connectivity, small islands and large distances: the *Cellana strigilis* limpet complex in the Southern Ocean. Mol. Ecol. 20: 3399–3413.

Rothäusler, E., I. Gómez, I.A. Hinojosa, U. Karsten, F. Tala and M. Thiel. 2009. Effect of temperature and grazing on growth and reproduction of floating *Macrocystis* spp. (Phaephyceae) along a latitudinal gradient. J. Phycol. 45: 547–559.

Rothäusler, E., I. Gómez, U. Karsten, F. Tala and M. Thiel. 2011. UV-radiation versus grazing pressure: long-term floating of kelp rafts (*Macrocystis pyrifera*) is facilitated by efficient photoacclimation but undermined by grazing losses. Mar. Biol. 158: 127–141.

Rothäusler, E., L. Gutow and M. Thiel. 2012. The ecology of floating seaweeds. Ecol. Studies 219: 359–380.

Rouch, J. 1954. Les Îles flottantes. Arch. Meteor. Geophy. A 7: 528–532.

Sano, M., M. Omori and K. Taniguchi. 2003. Predator-prey systems of drifting seaweed communities off the Tohoku coast, northern Japan, as determined by feeding habit analysis of phytal animals. Fisheries Sci. 69: 260–268.

Segawa, S., T. Sawada, M. Higaki, T. Yoshida, H. Ohshiro and F. Hayashida. 1962. Some comments on the movement of the floating seaweeds. Rec. Oceanogr. Work. Jpn. 6: 153–159.

Smith, S.D.A. 2002. Kelp rafts in the Southern Ocean. Global Ecol. Biogeogr. 2: 67–69.

Smith, J.M.B., H. Heatwole, M. Jones and B.M. Waterhouse. 1990. Drift disseminules on cays of the Swain Reefs, Great Barrier Reef, Australia. J. Biogeogr. 17: 5–17.

Steinke, T.D. 1986. A preliminary study of buoyancy behaviour in *Avicennia marina* propagules. S. Afr. J. Bot. 52: 559–565.

Steinke, T.D. and C.J. Ward. 2003. Use of plastic drift cards as indicators of possible dispersal of propagules of the mangrove *Avicennia marina* by ocean currents. Afr. J. Mar. Sci. 25: 169–176.

Stewart, H.L. 2008. The role of spatial and ontogenetic morphological variation in the expansion of the geographic range of the tropical brown alga, *Turbinaria ornata*. Integr. Comp. Biol. 48: 713–719.

Svavarsson, J., M.K.W. Osore and E. Ólafsson. 2002. Does the wood-borer *Sphaeroma terebrans* (Crustacea) shape the distribution of the mangrove *Rhizophora mucronata*? Ambio 31: 574–579.

Tala, F., I. Gómez, G. Luna-Jorquera and M. Thiel. 2013. Morphological, physiological and reproductive conditions of rafting bull kelp (*Durvillaea antarctica*) in northern-central Chile (30°S). Mar. Biol. 160: 1339–1351.

Thiel, M. 2003. Rafting of benthic macrofauna: important factors determining the temporal succession of the assemblage on detached macroalgae. Hydrobiologia 503: 49–57.

Thiel, M. and L. Gutow. 2005a. The ecology of rafting in the marine environment. I. The floating substrata. Oceanogr. Mar. Biol. Annu. Rev. 42: 181–263.

Thiel, M. and L. Gutow. 2005b. The ecology of rafting in the marine environment. II. The rafting organisms and community. Oceanogr. Mar. Biol. Annu. Rev. 43: 279–418.

Thiel, M. and P.A. Haye. 2006. The ecology of rafting in the marine environment. III. Biogeographical and evolutionary consequences. Oceanogr. Mar. Biol. Annu. Rev. 44: 323–429.

Thomson, A.C., P.H. York, T.M. Smith, C.D. Sherman, D.J. Booth, M.J. Keough, D.J. Ross and P.L. Macreadie. 2014. Seagrass viviparous propagules as a potential long-distance dispersal mechanism. Estuar. Coast. DOI 10.1007/s12237-014-9850-1.

van den Hoek, C. 1987. The possible significance of long-range dispersal for the biogeography of seaweeds. Helgoländer Meeresunters. 41: 261–272.

Vandendriessche, S., G. De Keersmaecker, M. Vincx and S. Degraer. 2006a. Food and habitat choice in floating seaweed clumps: the obligate opportunistic nature of the associated macrofauna. Mar. Biol. 149: 1499–1507.

Vandendriessche, S., M. Vincx and S. Degraer. 2006b. Floating seaweed in the neustonic environment: a case study from Belgian coastal waters. J. Sea Res. 55: 103–112.

Vandendriessche, S., M. Vincx and S. Degraer. 2007a. Floating seaweed and the influences of temperature, grazing and clump size on raft longevity—A microcosm study. J. Exp. Mar. Biol. Ecol. 343: 64–73.

Vandendriessche, S., E.W.M. Stienen, M. Vincx and S. Degraer. 2007b. Seabirds foraging at floating seaweeds in the Northeast Atlantic. Ardea 95: 289–298.

Vandendriessche, S., M. Messiaen, S. O'Flynn, M. Vincx and S. Degraer. 2007c. Hiding and feeding in floating seaweed: floating seaweed clumps as possible refuges or feeding grounds for fishes. Est. Coast. Shelf Sci. 71: 691–703.

Vásquez, J.A. 1993. Effects on the animal community of dislodgement of holdfasts of *Macrocystis pyrifera*. Pac. Sci. 47: 180–184.

Viejo, R.M. and P. Åberg. 2003. Temporal and spatial variation in the density of mobile epifauna and grazing damage on the seaweed *Ascophyllum nodosum*. Mar. Biol. 142: 1229–1241.

Wares, J.P. 2001. Intraspecific variation and geographic isolation in *Idotea balthica* (Isopoda: Valvifera). J. Crust. Biol. 21: 1007–1013.

Waters, J.M., C.I. Fraser and G.M. Hewitt. 2013. Founder takes all: density-dependent processes structure biodiversity. Trends Ecol. Evol. 28: 78–85.

Wernberg, T., M.S. Thomsen, F. Tuya, G.A. Kendrick, P.A. Staehr and B.D. Toohey. 2010. Decreasing resilience of kelp beds along a latitudinal temperature gradient: potential implications for a warmer future. Ecol. Lett. 13: 685–694.

Wernberg, T., B.D. Russell, M.S. Thomsen, C.F.D. Gurgel, C.J.A. Bradshaw, E.S. Poloczanska and S.D. Connell. 2011. Seaweed communities in retreat from ocean warming. Curr. Biol. 21: 1828–1832.

Wheeler, W.M. 1916. Ants carried in a floating log from the Brazilian coast to San Sebastian Island. Psyche 23: 180–183.

Witherington, B., S. Hirama and R. Hardy. 2012. Young sea turtles of the pelagic *Sargassum*-dominated drift community: habitat use, population density, and threats. Mar. Ecol. Prog. Ser. 463: 1–22.

Worcester, S.E. 1994. Adult rafting versus larval swimming—dispersal and recruitment of a botryllid ascidian on eelgrass. Mar. Biol. 121: 309–317.

5

The Role of Drifting Algae for Marine Biodiversity

Nina Larissa Arroyo[1,]* and *Erik Bonsdorff*[2]

Introduction

Drifting algal mats are conglomerates of (usually) ephemeral filamentous algae that are detached from their parent algal areas and end up forming clumps or mats of different sizes. They drift moved by winds or hydrodynamic forces until they disintegrate or aggregate covering wide extensions of the sea floor, macrophytic stands or coastal structures such as coral reefs, constituting the main benthic species implicated in nuisance blooms worldwide. The massive appearance of these algae is often the result of increased nutrient loading leading to eutrophication in coastal areas, the algae being frequently opportunistic, filamentous species that take advantage of increased nutrient inputs and develop in large amounts. If swept by strong currents they may reach far away areas affecting the dynamics of various habitats. More often they are gathered close to their areas of origin, where they have major impacts on sea-floor dynamics, causing hypoxia and considerable losses to benthic life in areas in which they cover large surfaces for long periods of time. When occurring intermittently in small to moderate amounts however, they provide a new and rich habitat for a variety of phytal and sediment species, increasing structural and functional diversity and production. They may also act as nursery grounds and shelter for juvenile fish. Moreover, they act as a means of transport for various species of invertebrates, enhancing colonization of new areas and recovery of previously disturbed patches by often the same opportunistic invertebrate species that can tolerate the conditions set forth by the algal accumulations.

[1] c/Honduras 2, 6°dcha, Santander 39005, Cantabria, Spain.
[2] Environmental and Marine Biology, Åbo Akademi University, BioCity, Artillerigatan 6, FI-20520 Turku Finland.
  Email: ebonsdor@abo.fi
* Corresponding author: nlarroyo@gmail.com

What are and what causes drifting algae?

Bloom-forming species are typically characterized by finely structured thalli with large surface area-to-volume ratios (either thin and blade-like or filamentous and repeatedly branched) with a high capacity for nutrient uptake and growth (Wallentinus 1984, Lotze and Schramm 2000). They most commonly consist of green algae such as *Enteromorpha*,[1] *Ulva, Chaetomorpha* or *Cladophora* (Valiela et al. 1997, Raffaelli et al. 1998), though other green, red (e.g., *Ceramium, Porphyra, Spyridia, Gracilaria, Hypnea* (Bell and Hall 1997, Arroyo et al. 2006, Thomsen et al. 2012a)) and brown species of macroalgae (e.g., *Pilayella, Ectocarpus*) can be bloom forming as well (Valiela et al. 1997, Berger et al. 2003, Arroyo et al. 2006, Lyons et al. 2009).

Due to the important differences in their growth form and ecology, different species may cause a variety of effects. For example, while *Ulva* is usually loosely associated with the sediment surface, in its *Enteromorpha* morph it is, like *Chaetomorpha*, more intimately associated with the sediment, often creating a dense matrix of filaments extending several cm beneath the surface. Also, different bloom forming species will show varying strategies and respond differently to environmental factors which will condition their bloom forming capability and the seasonality of the phenomena. Different *Ulva* species, for instance, have shown varying tolerances to stress (Wang et al. 2012), and while *U. prolifera* has a higher capacity, plasticity and more efficient nutrient absorption and assimilation systems, which allows it to successfully compete with *U. intestinalis* under optimal conditions, the latter is more stable and has a higher tolerance to stress provided by its antioxidant system, being more successful and resistant under limiting conditions (Wang et al. 2012).

The most common explanation for the increase in algal bloom occurrence is eutrophication and more particularly, increased availability of nitrogen and phosphorous originating from land—derived anthropogenic sewage or fertilization runoff (Raffaelli et al. 1998, Österling and Pihl 2001, Wang et al. 2012). NH_4^+ inputs from raw sewage, aquaculture activities, and dissolved organic nitrogen from benthic nutrient regeneration are increasingly more common, and also support macroalgal production (Sundbäck et al. 2003, Tsai et al. 2005). While nitrogen supply seems to control the peak seasonal rates of macroalgal growth in most coastal systems, phosphorous alone may limit production of macrophytes at certain times of the year and is often more limiting in tropical and carbonate-rich, than in temperate waters (Lapointe et al. 1992, Valiela et al. 1997, and references therein). Different species may show varying levels of nutrient limitation and diverse storage capacities, hence responding differently to N inputs and blooming accordingly, at various nitrogen levels (Teichberg et al. 2008). Moreover, differences in nutrient supply may cause changes in the morphology of macroalgae (Valiela et al. 1997), while different morphologies will respond differently to nutrient inputs (Teichberg et al. 2008, and references therein).

Besides, a species may be limited by different elements depending on nutrient levels; for example, as coastal waters increasingly undergo nutrient enrichment

---

[1] Despite findings reported in Hayden et al. (2003), regarding the conspecificity of both species, for clarity and consistency with previous publications, we will keep the denominations, *Ulva* and *Enteromorpha* as referenced by the respective authors in their contributions.

the limiting element supporting growth of *Ulva* spp. is initially nitrogen, shifting to phosphorous in those waters subject to the highest nitrogen loadings (Teichberg et al. 2010). Availability of other elements may also be determinant, for instance, iron depletion might be critical in *Ulva* growth during bloom formation, limiting N acquisition by the algae (Viaroli et al. 2005), while iron-sulphur-phosphorous interactions may condition macroalgal development, especially at the later stages of the bloom, when anoxic conditions, and high sulphide production occur (Viaroli et al. 2005). Changes in water circulation have also been invoked as a likely cause and wind-driven upwelling or tidal events, which replenish nutrients have also been linked to occurrences of blooms (Kiirikki and Blomster 1996, Schramm 1999). The conspicuous mass development of floating green algae along the open Atlantic coast of Brittany (France), known as "green tides", was explained by a specific hydrological regime in combination with high nutrient loads deriving from agriculture (Piriou and Menesguen 1992). In China, the large macroalgal blooms occurring along the coast of Qingdao have been attributed to cultivation of another alga, *Porphyra yezoensis* far away from the Qingdao area (Liu et al. 2009), indicating that the source for green tides may be found at long distances from their actual occurrence. More recently increasing evidence of a relationship between increasing eutrophication and the removal of top predators has been pointed out (Eriksson et al. 2009, Sieben et al. 2011a,b), authors suggesting cascading effects of top-predator removal are causing a decrease in grazers as a result of the release of intermediate predators, and this might be contributing to the expansion and overgrowth of opportunistic filamentous algae. This mechanism still needs both experimental and field evidence, however.

Factors favouring drift algal accumulation

The effects of drifting algal mats depend largely on their residence time in a specific location, which is mostly conditioned by the flow velocities of currents at that site and the roughness of the substrate (Berglund et al. 2003, Biber 2007). However, a number of physical, chemical and biological factors may also have a bearing on their persistence in a particular place, hampering our capability to predict their evolution and impacts. In general, factors such as nutrient supply, temperature, turbidity, bed stability, hydrography and the amount and type of substratum suitable for algal growth have been found to be important limiting factors for macroalgal bloom occurrence (Tagliapietra 1998, Biber 2007, Scanlan et al. 2007).

For instance, Scanlan et al. (2007) pointed out that since opportunistic algae cannot exist above the high tide limit nor grow at depths where turbidity levels limit light intensity, the standing stock in an estuary must be limited by the total available intertidal area. CEFAS (2004) found no evidence that factors such as attachment points, grazing or bed biogeochemistry played a significant part in determining macroalgal growth, though they reckoned that at a local level these may play a modifying role. On the other hand, Bell and Hall (1997) found that salinity and prevailing wind direction did not adequately predict large scale differences in drift macroalgal abundance among sites, but rather local energy regimes as shown by percent silt-cay better explained these variations. They suggested that less-active environments may tend to capture/accumulate algae at a greater rate than those sites with high current velocity or extensive exposure to waves. This was also found in the Baltic by Berglund et al. (2003), who concluded that occurrence of drift algal accumulations was affected by exposure, localities with higher exposure being more prone to algal accumulations, despite the algae being mainly originated at sheltered sites, where they also built up extensive biomasses.

World occurrence of drifting algae

Bloom forming algae are an increasing problem worldwide (see tables with compilations of cases and/or their effects in for example: Fletcher 1996, Morand and Briand 1996, Valiela et al. 1997, Morand and Merceron 2004, Teichberg et al. 2010), their extended and prolonged occurrence being reported more and more often from nearly every corner of the planet, especially since the 1980's. They occur mainly in the North Temperate Zone, America, Europe and Asia-Pacific being the most affected areas. They are typically found in eutrophic estuaries and inlets, where water residence times are long enough for macroalgae to absorb large pulses of nutrients and form blooms (Valiela et al. 1997), though they can also occur along well-flushed open coastlines (Piriou and Menesguen 1992, Lyons et al. 2009) if transported to beaches via coastal currents or upwelling phenomena (Kiirikki and Blomster 1996, Liu et al. 2009).

Once bloom-forming algae establish at a site, they may facilitate future blooms, becoming a recurrent problem in a specific area (Raffaelli et al. 1998). They may continue to grow even under low N concentrations, persisting over time also in places far away from nutrient input sources (Thornber et al. 2008).

One of the areas where drifting algae have become more persistent and caused larger problems in coastal ecosystems is the Baltic Sea (Fig. 1A). On the Swedish Skagerrak coast, shallow (0-1 m) soft-bottom bays were generally free from vegetation in the 1970's, the situation having changed dramatically over the past decades. Filamentous macroalgae appeared in these areas in the 1980's and have since increased, covering up to 30–50% of the shallow soft bottom areas from May to September (Pihl et al. 1999). Here, the increase in macroalgae is thought to be due to a fourfold increase in anthropogenic nitrogen to these areas over the last century (Österling and Pihl 2001).

Large patches (ca. 30 ha) of decomposing algae of varying thickness and composition have also been recorded in the Archipelago Sea, northern Baltic Sea (Vahteri et al. 2000), while in the Gulf of Finland, green tides occurred regularly every summer since 1992 and 2000 (Blomster et al. 2002).

Along the coast of Brittany, already in 1986, 25,000 m^3 wet weight of *Ulva* spp. accumulated in the Lannion Bay (Charlier et al. 2007), while nearly 100, 000t ww of algae are collected annually in the area (Dion and Le Bozec 1996, Fig. 1B). Long-term studies and on-going research indicates that the effects on the intertidal (tidal range: > 6 m) sandy beaches are both structurally and functionally fundamental (Bolam et al. 2000, Bohorquez et al. 2013). Other European areas include the notorious case

Figure 1. (A) Drifting algae being washed ashore on a sandy beach in Finland, northern Baltic Sea (photo: Erik Bonsdorff). (B) An *Ulva* bloom in Brittany (Photo: Nolwenn Quillien).

of the Venice and other north Adriatic lagoons and beaches, which were repeatedly inundated with *Ulva* spp. especially during the late 1980's and early 1990's (Vukadin 1991, Ravera 2000, Viaroli et al. 2005).

In the USA, the development of large algal blooms along the south-eastern coast of Florida has been occurring repeatedly (Virnstein and Carbonara 1985). Other examples are the massive ephemeral blooms of *Cladophora vagabunda* in Waquoit Bay, Massachusetts (Valiela et al. 1992), or nuisance blooms of *C. prolifera* in Bermuda (Lapointe and O'Connell 1989). Persistent blooms of *Cladophora sericea* have also been reported from West Maui Hawai'i, where the filamentous algae has been appearing seasonally since the mid 1980's, growing from the shore to depths greater than 30 m (Smith et al. 2005).

In Australia, algal blooms have been reported repeatedly from Peel Harvey estuary, where a bloom of *Cladophora* lasted for more than 10 years (Gordon and McComb 1989). Since 2002, the usually uncommon endemic filamentous brown alga *Hincksia sordida* has formed nuisance blooms annually during spring/early summer at Main Beach, Noosa on the subtropical east Australian coast (Phillips 2006).

Large scale green tides in China's Yellow Sea have broken out continuously from 2007 to 2011, posing serious problems to navigation, fisheries and tourism (Wang et al. 2012). The so-called "largest macroalgal bloom" ever reported was a massive green tide of about 600 km^2 occurring along the coast of Qingdao in 2008, just weeks before the Beijing Olympic games (Liu et al. 2009).

Seasonality and growth patterns

Drift macroalgal blooms have typically been described in the literature as following broadly consistent temporal trends in abundance, which may moderate their impact, shortening exposure times for both attached macrophytes and benthic fauna (Pihl et al. 1996, Rasmussen et al. 2013). In the Baltic, blooms typically peak in late summer (e.g., Berglund et al. 2003) due to light and nutrient levels, water temperature, and other factors associated with seasonality. The proliferation of algal mats from early spring to late summer has also been identified as a clear sign of nutrient enrichment in certain areas such as the Mondego estuary (Cabral et al. 1999), since during this period high temperatures, and river runoff due to ice melting and spring rains are maximal. However, in Florida, several authors (e.g., Benz et al. 1979, Virnstein and Carbonara 1985) noted high peaks of macroalgae in spring months, and a decline in summer. Moreover, persistence of macroalgal mats through winter has been reported in the Baltic Sea (Bäck et al. 2000, Salovius and Bonsdorff 2003), while somatic cells of *Ulva prolifera* lying in the sediment have been found to act as overwintering stages for the annual spring blooms occurring later in the year (Zhang et al. 2010). Temporal abundance patterns may be related to a combination of salinity, temperature and light conditions, not only within the locations where they accumulate but also at off site sources (Bell and Hall 1997, Höffle et al. 2012). Gubelit and Berezina (2010) described two peaks of *Cladophora glomerata* growth in the Neva estuary, in July and autumn which they attributed to maximal growth rates during summer and growth from summer zoospore development during autumn, respectively. Subtidal stocks of several thousand tons of size on average have also been detected lying at depths down to 15 m in some sites of the Brittany coast, and suspected to be the source of inoculum for the intertidal Spring blooms occurring intertidally (Merceron and Morand 2004).

The beginning of macroalgal mass development is usually determined by the ecophysiological background for timing and the rate of germination, which differs between species (Schramm 1999). This may condition dominance by a specific species and their evolution throughout the blooming season. For instance, field observations and laboratory experiments have shown the importance of the mode of resting stages and overwintering, as well as of abiotic environmental factors, in particular light, temperature, and nutrient supply, in stimulating germination (Lotze et al. 1999). Lotze et al. (1999) revealed pronounced differences between *Pilayella* and *Enteromorpha* due to species-specific ecophysiological requirements for germination. There are also differences in the appetence for germlings or adult stages of different species by mesograzers (Lotze et al. 1998), which may indicate the importance of grazer diversity for an effective control of bloom-forming algae (Myers and Heck 2013). Temperature seems to control algal growth rate also while the plants are floating and several bloom-forming species have been found to do best at specific temperature ranges when their increase is maximal (Liu et al. 2009). When sufficient light, nutrient availability and optimal temperature ranges occur together, algal proliferation is guaranteed and the formation and persistence of floating algal conglomerates can be expected.

Sessile filamentous algae gradually detach from rocky areas towards the end of their life cycles. Loose filaments are either washed up on the shore, or sink to the bottom, often below the photic layer, where algal degradation proceeds. The degradation process is strongly temperature dependent and seems to occur at a lower rate under anaerobic conditions (García-Robledo et al. 2008), which causes a higher degree of organic carbon preservation in anoxic sediments. Decomposition of the algal detritus is known to be mediated by bacteria and fungi, though it initially involves the rapid leaching of soluble materials in a cellular autolytic process that is independent of microbial activity (García Robledo et al. 2008). This is followed by a phase in which microbial activity continues the degradation process (García-Robledo et al. 2008). Algal respiration and bacterial activity during decomposition induce oxygen deficiency and anoxia in the interstitial (Hull 1987) and the algal—sediment interface (Sundbäck et al. 1990, Norkko and Bonsdorff 1996a,b). Bacterial activity is often intense during hypoxia and anoxia, in particular by sulphur bacteria of the genus *Beggiatoa*, which can be seen as a white cover on the sediment (Rosenberg and Diaz 1993) and on the surface of the algal mat (Holmström 1998, Vahteri et al. 2000). Highly toxic hydrogen sulphide is further produced by sulphate respiring bacteria intervening in organic matter decomposition and to a lesser extent also by the fermentative degradation of proteins (Fenchel et al. 1998). Moreover, the anaerobic state at the bottom induces the release of nutrients (Norkko and Bonsdorff 1996a,b,c, Gubelit and Berezina 2010) which are further utilised by the next generation of opportunistic micro- and macroalgal species (García-Robledo et al. 2008, Corzo et al. 2009), in what has been termed a self-regeneration process of the algal mats. On the other hand, bacterial enrichment of the algal mass converts it into a large food source for many organisms, as illustrated by Vetter (1994), who recorded extremely high levels of benthic secondary production in connection to macroalgal debris.

Decomposition rates are initially rapid but decline with time, when the more labile constituents have decomposed. The process is variable depending on differences in chemical composition of the particulate organic matter, particle size and nutrient availability, among other factors, it is species-specific, and may last from months to years. Holmström (1998) studied the leakage of nutrients from drifting filamentous algae and concluded that most of the phosphorous leakage (mainly as phosphate) occurred within 1 week, while that of nitrogen (mainly as ammonia) was less concentrated. In general, degradability of organic matter is inversely related to its C:N ratio, and therefore macroalgal detritus is more refractory than that of, for example microphytobenthos. Salovius and Bonsdorff (2003) found that algal decomposition was always more rapid close to the sediment than in the water column, and at shallow vs. deeper levels within the water column. They attributed these differences to variations in temperature between depths, but mainly to the larger bacterial populations and metabolic activity occurring close to the sediment as compared to those in the water column. They also suggested an active decomposition by grazers (mainly gammarid amphipods), which would thrive in the decomposing algal habitat, with plenty food resources, abundant diatoms and bacteria, and shelter from predators.

Algae may also drift off-shore reaching other coastal areas or float alive in the water column for extended periods of time. Lyons et al. (2009) found a significant negative relationship between algal density and NO-concentration, and suggested that algae were likely using nutrients to proliferate while drifting, their growth maybe being nutrient-limited in dense aggregations (Escartin and Aubrey 1995). Fragments of *Ulva prolifera* released from nets and rafts used for the aquaculture of *Porphyra yezoensis* were hypothesized to keep growing in the water and transported over long distances, eventually forming the green tide originated along the Qingdao coast in 2008 (Liu et al. 2009). Moreover, Vermeij et al. (2009), showed that increased nutrient availability increased the dispersal potential of the invasive alga *Hypnea musciformis*. While growth rates were reduced, its recruitment success increased since it was able to attach more easily to other algae the smaller their fragment size. This species smothers native reef communities and forms localized blooms. The authors suggested that nutrient reductions should reduce their dispersive ability through a reduction in the life-span of their fragments and highlighted that drifting fragments of this and other invasive species form a robust life-phase which allows them to maximize their dispersive range and their success as invasive species.

Macroalgal blooms have sometimes terminated or suffered unexplained collapses after increases in nutrient loads through a certain number of years, the explanation possibly being the increased stimulation of phytoplankton and consequent light interception in the water column, which would eventually cause shading of the macrophytes lying below (Lavery et al. 1991, Valiela et al. 1997). Conversely, other studies have reported a strong sensitivity of *Ulva* species to light, and a halt in their growth every year at the beginning of summer, with increasing light strengths and periods (Piriou and Duval 1990).

Implications for biodiversity—effects of drifting algal mats on macrophytic communities and benthic fauna

Macroalgal blooms in almost all cases end up displacing seagrasses, corals or brown and red algae of the areas which they occupy (see table of effects in Valiela et al. 1997). The degree to which these effects cause changes in the affected communities is also varied, and may depend on a series of factors as the already mentioned residence time or the disturbance history of the locations where they occur, more perturbed sites being less resistant to the additional impacts caused by the algae (Cardoso et al. 2004). For example, Tagliapietra et al. (1998) concluded that the benthic community at Palude della Rosa was led by the overgrowth of the green macroalga *Ulva rigida*, its evolution causing notable differences in benthic community structure. While in low amounts, *Ulva* contributed to the water column oxygenation and became a suitable substratum for a rich community, when surpassing its carrying capacity it acquired a very noxious character, devastating the benthos of the lagoon.

Macroalgal blooms in coastal areas may be large enough to create significant biogeochemical changes (Valiela et al. 1997). In particular, they seem to inhibit the photosynthetic activity of the microphytobenthos (Corzo et al. 2009, García-Robledo and Corzo 2011), shifting community structure from a diatom—to a cyanobacteria dominated one, and reducing sediment metabolism (which turns heterotrophic), with reduced coupled nitrification-denitrification rates (García-Robledo and Corzo 2011). Benthic cyanobacteria are better adapted to the light conditions prevailing under the algal mats, due to their photosynthetic pigments (phycocyanin and phycoerythrin), and will have higher photosynthetic efficiencies compared with diatoms below the mats (García-Robledo et al. 2012). This community photo-acclimation may provide sediments with increased resilience to the detrimental effects of the macroalgal mats (García-Robledo et al. 2012). Moreover, cyanobacteria are more resistant to the anoxic conditions occurring under the algal mats and have a higher capacity for using organic substrates, which may favour their growth under conditions prevailing under the algae (García-Robledo et al. 2008, and references therein). Even at low densities, shading by macroalgal canopies can negatively affect the primary production of microphytobenthos, and the redox status of surface sediments, though inhibition of the autotrophic metabolism is quickly re-established with macro-algal detritus disintegration (García-Robledo et al. 2008). Under heavy algal mat colonization however, the inhibition of the microphytobenthos photosynthetic activity might have cascading effects on sediment biogeochemistry, by altering rates of carbon and nitrogen biogeochemical cycles (Corzo et al. 2009, García-Robledo et al. 2008).

Sundbäck et al. (1990) found no significant flow of organic material from the drifting algae to the sediment, concluding that the mat constituted an apparently independent habitat in terms of particulate organic carbon inputs to the sediment. In their study, most of the macroalgal carbon was mineralized within 3 weeks in the algal mat itself, and was released as CO_2 and as dissolved organic matter (DOM). However, they reckoned that carbon might be transferred when algal filaments are planted into the sediment or when grazers fragment and graze the algae. These authors also found inorganic nutrients (mainly ammonium) to be released only under very low oxygen levels. Conversely, Norkko and Bonsdorff (1996c) found a quick deposition of organic material from the algae to the sediment, while Kautsky (1995), calculated that the detached filamentous algae contributed up to 50% of the annual input of POC to coastal sediment communities. Corzo et al. (2009) found that macroalgae increased the content of inorganic nutrients and the mean annual contents of C and N in the sediment. They reflected that oxygen availability might be determinant in establishing which fraction of mineralised inorganic nutrient within the sediment is released to the water column and what amount is used for nitrification and de-nitrification. As mentioned before, part of the nutrients released into the water column might be re-utilized by the macroalgae, which could be one of the mechanisms explaining how the macroalgae manage to perpetuate dense blooms. In this sense, Sundbäck et al. (2003) estimated that inorganic nutrient regeneration could supply up to 55–100% of the nitrogen demand of macroalgae and up to 30–70% of the phosphorous demand. These amounts, coupled by the nutrients excreted by the macro- and meiofauna associated to the algae (Bracken et al. 2007), could well account for a self-sustaining system of filamentous macroalgal mats.

Impacts on resident attached macrophytes and other benthic habitats

The bulk of studies addressing the impacts of filamentous bloom forming algae on attached macrophytes have been conducted on seagrass systems, where their effects appear to be variable and context-dependent (Holmquist 1997, Hauxwell et al. 2001, Rasmussen et al. 2012, 2013). One of the main effects caused by the algal mats seems to be shading, which affects the seagrasses both structurally and functionally (Gustafsson and Boström 2013). In their recent meta-analysis on seaweed impacts on seagrasses, Thomsen et al. (2012), suggested that unattached drifting algae cause more harm to seagrasses than attached macroalgae due to their horizontal position at the sediment surface, causing shading on small seagrasses (also shown by Holmer et al. 2011) and that seagrasses with low light requirements will probably be more tolerant to shading induced by the drifting algae (Thomsen et al. 2012b). Similarly, seedlings seem to be more sensitive to shading than adults, even at low coverage densities, and particularly their growth rates seem to be affected by the algal mats (Rasmussen et al. 2012).

Declines in density, recruitment, growth-rate and production of seagrasses have also been attributed to sediment instability (McGlathery 2001), hypoxic conditions and potentially toxic concentrations of ammonium (Hauxwell et al. 2001) or sulphide (Holmer et al. 2011, Thomsen et al. 2012b). As regards the latter, respiration of macroalgae during night (Krause-Jensen et al. 1999), burial of macroalgae organic matter (Holmer et al. 2004) and increasing temperatures can stimulate sulphide concentrations in the sediments through stimulated microbial sulphate reduction. A few studies have also analysed the concomitant effect of increasing temperature and drift algae, finding additive effects of these two stressors and a significant reduction in the main morphological attributes of seagrasses (Holmer et al. 2011), coupled by increased shoot mortality and leaf loss rate (Höffle et al. 2012).

Drift algae can originate from hard substrata located outside of seagrass beds, or alternatively break off *in situ* after epiphytic growth on the plants. Macroalgae display highly aggregated distributions in seagrass beds, sometimes accumulating in the bare patches within seagrass vegetation, though there is often a high variability in density on the scale of meters (e.g., Josselyn 1977). Moreover, some seagrass species may retain more drift algae than others because of morphology (Virnstein and Carbonara 1985); thus spatial distribution of drift algae may be closely linked to seagrass species composition and zonal distribution (Bell and Hall 1997).

Drifting macroalgae also become entangled among *Spartina alterniflora* stalks, accumulating on intertidal flats dominated by cordgrass. As with seagrasses, thick mats of these filamentous algae may cause shading of lower marsh plants, decreasing light availability and inhibiting photosynthetic ability. In these systems, though macroalgal decomposition may have a positive impact on sediment nitrogen concentrations, and thus on cordgrass growth rates, accumulations of large algal biomasses tend to smother and break the culms, large stands of *S. alterniflora* having been found to become flattened under the mats (Newton and Thornber 2013).

Charophytes, have also been found to be negatively affected by increased shading from phytoplankton and epiphyton as a result of increased eutrophication (Phillips et al. 1978), and may also compete with filamentous algae on sheltered soft-bottoms. In the Baltic Sea, these macrophytes are sometimes outcompeted by the algal mats, causing a strong impact on the structure and function of the vegetated bottoms in shallow areas (Blindow 2000).

Shading and scouring of germlings also negatively affect perennial large macroalgal species such as *Fucus vesiculosus* (Worm et al. 1999, 2000, Berger et al. 2003), whose settlement and germling early survival might be compromised by the presence of the algae.

Large-scale impact on attached perennial macrophytes over extended periods of time may result in their disappearance from the sea bottom, limiting primary production only to the pelagic environment, where phytoplankton will continue to bloom whenever nutrient inputs are guaranteed (Valiela et al. 1997, Schramm 1999). Moreover, since carbon fixed by these soft, ephemeral species moves faster through trophic webs than that fixed by seagrasses or vascular plants, the substitution of these perennial forms will propagate through the entire food web (Valiela et al. 1997).

Despite the fact that reports of macroalgal ephemeral blooms on coral reefs are less common, the few studies investigating their effects on corals have also found adverse effects due to coral overgrowth by seaweed and a combination of shading, oxygen depletion and toxins resulting from the decaying algae and the anoxic sediments (Riegl and Velmirov 1991, Lapointe 1997, 1999, Fabricius 2005). Competition for nutrients by their symbiotic zooxanthelae with the opportunistic algae seems to be particularly damaging (Marszalek et al. 1981), while gas exchange restriction, smothering, sediment retention and oxygen depletion during nights as a result of the algal colonization have also been identified as responsible for damaging large coral areas (Fabricius 2005). Loya et al. (2004) reported a 50% reduction in coral cover in the Gulf of Aqaba (Eilat, Red Sea) following a 50% increase in nutrients and the consequent formation of blooms in the area, while Abelson et al. (2005) suggested biofouling as a result of the increase in nutrients in the same area, compromised coral recruitment, explaining the general decline and the slow recovery of the corals in those sites affected by eutrophication.

Impacts on benthic fauna

Apart from the concomitant effects that degradation of benthic habitats may have on the associated communities, and the overall degradation of the coastal system as a result of the disappearance of the macrophytes or coral reefs (Benedetti-Cecchi et al. 2001), impacts of algal mats on underlying sediment-dwelling macro- and meiofauna are many, and again very much dependent on the amount and extension of the accumulations of drifting algae affecting a specific area. Most authors coincide in the fact that small amounts of filamentous algae provide an alternative and rich in resources habitat for the benthic fauna, providing refuge for small fish, crustaceans and gastropods (Bonsdorff 1992, Everett 1994, Pihl et al. 1996, Holmquist 1997, Raffaelli et al. 1998, Norkko et al. 2000, Arroyo et al. 2006), while larger amounts may provoke dramatic losses of benthic life (e.g., Hull 1987, Bonsdorff 1992, Raffaelli et al. 1998, Bolam et al. 2000, Cummins et al. 2004), altered species compositions and even local extinctions (Pardal et al. 2000). The responses of benthic macrofauna to drifting algae are also highly species specific, some species (or life stages) being adversely affected and others being promoted by the algae. These different effects probably reflect different environmental tolerances of different animal species (e.g., tolerant vs. sensitive to anoxia) (Norkko and Bonsdorff 1996c, Cummins et al. 2004), different algal "health levels" (e.g., decomposed or not) (Cummins et al. 2004, Salovius and Kraufvelin 2004, Ólafsson et al. 2013) or different algal densities (Thomsen 2010).

Fauna found in the drift algae consists both of species escaping from the adverse conditions occurring under the algal mat and of species actively colonising the algae, utilising them as a food source or refuge (Salovius et al. 2005). The community structure is often different from that prevailing in the sediments below, and dominated more by typical phytal taxa, and other opportunistic ones (Arroyo et al. 2006, Bohórquez et al. 2013). Among macrofauna, hydrobiid snails, bivalves, oligochaetes and chironomids are usually the dominant taxa (Norkko et al. 2000, Salovius and Kraufvelin 2004), while meiobenthic assemblages are mainly comprised of harpacticoids, ostracods and the nematode *Adoncholaimus thalassophygas* (Ólafsson et al. 2013). The algae seem to facilitate many invertebrates already living in seagrass beds (Holmquist 1997, Thomsen 2010), and other benthic systems (Martinetto et al. 2010), in what has been termed a "habitat cascade" (Thomsen et al. 2012a), increasing species richness and animal abundances (Salovius et al. 2005). Additionally, drifting algae may provide protection against predators, enabling many invertebrates to disperse without swimming, thereby impeding any predation efficiency which is dependent upon the detection of prey movements. The net effect of macroalgae on benthic systems might thus depend on their density, and in those cases in which unfavourable conditions such as anoxic or hypoxic events are not reached, the eutrophication processes triggering algal blooms might be beneficial, supporting high densities of invertebrates and algae, and increasing food availability for higher trophic levels (Nordström et al. 2009, Martinetto et al. 2010, Bohórquez et al. 2013).

Moderate amounts of drifting algae may thus increase habitat complexity and the diversity of the benthic faunal assemblages in otherwise poorly vegetated coastal areas. However, habitat provision by filamentous algae, is usually far from compensating the disappearance of perennial macrophytes, which usually harbour a much more diverse and abundant fauna than that associated with the typical opportunistic species (Råberg and Kautsky 2007, Boström et al. 2010). Therefore, despite the fact that apparently these algae may provide a transient rich resource for benthic fauna, the overall balance of species abundances and biomasses should be considered taking into account the loss associated with macrophytic disappearance or degradation as a result of the drifting algae.

Response to hypoxia

As a consequence of reduced oxygen concentration in the sediment pore water, the redox potential discontinuity layer approaches the sediment surface, and macrofauna from deeper sediment layers are forced to concentrate in the narrow oxidized surface layer (Norkko and Bonsdorff 1996c). The number of available resources is then reduced and species richness decreases. When anaerobic conditions are reached, sulphide and toxic compounds occur (Isaksson and Pihl 1992), leading to a massive mortality of benthic organisms (Bonsdorff 1992, Tagliapietra et al. 1998), in particular juvenile stages of the macrobenthos (Fig. 3).

As a result of these changes deep burrowing forms, including orbinid and opheliid polychaetes worms, tellinid bivalves, some gastropods and infaunal amphipods are negatively affected, as well as other surface-feeding worms such as spionids whose abundances generally decline under the algal canopy (Ólafsson 1988, Everett 1994, Norkko and Bonsdorff 1996a, Cummins et al. 2004). Conversely, small epifaunal species such as the gastropod *Potampopyrgus antipodarium* or *Hydrobia ulvae* can be very abundant (Hull 1987, Cummins et al. 2004). *Hydrobia ulvae* in particular, has often been reported to be benefitted by the algal mats, which it normally inhabits (Norkko and Bonsdorff 1996c, Norkko et al. 2000), and where these deposit-feeding snails find an abundant food supply (Fig. 2). Moreover, Norkko et al. (2000) demonstrated that *Hydrobia* spp. were capable of moving up through the algal canopy to escape anoxic conditions. Oligochaetes can attain high abundances under drifting algal mats (Fletcher 1996, Norkko and Bonsdorff 1996c, Thiel and Watling 1998, Österling and Pihl 2001), algae possibly providing a refuge from predators (Norkko 1998). Significantly higher amounts of tubificid oligochaete egg capsules were found by Norkko and Bonsdorff (1996c), who attributed it to reduced competition and predation, together with greater food supplies for the worms under the mats. Within polychaetes, capitellids have generally been found to increase under the algal mats (e.g., Hull 1987, Bolam et al. 2000, Lopes et al. 2000), probably as a response to organic enrichment, but may be sensitive when extreme anoxia levels are reached. The same is applicable to *Hediste diversicolor*, another of the species which has been found to be more tolerant to the extreme hypoxic and high sulphidic levels induced by the algae (Norkko et al. 1996c, Norkko et al. 2000), but which may also be negatively affected after long exposure times (Lopes et al. 2000) or extreme anoxic levels (Arroyo et al. 2012). The invasive polychaetes *Marenzelleria* spp. have also been found to be tolerant to and abundant under drifting algal mats (Lauringson and Kotta 2006, Arroyo et al. 2012), probably as a response of increased food resources and refuge from predation under the algae.

Figure 2. Two examples of drifting algae suffocating attached macrophytes (A) *Fucus vesiculosus* in the Baltic Sea (photo: Erik Bonsdorff). (B) *Zostera marina* in the Baltic Sea. Mud snails (*Hydrobia* spp.) take advantage of the new habitat and food source (photo: Christoffer Boström).

Thus, algal mats may also benefit opportunistic and highly tolerant invasive species such as these polychaetes, hence facilitating their dispersal and the negative (or positive) effects caused by them in the habitats they colonise.

Regarding bivalves, adult specimens seem to avoid filamentous algae by horizontal migration across the sediment surface (Österling and Pihl 2001). Norkko et al. (2000) showed a higher survival of adult *M. balthica* compared to juveniles, despite the fact that it cannot migrate into or adapt to a life in drifting algae. Severe effects on the settling spat of *M. balthica* were also demonstrated by Bonsdorff (1992) and Bonsdorff et al. (1995a). Siphon cropping may also be a negative consequence of bivalves having to adapt to the hypoxic near-bottom waters (Bonsdorff et al. 1995b). Other studies have also shown *M. balthica* to be unable to move away horizontally

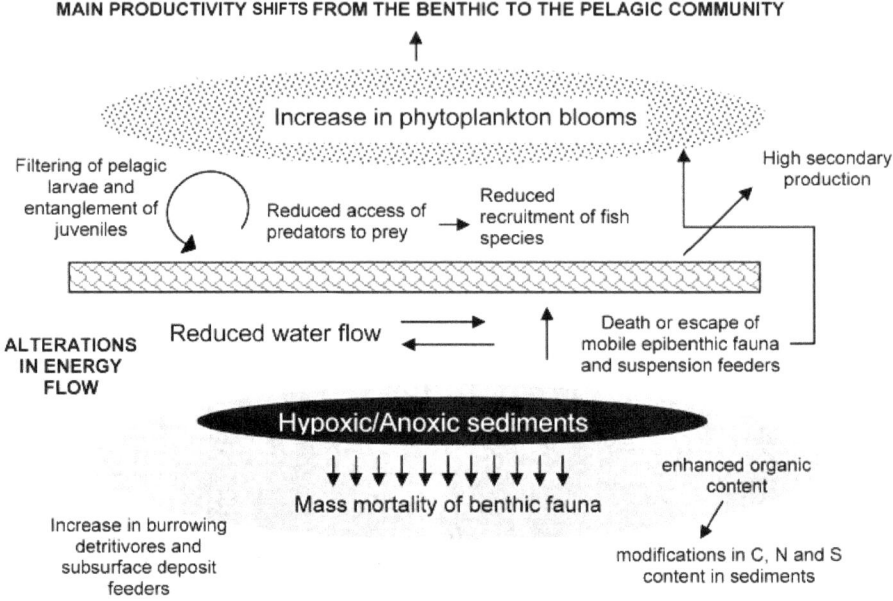

Figure 3. Schematic representation of the most common noxious effects of drifting algal mats on soft sediment benthic habitats.

from algal mats and that drift algae inhibit the bivalves from extending their siphons (Norkko 1998). Norkko and Bonsdorff (1996a) found 70% of the population of adult *Macoma balthica* to have emerged from the sediment after 17 days of algal cover. *M. balthica* has been found to be indifferent to algal mat presence or even increase under the drifting algae in areas other than the Baltic (Hull 1987, Everett 1994), the latter's hyposaline conditions probably posing further stress to the clams, and causing their decline under hypoxic events.

Crustaceans are normally sensitive to hypoxic conditions and several studies have shown their limited tolerance to the algal mats. The surface detritivore *Corophium volutator* seems to escape the algal mats and prefer adjacent algal-free sediment (Österling and Pihl 2001, Raffaelli et al. 1991), though it may use them as a refuge from predation and has shown a tendency to migrate from bare sand to loose patches of drifting brown algae even when predators are not present (Norkko 1998). High algal biomasses however, have been shown to have dramatic effects on the density of this amphipod, the mechanism for this interaction seemingly being related to interference by the algal filaments with the deposit-feeding behaviour of the amphipod at the sediment—water interface (Hull 1987, Raffaelli 2000).

Meiofauna have traditionally been considered more resistant to hypoxia/anoxia and disturbances caused in the sediment (Meyers et al. 1987, Modig and Ólafsson 1998), but their responses to drifting algal mats have not been as exhaustively explored as those of the macrofauna. The few studies investigating these effects however, seem to conclude that meiofauna can also be vulnerable to anoxia in the sediments and within the algal mat even after short exposure times (Franz and Friedman 2002, Arroyo et al. 2012) and do also show escaping responses into the algal mat (Arroyo et al. 2012) or the water column (Franz and Friedman 2002). Villano and Warwick (1995) investigated the effects of algal mats on meiofauna at the Venice lagoon and reported that they showed less vulnerability compared to the macrofauna, with changes in species composition but not a population crash as that shown for the macrofauna. On the other hand, macroalgal biomass may increase the organic matter content of the sediment, favouring meiobenthic detritivores and causing increases in some meiobenthic groups such as nematodes (Bohórquez et al. 2013). Generally, low doses of macroalgal mats will increase the abundance of harpacticoids, rotifers or foraminiferans, together with that of ostracods and bivalves, while increased amounts will shift community structure to a nematode dominated one, significantly decreasing the abundance of less tolerant groups which initially benefitted from the algae (Sundbäck et al. 1990, Arroyo et al. 2006, Bohórquez et al. 2013).

Taking into account that these changes in the abundance of major invertebrate fauna seem to occur from scales of less than 1 m^2 to those of entire estuaries or bays (Raffaelli 2000), it is obvious that the persistence of these mats may have strong implications for species distributions and diversity in affected areas. More so taking into account that mortality events are often followed by recolonization phases in which a few opportunistic species will take advantage of the impoverished sea floor (Arroyo et al. 2006). Often, polychaetes (e.g., *Capitella capitata, Streblopsio shrubsolii, Polydora ligni* and *Hediste diversicolor*) and oligochaetes, both tolerant to the conditions set forth by the drifting algae, are the first colonizers, followed by other infaunal species such as *Corophium* spp. or *Melita palmata* (Tagliapietra et al. 1998), or opportunistic species which are benefitted by the algae and tolerate the conditions prevailing under them (Norkko and Bonsdorff 1996b, Arroyo et al. 2006). If conditions prior to the "algal crisis" are not restored, the benthic communities may remain severely altered, with drastic changes in community structure and a dominance of opportunistic, highly tolerant taxa.

Larvae filtering—effects on recruitment

The effects on recruitment of benthic fauna seem to be also density and species dependent. Filamentous algae have been shown to filter out macrofaunal pelagic larvae from the water column during settlement (Ólafsson 1988, Bonsdorff 1992, Österling and Pihl 2001). As the production and spread of filamentous algae normally coincides with the time of larval recruitment, such filtering effects may lead to significant reductions in recruitment to macrofaunal populations, especially if algae cover a major part of a shallow area. Juvenile bivalves are also known to drift in the water column (Armonies 1996), and may be trapped in filamentous algae (Bonsdorff et al. 1995a).

However, enhanced recruitment may occur where disturbance caused by the algae is low (Raffaelli et al. 1998), and where particular species may find the algae as a suitable substratum to grow their juveniles. Hull (1987) reported increases in juvenile tellinid bivalves, which could have benefitted in their settlement by the reducing flow velocities under the algal mats. Arroyo et al. (2006) found juveniles of *Cerastoderma glaucum* and *Hydrobia* spp. to be the dominant macrofauna occurring among drifting algal mats, and Österling and Pihl (2001) found an apparent migration of juvenile bivalves into the algae. This may lead to a decline in young bivalves in the sediment, but *Cerastoderma glaucum* larvae normally settle in attached algal belts before the secondary spat fall to the bottom, so patches of drifting algae could also function as a "stepping stone" for recolonisation of sediments that have previously been defaunated by excessive amounts of drifting algae (Arroyo et al. 2006).

Migration into the algae, drifting algae as mobile corridors

As sediments become more anoxic, adult invertebrates are forced to migrate towards the sediment surface, where they are at a greater risk of predation (Norkko and Bonsdorff 1996a). Also, physical interference with feeding behaviour might cause migration of bivalves (Norkko and Bonsdorff 1996a) and spionid worms (Everett 1994) into the algae. In general, macrofauna seems to respond to stress induced by the algae by emerging into the water column or the drifting algae (Norkko et al. 2000), which may have important dispersal implications for the macro- (and meiofauna) inhabiting disturbed habitats. Several authors have stressed the potential role of drift macroalgae as contributors to colonisation of sediments (Cummins et al. 2004, Arroyo et al. 2006), or seagrasses (Holmquist 1994, Brooks and Bell 2001), sometimes covering distances of 0.5 km per day (Holmquist 1994). Drift algal dispersal by phytal and sediment fauna should facilitate species exchange within a landscape, thereby decreasing recovery time from disturbance events. This would effectively decrease the degree of isolation between fragments within a landscape, particularly for fauna lacking planktonic larval stages.

Colonisation of drifting algae and seagrasses by invertebrates is known to be rapid (Boström et al. 2010), the most likely sources of colonisation being ambient benthic fauna, fauna from attached algal belts in the littoral zone and pelagic larvae (e.g., *Cerastoderma glaucum*). However, the habitat provided by drift algae changes as the algae degrade, are colonised by microorganisms and invertebrates and become fragmented, which may make it more or less appetized by the various fauna inhabiting it (Ólafsson et al. 2013). Differences in the nutrient quality of the degrading algae may have cascading effects on benthic food webs, less mobile taxa being unable to shift to preferred food items and having to adapt to those existing in the degrading algae. Algal degradation seems to cause depletion in $\delta^{15}N$ which may be transferred through the food web, especially in those taxa feeding directly from the algal tissues (Ólafsson et al. 2013).

Effects of invertebrates on the algal mats

Macrofauna may effectively help in disintegrating the algal mats, not only through direct consumption but also as a result of the vertical redistribution of the mat, since algae in contact with the sediment are degraded faster than algae in the water column (Salovius and Bonsdorff 2003). *Hediste diversicolor* has been repeatedly found to be able to contribute to the redistribution and biomass loss of macroalgae on marine soft bottoms (Raffaelli 2000, Nordström et al. 2006). On the other hand, Reise (1983) argued that burrowing polychaetes like *Hediste diversicolor* might have a positive effect on the development of algal mats in the Wadden Sea, by anchoring filaments in the burrows. *Hydrobia* spp. may also play a positive role in the development of macroalgal mats by providing a suitable substrate for settlement and germination (Schories and Reise 1993, Raffaelli 2000).

This top-down effect of the macrofauna on the algae and on the macrophytic communities affected by them has been argued as one of the mechanisms impeding algal colonization of seagrass meadows (Whalen et al. 2013), and aiding in their conservation. Though the controlling role of mesograzers on epiphytic algae is now well established, their ability to disintegrate algal mats when they attain high densities is still not clear. As regards fish, Gacia et al. (1999) showed how herbivorous fish had only a limited ability to control populations of filamentous algae, and that it was restricted to green epiphytic forms on the seagrass *Thalassia testudinum*. They reasoned that these results showed how top-down interactions in maintaining this seagrass's growth were limited to the occasional reduction of shading by epiphytic green algae. They further suggested that fish—algal interactions have only limited capacity to buffer the consequences of eutrophication for seagrass growth.

Functional and Trophic implications of macroalgal blooms

As shown before, cumulative effects of drifting algal mats in benthic systems include modifications of the carbon, nitrogen and sulphide content of sediments, severe modifications of macro- and meiofaunal community composition and diversity, shifts in species structure toward smaller and/or (bigger) pollution-tolerant, opportunistic taxa, and escape responses and high mortality rates of larger bivalves, crustaceans and fish. This is likely to have strong impacts on the trophic structure of the benthic systems affected by the algae, and on the functional role of the major groups inhabiting them (Boström et al. 2010). Trophic relationships have been postulated to be simplified as the gradient of organic input to a particular environment increases (Lindeman 1942, Pearson and Rosenberg 1978), complex food-webs occurring in oligotrophic situations, gradually being transformed into simplified communities composed entirely of detritus feeders and omnivores (Tagliapietra et al. 1998). This seems to be the trend in environments heavily affected by macroalgal blooms. Norkko et al. (2000), predicted that if the amounts of algae continue to increase in the northern Baltic Sea, the benthic community on shallow sandy bottoms will change from being dominated by long-lived species such as bivalves to being mainly composed of opportunistic short-lived ones as chironomids, which are favoured by their tolerance of anoxic conditions and temporal disappearance of zoobenthos and fish, and can also actively take part in fragmentation of the algal material.

Further, losses of key species, such as the bivalve *Macoma balthica* or the amphipod *Corophium volutator*, might induce fundamental changes to community function and food-web dynamics by altering energy flow in the community (Norkko and Bonsdorff 1996c, Raffaelli 2000). The decrease in suspension feeders such as bivalves caused by the algae may have cascading effects in the overall well-being of coastal systems, since they are permanently present in the system and filter large amounts of phytoplankton, decreasing the probability of phytoplankton blooms as well as the sedimentation of detritus (Flindt et al. 1999). Moreover, other structurally important and habitat forming fauna such as bivalves may be negatively affected by the algal mats (Thomsen and McGlathery 2006), precluding reef formation and having large scale consequences for all the benthic organisms associated with them.

In general, changes in community structure will depend on tolerance, feeding mode and mobility of the macro- and meiobenthic species present (Norkko and Bonsdorff 1996c, Franz and Friedman 2002, Arroyo et al. 2006, Arroyo et al. 2012), and while non-severe disturbances may enhance the presence of certain groups such as burrowing detritivores or mobile epibenthic fauna (Isaksson and Pihl 1992, Norkko and Bonsdorff 1996b,c, Thiel and Watling 1998), extreme situations are likely to cause fundamental structural changes of the ecosystems involved: from grazing and/ or nutrient controlled stable systems to unstable detritus/mineralisation systems where the turnover of oxygen and nutrients is much more dynamic and oscillations between aerobic and anaerobic states frequently occur (Fig. 4).

Grazing is an important factor controlling biomass accumulation of free-floating algae, and it has been found to be negatively correlated with eutrophication (Valiela et al. 1997, Worm and Lotze 2006). In this way, excessive eutrophication can override top-down control in benthic systems, favouring the development of algal mats, which will in turn cause further damage to consumer trophic function (Myers and Heck 2013). Littorinid snails, amphipods and isopods can reduce or even prevent algal blooms through selective feeding on their early life-history stages, but their capability to control algal blooms is often hindered by increased eutrophication (Worm and Lotze 2006). Under high nitrogen loads the abundance of isopods and amphipods decreases (probably due to more frequent hypoxic conditions), and their capability to control macroalgal blooms and growth of epiphytic algae is reduced (Valiela et al. 1997, Flindt et al. 1999). Österling and Pihl (2001) found a strong decrease in juvenile suspension feeders, which declined most under algae compared to undisturbed controls. These systems generally become bottom-up controlled, since grazers are incapable of controlling macroalgal proliferation, which is further triggered by continuous nutrient-supply, though in situations in which macroalgae don't cover wide surfaces over long periods of time or oxygen is replenished frequently, consumers may remain highly abundant, top-down forces continuing to be able to control algal growth (Martinetto et al. 2011). Nevertheless, grazers may modify their pressure over algae under increased nutrient inputs, eating less enriched algal material compared to when macrophytes are not nutrient enriched, thus diminishing their impact over the algal mass (Korpinen et al. 2007, Martinetto et al. 2011).

Reduced water flow under the algae may enhance food levels for burrowing detritivores (Norkko and Bonsdorff 1996b,c, Thiel and Watling 1998), as long as disturbance is not severe. Thiel and Watling (1998) found increased abundances of subsurface deposit feeders due to enhanced organic content in the sediment. Moreover, Isaksson and Pihl (1992), found that mobile epibenthic fauna increased in biomass at moderate (30–50%) algal cover, towards species normally associated with vegetation, whereas when algae increased to 90% cover, the biomass of mobile epibenthic fauna declined. Both suspension feeders and surface detritivores seem to be negatively

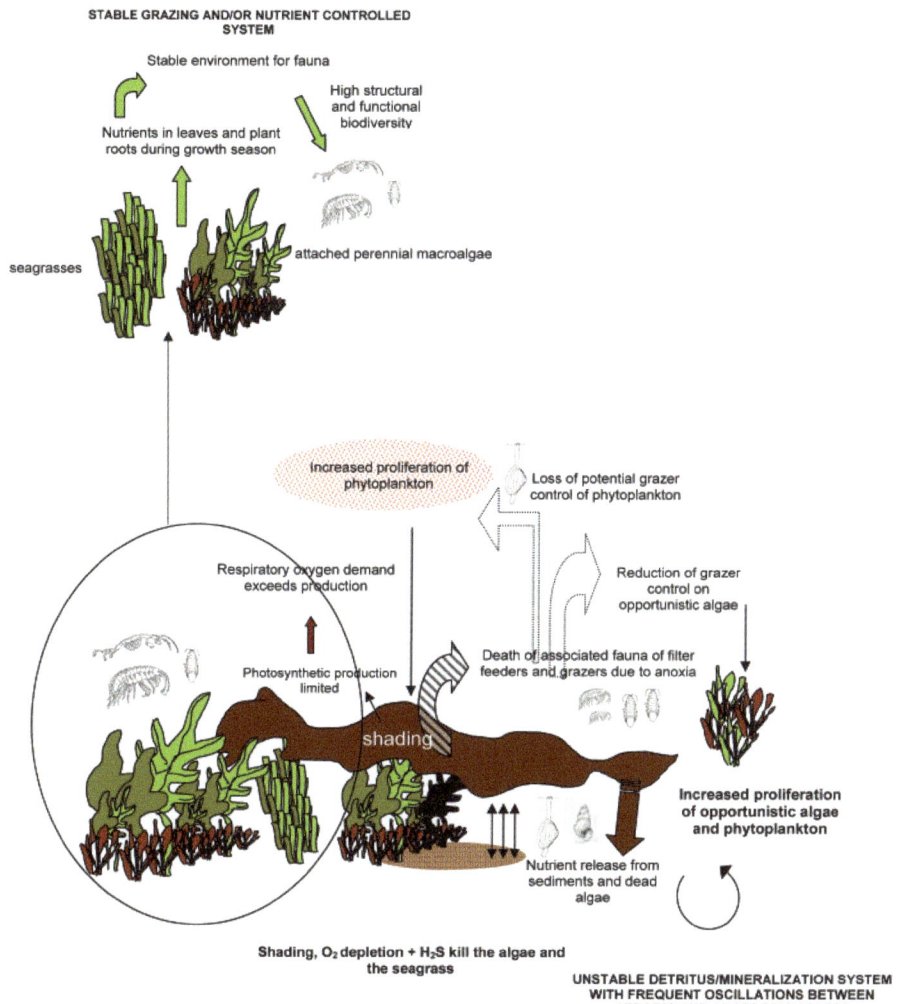

Figure 4. Schematic representation of the most common noxious effects of drifting algal mats on vegetated benthic habitats, which pass from being stable grazing and/or nutrient controlled systems to more unstable and detritus or mineralization based ones.

affected by filamentous algae, but different degrees of physical disturbance induced by the algae on the sediment result in different degrees of reduction in abundance for these functional feeding groups (Österling and Pihl 2001).

The decrease in benthic suspension feeders promoted by the drifting algal mats may seriously compromise pelagic-benthic coupling and phytoplankton control, since these bivalves (mostly) are responsible for converting the bulk of the primary production coming from phytoplankton blooms, while they may also control its production through nutrient release into the water column, and facilitate POM sedimentation in the form of pseudofaeces, increasing the overall sedimentation rate (Grall and Chauvaud 2002). Moreover, the high amounts of detritus provided by the macroalgal blooms are quite refractory, and thus will probably be used by macrofaunal deposit feeders rather than by the suspension and surface deposit feeders, microorganisms and meiobenthic copepods which normally mineralize living phytoplankton (Widbom and Elmgren 1988).

Altered trophic relationships between predators and macrofaunal prey also occur as a consequence of drifting macroalgal mats (Norkko and Bonsdorff 1996a, Norkko 1998). In the Baltic Sea, spawning and feeding grounds for flatfish are lost with drifting algae, with potential cascading effects (Österling and Pihl 2000, Nordström and Booth 2007). For instance, the presence of algae may significantly reduce the predation efficiency of juvenile flounders for three of its preferred prey items (i.e., *Macoma balthica*, ostracods, *Hydrobia* spp.) which would find refuge among and under the algal mat (Aarnio and Mattila 2000). Since a 50% cover of sandy bottoms is normal in the northern Baltic Sea, this may have severe effects on juvenile flounders, which normally utilize and forage in uncovered areas, and eventually entire flounder populations in these areas may be affected negatively. Flounders have also been found dead entangled among the algal mat (Norkko and Bonsdorff 1996b), evidencing further the habitat loss experienced by fish in areas affected by the algae. The mortality under algal mats of, e.g., juvenile bivalves such as *Macoma balthica* (Bonsdorff 1992) may lead to compensatory feeding on non-preferred prey by juvenile flounder, which in turn may lead to starvation of the fish if the alternative food (e.g., mudsnails; *Hydrobia* spp.) is unpalatable (Aarnio and Bonsdorff 1997). In the Ythan estuary, the pronounced decline of *Corophium volutator* under drifting algal mats was postulated to have strong implications for higher trophic levels' access to food, since this species occurs in the diets of many shorebirds, fish and epibenthic crustaceans (Hall and Raffaelli 1991), and its decline over large areas (40%) of the intertidal flats was registered by Raffaelli (2000). Cabral et al. (1999) also found waders to avoid parts of the Mondego estuary covered by drifting algal mats, suggesting that some species may find it more difficult to forage in these areas. However, those birds eating from the algal mats seemed to do it more frequently, indicating that they were probably feeding on prey items which were abundant there. Actually, drifting algae, invertebrates living in the algae and/or microorganisms growing on the algae can provide a valuable food resource for higher trophic levels (Ólafsson et al. 2013), and predation might also be enhanced when the disturbed fauna emerges to oxic layers in the sediment surface or the algal mat itself (Norkko and Bonsdorff 1996c, Norkko 1998, Österling and Pihl 2001, Franz and Friedman 2002, Arroyo et al. 2012). Norkko and Bonsdorff (1996c) observed that when removing the algae during their experiments, epibenthic predators such as gobiids and brown shrimps *Crangon crangon* (L.) accumulated in the vicinity of the exposed plots, probably to exploit the stressed infauna at the sediment surface. Lenanton et al. (1985), also found that an increase in fish catches followed a rise in the biomass of *Cladophora* and *Chaetomorpha*.

Trends in drifting algae evolution under current threats to coastal ecosystems

Eutrophication, climate change and invasive species lie among the most invoked threats to coastal ecosystems worldwide. Eutrophication is an increasing and more generalized problem throughout the globe, where algal blooms are changing from stochastic to more predictable, and their impacts may significantly alter the structure and function of the benthic systems affected by them.

Rising temperatures, changes in nutrient availability and ultraviolet radiation, along with stronger and more frequently occurring storms will probably increase the number and intensity of macroalgal blooms (IPCC 2007, Lotze and Worm 2002), their effects acting additively and synergistically with invasive species, thus causing a more pronounced decline in benthic life and alterations to the ecosystems than previously anticipated. Many exotic or non-native species are themselves opportunistic and filamentous and may eventually become seasonal blooming phenomena in the areas to which they arrive (Lapointe and O'Connell 1989, Vermeij et al. 2009, Irigoyen et al. 2010). Elevated temperature and eutrophication interact synergistically to increase bacterial metabolism in the sediment, usually by a factor of 2 to 3 for each 10°C increment (Thamdrup et al. 1998), thereby stimulating oxygen consumption in the sediments and the accumulation of sulphide in the sediment pore water. As mentioned earlier, the deleterious effects of drifting algae on seagrasses have been found to increase with ascending temperatures (Holmer et al. 2011, Höffle et al. 2012), while all these effects may further interact with other stressors such as augmenting sedimentation as a result of urbanization of coastal areas, increasing the pressures exerted over their already distressed and fragile ecosystems. There is an urgent need to investigate the commonality of all these factors in order to make predictions of their future impacts and propose realistic management practices to try to prevent further disasters.

Moreover, considering recent evidences of trophic downgrading (Estes et al. 2011), then processes as those described by Eriksson et al. (2009), in which the liberation of secondary consumers due to top-predator declines will limit the controlling function of mesograzers over algal mats, are prone to promote algal growth and the development of blooms. Thus, managing a high trophic diversity of fish communities is important for water quality in near shore environments, and synergistic effects of bottom-up and top-down processes need to be incorporated into future ecosystem management of marine resources (Eriksson et al. 2009).

Management

As regards prevention and managerial recommendations, an assessment of whether an estuary is at risk might focus on issues of availability of intertidal area, turbidity, nutrient concentrations and bed stability (Scanlan et al. 2007). Several recommendations for monitoring based on % cover and biomass were provided by Scanlan et al. 2007. Aerial photography is a rapid and applicable method for studying coverage and occurrence of macrophytes in shallow bays (Pihl et al. 1999), and has proved more efficient than estimation of coverage by ground-truth surveys (Berglund et al. 2003). More recently, satellite monitoring has been used and seems to be an efficient way of providing an early detection of algal patches at first appearance, thus being useful for taking preventive measures (Cui et al. 2012).

Early detection is becoming a more necessary tool, invoked by coastal managers and stakeholders, since far from being solely an ecological problem, drifting algal mats can also negatively affect the recreational use of the shores, causing unpleasant smells (Villares et al. 1999, Charlier et al. 2007, Lyons et al. 2009), and the need to have beaches cleaned at regular intervals, thus increasing maintenance costs. Given the very high revenues provided by coastal areas, particularly highly touristic ones, the need to locate and control macroalgal blooms is becoming more and more a managerial tool which will need precise monitoring strategies to be successful (Charlier et al. 2007). Algal mats may also pose a health hazard in coastal areas due to the development of pathogenic bacteria such as *Salmonella* and enterobacteria which may not only grow within the algal mat, but also find a means of dispersal (Gubelit and Vainshtein 2011) along coastal areas. Additionally, they also foul fishing gear and represent a threat to power plants which use water in their cooling systems (Lehvo and Bäck 2000). Moreover, green tides have already caused massive losses in aquaculture activities in China (Rosenberg et al. 1990, Ye et al. 2011), while mass cultivation of macroalgae (which was considered a viable solution to eutrophication problems in coastal areas), has also been found to pose possible problems of opportunistic algae proliferation and eventual blooms even in areas far away from the cultivation area (Liu et al. 2009).

Due to these increasing trends in algal accumulation, harvesting has become increasingly common. It is considered as a short-term tool to improve the cosmetic state of normally exposed beaches (Lavery et al. 1999, Berglund et al. 2003). Several initiatives have tried to put these algal masses into value, the most profitable and commercially viable one apparently being that of their use as fertilizer (Charlier et al. 2007, Ye et al. 2011).

However, a more permanent solution to the problem can only be achieved by significantly reducing the present level of eutrophication in the affected areas, which can only be attained by dropping the levels of nutrient effluents into coastal systems considerably or by creating natural filters that prevent abnormal levels from reaching them. In this sense, saltmarsh fringes have been found to be able to mitigate macroalgal bloom formation through nutrient availability reduction by de-nitrification (Lyons et al. 1995, Valiela et al. 1997). Recent field surveys indicate the recovery of *F. vesiculosus* populations and associated fauna in the Baltic Sea following the reduction of nutrient loads (Nilsson et al. 2004). In other cases, legislated reductions in chemical and nutrient inputs have also successfully diminished algal growth (Bushaw-Newton and Sellneri 1999), which may be interpreted as a hopeful sign that wise management actions can reverse the deleterious trends discussed in this paper even over large spatial scales.

Acknowledgements

The authors would like to acknowledge Emil Ólafsson for inviting them to write this chapter. We thank Christoffer Boström and Nolwenn Quillien for providing photographs. Åbo Akademi University is thanked for financial support (sabbatical; EB).

References

Aarnio, K. and E. Bonsdorff. 1997. Passing the gut of juvenile flounder, *Platichthys flesus* (L.)—differential survival of zoobenthic prey species. Mar. Biol. 129: 11–14.

Aarnio, K. and J. Mattila. 2000. Predation by juvenile *Platichthys flesus* (L.) on shelled prey species in a bare sand and a drift algae habitat. Hydrobiologia 440: 347–355.

Abelson, A., R. Olinky and S. Gaines. 2005. Coral recruitment to the reefs of Eilat, Red Sea: temporal and spatial variation, and possible effects of anthropogenic disturbances. Mar. Poll. Bull. 50: 576–582.

Armonies, W. 1996. Changes in distribution patterns of 0-group bivalves in the Wadden sea: byssus-drifting releases juveniles from the constraints of hydrography. J. Sea Res. 35(4): 323–334.

Arroyo, N.L., K. Aarnio and E. Bonsdorff. 2006. Drifting algae as a means or re-colonizing defaunated sediments in the Baltic Sea. Hydrobiologia 553: 83–95.

Arroyo, N.L., K. Aarnio, M. Mäensivu and E. Bonsdorff. 2012. Drifting filamentous algal mats disturb sediment fauna: impacts on macro-meiofaunal interactions. J. Exp. Mar. Biol. Ecol. 420-421: 77–90.

Bäck, S., A. Lehvo and J. Blomster. 2000. Mass occurrence of unattached *Enteromorpha intestinalis* on the Finnish Baltic coast. Annales Botanici Fennici 37: 155–161.

Bell, S.S. and M.O. Hall. 1997. Drift macroalgal abundance in seagrass beds: investigating large-scale associations with physical and biotic attributes. Mar. Ecol. Prog. Ser. 147: 277–283.

Benedetti-Cecchi, L., F. Pannacciulli, F. Bulleri, P.S. Moschella, L. Airoldi, G. Relini and F. Cinelli. 2001. Predicting the consequences of anthropogenic disturbance: large-scale effects of loss of canopy algae on rocky shores. Mar. Ecol. Prog. Ser. 214: 137–150.

Benz, M., N.J. Eiseman and E.E. Gallaher. 1979. Seasonal occurrence and variation in standing crop of a drift algae community in the Indian River, Florida. Bot. Mar. 22: 413–420.

Berger, R., E. Henriksson, L. Kautsky and T. Malm. 2003. Effects of filamentous algae and deposited matter on the survival of *Fucus vesiculosus* L. germlings in the Baltic Sea. Aquat. Ecol. 37: 1–11.

Berglund, J., J. Mattila, O. Rönnberg, J. Heikkilä and E. Bonsdorff. 2003. Seasonal and inter-annual variation in occurrence and biomass of rooted macrophytes and drift algae in shallow bays. Estuar. Coast. Shelf Sci. 56: 1167–1175.

Biber, P.D. 2007. Transport and persistence of drifting macroalgae (Rhodophyta) are strongly influenced by flow velocity and substratum complexity in tropical seagrass habitats. Mar. Ecol. Prog. Ser. 343: 115–122.

Blindow, I. 2000. Distribution of Charophytes along the Swedish Coast in Relation to Salinity and Eutrophication. Internat. Rev. Hydrobiol. 85: 707–717.

Blomster, J., S. Bäck, D.P. Fewer, M. Kiirikki, A. Lehvo, C.A. Maggs and M.J. Stanhope. 2002. Novel morphology in *Enteromorpha* (Ulvophyceae) forming green tides. Am. J. Bot. 89: 1756–1763.

Bohórquez, J., S. Papaspyrou, M. Yúfera, S.A. van Bergeijk, E. García-Robledo, J.L. Jiménez-Arias, M. Bright and A. Corzo. 2013. Effects of green macroalgal blooms on the meiofauna community structure in the Bay of Cádiz. Mar. Poll. Bull. 70: 10–17.

Bolam, S.G., T.F. Fernandes, P. Read and D. Raffaelli. 2000. Effects of macroalgal mats on intertidal sandflats: an experimental study. J. Exp. Mar. Biol. Ecol. 249: 123–137.

Bonsdorff, E. 1992. Drifting algae and zoobenthos—effects on settling and community structure. Neth. J. Sea Res. 30: 57–62.

Bonsdorff, E., A. Norkko and C. Boström. 1995a. Recruitment and population maintenance of the bivalve *Macoma balthica* (L.)—factors affecting settling success and early survival on shallow sandy bottoms. pp. 253–260. *In*: Eleftheriou, A., A.D. Ansell and C.J. Smith (eds.). Biology and Ecology of Shallow Coastal Waters. Proc. 28th European Marine Biology Symposium, Olsen & Olsen, Fredensborg.

Bonsdorff, E., A. Norkko and E. Sandberg. 1995b. Structuring zoobenthos: the importance of predation, cropping and physical disturbance. J. Exp. Mar. Biol. Ecol. 192: 125–144.

Boström, C., A. Törnroos and E. Bonsdorff. 2010. Invertebrate dispersal and habitat heterogeneity: expression of biological traits in a seagrass landscape. J. Exp. Mar. Biol. Ecol. 390: 106–117.

Bracken, M.E., C.A. González-Dorantes and J.J. Stachowicz. 2007. Whole-community mutualism: associated invertebrates facilitate a dominant habitat-forming seaweed. Ecology 88(9): 2211–9.

Brooks, R.B. and S.S. Bell. 2001. Mobile corridors in marine landscapes: enhancement of faunal exchange at seagrass/sand ecotones. J. Exp. Mar. Biol. Ecol. 264: 67–84.

Bushaw-Newton, K.L. and K.G. Sellner. 1999. Harmful algal blooms. *In*: NOAA's State of the Coast Report. National Oceanic and Atmospheric Administration, Silver Spring, MD.

Cabral, J.A., M.Â. Pardal, R.J. Lopes, T. Múrias and J.C. Marques. 1999. The impact of macroalgal blooms on the use of the intertidal area and feeding behaviour of waders (Charadrii) in the Mondego estuary (west Portugal). Acta Oecol. 20(4): 417–427.

CEFAS. 2004. Investigation of factors controlling the presence of macroalgae in some estuaries of South East England. Report to the Environment Agency, Southern Region.

Charlier, R.H., P. Morand, C.W. Finkl and A. Thys. 2007. Green tides on the Brittany coasts. Environ. Res. Eng. Man. 3(41): 52–59.

Corzo, A., S.A. van Bergeijk and E. García-Robledo. 2009. Effects of green macroalgal blooms on intertidal sediments: net metabolism and carbon and nitrogen contents. Mar. Ecol. Prog. Ser. 380: 81–93.

Cui, T.-W., J. Zhang, L.-E. Sun, Y.-J. Jia, W. Zhao, Z.-L. Wang and J.-M. Meng. 2012. Satellite monitoring of massive green macroalgae bloom (GMB): imaging ability comparison of multi-source data and drifting velocity estimation. Int. J. Remote Sens. 33(17): 5513–5527.

Cummins, S.P., D.E. Roberts and K.D. Zimmerman. 2004. Effects of the green macroalga *Enteromorpha intestinalis* on macrobenthic and seagrass assemblages in a shallow coastal estuary. Mar. Ecol. Prog. Ser. 266: 77–87.

Dion, P. and S. Le Bozec. 1996. The French Atlantic coasts. pp. 251–264. *In*: Schramm, S.W. and P.H. Nienhuis (eds.). Marine Benthic Vegetation. Recent Changes and the Effects of Eutrophication, Ecol. Stud., Vol. 123. Springer-Verlag, Berlin.

Eriksson, B.K., L. Ljunggren, A. Sandstrom, G. Johansson, J. Mattila, A. Rubach and M. Snickars. 2009. Declines in predatory fish promote bloom-forming macroalgae. Ecol. Appl. 19: 1975–1988.

Escartin, J. and D.G. Aubrey. 1995. Flow structure and dispersion within algal mats. Estuar. Coast Shelf Sci. 40: 451–472.

Estes, J.A., J. Terborgh, J.S. Brashares, M.E. Power, J. Berger, W.J. Bond, S.R. Carpenter, T.E. Essington, R.D. Holt, J.B.C. Jackson, R.J. Marquis, L. Oksanen, T. Oksanen, R.T. Paine, E.K. Pikitch, W.J. Ripple, S.A. Sandin, M. Scheffer, T.W. Schoener, J.B. Shurin, A.R.E. Sinclair, M.E. Soulé, R. Virtanen and D.A. Wardless. 2011. Trophic downgrading of planet earth. Science 333: 301–306.

Everett, R.A. 1994. Macroalgae in marine soft-sediment communities: effects on benthic faunal assemblages. J. Exp. Mar. Biol. Ecol. 150: 223–247.

Fabricius, K.E. 2005. Effects of terrestrial runoff on the ecology of corals and coral reefs: review and synthesis. Mar. Poll. Bull. 50: 125–146.

Fenchel, T.M., T.M. King and T.H. Blackburn. 1998. Bacterial Biochemistry: the Ecophysiology of Mineral Cycling. Academic Press, San Diego, p. 307.

Fletcher, R.L. 1996. The occurrence of green tides. pp. 7–43. *In*: Schramm, W. and P.H. Nienhuis (eds.). Ecological Studies, Vol. 123. Marine Benthic Vegetation. Recent Changes and the Effects of Eutrophication. Springer-Verlag. Berlin, Heidelberg.

Flindt, M.R., M.A. Pardal, A.I. Lillebø, I. Martins and J.C. Marques. 1999. Nutrient cycling and plant dynamics in estuaries: a brief review. Acta Oecologica 20(4): 237–248.

Franz, D.R. and I. Friedman. 2002. Effects of a macroalgal mat (*Ulva lactuca*) on estuarine sand flat copepods: an experimental study. J. Exp. Mar. Biol. Ecol. 271: 209–226.

Gacia, E., M.M. Littler and D.S. Littler. 1999. An experimental test of the capacity of food web interactions (Fish–Epiphytes–seagrasses) to offset the negative consequences of eutrophication on seagrass communities. Estuar. Coast. Shelf Sci. 48: 757–766.

García-Robledo, E., A. Corzo, J. García de Lomas and S.A. van Bergeijk. 2008. Biogeochemical effects of macroalgal decomposition on intertidal microbenthos: a microcosm experiment. Mar. Ecol. Prog. Ser. 356: 139–151.

García-Robledo, E. and A. Corzo. 2011. Effects of macroalgal blooms on carbon and nitrogen biogeochemical cycling in photoautotrophic sediments: en experimental mesocosm. Mar. Poll. Bull. 62: 1550–1556.

García-Robledo, E., A. Corzo, S. Papaspyrou and E.P. Morris. 2012. Photosynthetic activity and community shifts of microphytobenthos covered by green macroalgae. Environ. Microbiol. Rep. 4: 316–325.

Grall, J. and L. Chavaud. 2002. Marine eutrophication and benthos: the need for new approaches and concepts. Global Change Biology 8: 813–830.

Gross, V.A. 1994. Biological and oceanographic factors controlling the nuisance algal bloom of free-living *Pilayella littoralis* in Nahant Bay, Massachusetts. M.Sc. Thesis. Northeastern University, Boston, MA., 133 pp.

Gubelit, Y.I. and N.A. Berezina. 2010. The causes and consequences of algal blooms: the *Cladophora glomerata* bloom and the Neva estuary (eastern Baltic Sea). Mar. Poll. Bull. 61: 183–188.

Gubelit, Y.I. and M.B. Vainshtein. 2011. Growth of enterobacteria on algal mats in the eastern part of the Gulf of Finland. Inland Water Biol. 4(2): 132–136.

Gustafsson, C. and C. Boström. 2013. Influence of Neighboring Plants on Shading Stress Resistance and Recovery of Eelgrass, *Zostera marina* L. PLoS ONE 8(5): e64064. doi: 10.137/journal.pone.0064064.

Hall, S.J. and D. Raffaelli. 1991. Food-web patterns—lessons from a species-rich web. J. Anim. Ecol. 60(3): 823–842.

Hauxwell, J., J. Cebrián, C. Furlong and I. Valiela. 2001. Macroalgal canopies contribute to eelgrass (*Zostera marina*) decline in temperate estuarine ecosystems. Ecology 82: 1007–1022.

Hayden, H.S., J. Blomster, C.A. Maggs, P.C. Silva, M.J. Stanhope and J.R. Waaland. 2003. Linnaeus was right all along: *Enteromorpha* and *Ulva* are not distinct genera. Eur. J. Phycol. 38: 277–293.

Holmer, M., C.M. Duarte, H.T.S. Boschker and C. Barrón. 2004. Carbon cycling and bacterial carbon sources in pristine and impacted Mediterranean seagrass sediments. Aquat. Microb. Ecol. 36: 227–237.

Holmer, M., P. Wirachwong and M.S. Thomsen. 2011. Negative effects of stress-resistant drift algae and high temperature on a small ephemeral seagrass species. Mar. Biol. 158: 297–309.

Holmquist, J.G. 1994. Benthic macroalgae as a dispersal mechanism for fauna: influence of a marine tumbleweed. J. Exp. Mar. Biol. Ecol. 180: 235–251.

Holmquist, J.G. 1997. Disturbance and gap formation in a marine benthic mosaic; influence of shifting macroalgal patches on seagrass structure and mobile invertebrates. Mar. Ecol. Prog. Ser. 158: 121–130.

Holmström, A. 1998. Occurrence, species composition, biomass and decomposition of drifting algae in the northwestern Åland archipelago, Northern Baltic Sea. M.Sc. Thesis. Åbo Akademi Uni., Dept. of Biology, p. 69.

Höffle, H., T. Wernberg, M.S. Thomsen and M. Holmer. 2012. Drift algae, an invasive snail and elevated temperature reduce ecological performance of a warm-temperate seagrass, through additive effects. Mar. Ecol. Prog. Ser. 450: 67–80.

Hull, S.C. 1987. Macroalgal mats and species abundance: a field experiment. Estuar. Coast. Shelf Sci. 25: 519–532.

IPCC. 2007. Climate Change 2007: Synthesis Report. Valencia, Spain.

Irigoyen, A.J., C. Eyras and A.M. Parma. 2010. Alien algae *Undaria pinnatifida* causes habitat loss for rocky reef fishes in north Patagonia. Biol. Invasions, DOI 10.1007/s10530-010-9780-1.

Isaksson, I. and L. Pihl. 1992. Structural changes in benthic macrovegetation and associated epibenthic faunal communities. Neth. J. Sea Res. 30: 131–140.

Kautsky, U. 1995. Ecosystem processes in coastal areas of the Baltic Sea. PhD Thesis, Stockholm University.

Kiirikki, M. and J. Blomster. 1996. Wind induced upwelling as a possible explanation for mass occurrence of epiphytic *Ectocarpus siliculosus Phaeophyta* in the northern Baltic proper. Mar. Biol. 127: 353–358.

Krause-Jensen, D., P.B. Christensen and S. Rysgaard. 1999. Oxygen and nutrient dynamics within mats of the filamentous macroalga *Chaetomorpha linum*. Estuaries 22: 31–38.

Lapointe, B.E. 1997. Nutrient thresholds for bottom-up control of macroalgal blooms on coral reefs in Jamaica and southeast Florida. Limnol. Oceanogr. 42: 1119–1131.

Lapointe, B.E. 1999. Simultaneous top-down and bottom-up forces control macroalgal blooms on coral reefs (Reply to the comment by Hughes et al.). Limnol. Oceanogr. 44: 1586–1592.

Lapointe, B.E. and J. O'Connell. 1989. Nutrient-enhanced growth of *Cladophora prolifera* in Harrington Sound, Bermuda: eutrophication of a confined, phosphorus-limited ecosystem. Estuar. Coast. Shelf Sci. 28: 347–360.

Lapointe, B.E., M.M. Littler and D.S. Littler. 1992. Nutrient availability to marine macroalgae in siliciclastic versus carbonate-rich coastal waters. Estuaries 15: 75–82.

Lauringson, V. and J. Kotta. 2006. Influence of the thin drift algal mats on the distribution of macrozoobenthos in Kõiguste Bay, NE Baltic Sea. Hydrobiologia 554: 97–105.

Lavery, P.S. and A.J. McComb. 1991. Macroalgal-sediment nutrient interactions and their importance to macroalgal nutrition in a eutrophic estuary. Estuarine, Coastal Shelf Sci. 32: 281–295.

Lehvö, A. and S. Bäck. 2001. Survey of macroalgal mats in the Gulf of Finland, Baltic Sea. Aquat. Conserv. 11: 11–18.

Lindeman, R.L. 1942. The Trophic-Dynamic Aspect of Ecology. Ecology 23(4): 399–417.

Liu, D., J.K. Keesing, Q. Xing and P. Shi. 2009. World's largest macroalgal bloom caused by expansion of seaweed aquaculture in China. Mar. Poll. Bull. 58: 888–895.

Lopes, R.J., M.A. Pardal and J.C. Marques. 2000. Impact of macroalgal blooms and wader predation on intertidal macroinvertebrates: experimental evidence from the Mondego estuary (Portugal). J. Exp. Mar. Biol. Ecol. 249: 165–179.

Lotze, H.K. and W. Schramm. 2000. Ecophysiological traits explain species dominance patterns in macroalgal blooms. J. Phycol. 36: 287–295.

Lotze, H.K. and B. Worm. 2002. Complex interactions of climatic and ecological controls on macroalgal recruitment. Limnol. Oceanogr. 47: 1734–1741.

Lotze, H.K., W. Schramm, D. Schories and B. Worm. 1999. Control of macroalgal blooms at early developmental stages: *Pilayella littoralis* versus *Enteromorpha* spp. Oecologia 119: 46–54.

Loya, Y., H. Lubinevsky, M. Rosenfeld and E. Kramarsky-Winter. 2004. Nutrient enrichment caused by *in situ* fish farms at Eilat, Red Sea is detrimental to coral reproduction. Mar. Poll. Bull. 49(4): 344–353.

Lyons, P., C. Thornber, J. Portnoy and E. Gwilliam. 2009. Dynamics of macroalgal blooms along the Cape Cod National Seashore. Northeast. Nat. 16(1): 53–66.

Marszalek, D.S. 1981. Effects of sewage on reef corals. Proceedings of the 4th International Coral Reef Symposium Manila, pp. 18–22.

Martinetto, P., P. Daleo, M. Escapa, J. Alberti, J.P. Isacch, E. Fanjul, F. Botto, M.L. Piriz, G. Ponce, G. Casas and O. Iribarne. 2010. High abundance and diversity of consumers associated with eutrophic areas in a semi-desert macrotidal coastal ecosystem in Patagonia, Argentina. Estuar. Coast. Shelf Sci. 88: 357–364.

Martinetto, P., M. Teichberg, I. Valiela, D. Montemayor and O. Iribarne. 2011. Top-down and bottom-up regulation in a high nutrient-high herbivory coastal ecosystem. Mar. Ecol. Prog. Ser. 432: 69–82.

Merceron, M. and M. Morand. 2004. Existence of a deep subtidal stock of drifting *Ulva* in relation to intertidal algal mat developments. J. Sea. Res. 52: 269–280.

Meyers, M.B., H. Fossing and E.N. Powell. 1987. Microdistribution of interstitial meiofauna, oxygen and sulphide gradients and the tubes of macro-infauna. Mar. Ecol. Prog. Ser. 35: 223–241.

Modig, H. and E. Ólafsson. 1998. Responses of Baltic benthic invertebrates to hypoxic events. J. Exp. Mar. Biol. Ecol. 229: 133–148.

Morand, P. and X. Brian. 1996. Excessive growth of macroalgae: a symptom of environmental disturbance. Bot. Mar. 39: 491–516.

Morand, P. and M. Merceron. 2004. Coastal eutrophication and excessive growth of macroalgae. pp. 395–449. *In*: Pandai, S.G. (ed.). Recent Research Developments in Environmental Biology, Vol. 1(2). Research Signpost, Trivandrum, Kerala, India.

Morand, P. and M. Merceron. 2005. Macroalgal population and sustainability. J. Coast. Res. 21(5): 1009–1020.

Myers, J.A. and K.L. Heck, Jr. 2013. Amphipod control of epiphytic load and its concomitant effects on shoalgrass *Halodule wrightii* biomass. Mar. Ecol. Prog. Ser. 483: 133–142.

Newton, C. and C. Thornber. 2013. Ecological impacts of macroalgal blooms on salt marsh communities. Estuar. Coast. 36: 365–376.

Nordström, M. and D.M. Booth. 2007. Drift algae reduce foraging efficiency of juvenile flatfish. J. Sea Res. 58: 335–341.

Nordström, M., E. Bonsdorff and S. Salovius. 2006. The impact of infauna (*Nereis diversicolor* and *Saduria entomon*) on the redistribution and biomass of macroalgae on marine soft bottoms. J. Exp. Mar. Biol. Ecol. 333: 58–70.

Nordström, M., K. Aarnio and E. Bonsdorff. 2009. Temporal variability of a benthic food web: patterns and processes in a low-diversity system. Mar. Ecol. Prog. Ser. 378: 13–26.

Norkko, A. 1998. The impact of loose-lying algal mats and predation by the brown shrimp *Crangon crangon* (L.) on infaunal prey dispersal and survival. J. Exp. Mar. Biol. Ecol. 221: 99–116.

Norkko, A. and E. Bonsdorff. 1996a. Altered benthic prey-availability due to episodic oxygen deficiency caused by drifting algal mats. Mar. Ecol. 17: 355–372.

Norkko, A. and E. Bonsdorff. 1996b. Rapid zoobenthic community responses to accumulations of drifting algae. Mar. Ecol. Prog. Ser. 131: 143–157.

Norkko, A. and E. Bonsdorff. 1996c. Population responses of coastal zoobenthos to stress induced by drifting algal mats. Mar. Ecol. Prog. Ser. 140: 141–151.

Norkko, J., E. Bonsdorff and A. Norkko. 2000. Drifting algal mats as an alternative habitat for benthic invertebrates: species specific responses to a transient resource. J. Exp. Mar. Biol. Ecol. 248: 79–104.

Ólafsson, E.B. 1988. Inhibition of larval settlement to a soft bottom benthic community by drifting algal mats—an experimental test. Mar. Biol. 97: 571–574.

Ólafsson, E., K. Aarnio, E. Bonsdorff and N.L. Arroyo. 2013. Fauna of the green alga *Cladophora glomerata* in the Baltic Sea: density, diversity and algal decomposition stage. Mar. Biol. 160: 2353–2362.

Österling, M. and L. Pihl. 2001. Effects of filamentous green algal mats on benthic macrofaunal functional feeding groups. J. Exp. Mar. Biol. Ecol. 263: 159–183.

Pardal, M.A., J.C. Marques, I. Metelo, A.I. Lillebø and M.R. Flindt. 2000. Impact of eutrophication on the life cycle, population dynamics and production of *Ampithoe valida* (Amphipoda) along an estuarine spatial gradient (Mondego estuary, Portugal). Mar. Ecol. Prog. Ser. 196: 207–219.

Pearson, T.H. and R. Rosenberg. 1978. Macrobenthic succession in relation to organic enrichment and pollution of the marine environment. Ocean. Mar. Biol. Ann. Rev. 16: 229–311.

Phillips, G.L., D. Eminson and B. Moss. 1978. A mechanism to account for macrophyte decline in progressively eutrophicated freshwater. Aquat. Bot. 4: 103–126.

Phillips, J.A. 2006. Drifting blooms of the endemic filamentous brown algae Hincksia sordida at Noosa on the subtropical east Australian coast. Mar. Poll. Bull. 52(8): 962–968.

Pihl, L., A. Svenson, P.-O. Moksnes and H. Wennhage. 1999. Distribution of green algal mats throughout shallow soft bottoms of the Swedish Skagerrak archipelago in relation to nutrient sources and wave exposure. J. Sea Res. 41: 281–294.

Piriou, J.Y. and V. Duval. 1990. Photosynthèse et nutrition carbonée chez *Ulva* sp. Document DRO.EL n°90.31. IFREMER (Institut Français de Recherche pour l'Exploitation de la MER), BP 70, 29263 PLouzané, France, XI-13 p.

Piriou, J.Y. and A. Ménesguen. 1990. Environmental factors controlling the *Ulva* sp. blooms in Brittany (France). Proc. 25th EMBS. Marine Eutrophication and Population Dynamics. Olsen and Olsen, Fredensborg, Denmark, pp. 111–115.

Raffaelli, D. 2000. Interactions between macro-algal mats and invertebrates in the Ythan estuary, Aberdeenshire, Scotland. Helgol Mar. Res. 54: 71–79.

Raffaelli, D., J. Limia, S. Hull and S. Pont. 1991. Interactions between the amphipod *Corophium volutator* and macroalgal mats on estuarine mudflats. J. Mar. Biol. Assoc. U.K. 71: 899–908.

Raffaelli, D., J.A. Raven and L.J. Poole. 1998. Ecological impact of green macroalgal blooms. Oceanogr. Mar. Biol. Annu. Rev. 36: 97–125.

Rasmussen, J.R., B. Olesen and D. Kruse-Jensen. 2012. Effects of filamentous macroalgae mats on growth and survival of eelgrass, *Zostera marina*, seedlings. Aquat. Bot. 99: 41–48.

Rasmussen, J.R., M.F. Pedersen, B. Olesen, S.L. Nielsen and T.M. Pedersen. 2013. Temporal and spatial dynamics of ephemeral drift-algae in eelgrass, *Zostera marina*, beds. Estuar. Coast. Shelf Sci. 119: 167–175.

Ravera, O. 2000. The Lagoon of Venice: the result of both natural factors and human influence. J. Limnol. 59(1): 19–30.

Riegl, B. and B. Velmirov. 1991. How many damaged corals in the Red Sea reef systems? A quantitative survey. Hydrobiologia 216/217: 249–256.

Rosenberg, R. and R.J. Diaz. 1993. Sulphur bacteria (*Beggiatoa* spp.) mats indicate hypoxic conditions in the inner Stockholm archipelago. Ambio 22: 32–36.

Råberg, S. and L. Kautsky. 2007. A comparative biodiversity study of the associated fauna of perennial fucoids and filamentous algae. Estuar. Coast. Shelf Sci. 73: 249–258.

Salovius, S. and E. Bonsdorff. 2003. Effects of depth, sediment and grazers on the degradation of drifting filamentous algae (*Cladophora glomerata* and *Pilayella littoralis*). J. Exp. Mar. Biol. Ecol. 298: 93–109.

Salovius, S. and P. Kraufvelin. 2004. The filamentous green alga *Cladophora glomerata* as a habitat for littoral macrofauna in the Northern Baltic Sea. Ophelia 58(2): 65–78.

Salovius, S., M. Nyqvist and E. Bonsdorff. 2005. Life in the fast lane: macrobenthos use temporary drifting algal habitats. J. Sea Res. 53: 169–180.

Scanlan, C.M., J. Foden, E. Wells and M.A. Best. 2007. The monitoring of opportunistic macroalgal blooms for the water framework directive. Mar. Poll. Bull. 55: 162–171.

Schramm, W. 1999. Factors influencing seaweed responses to eutrophication: some results from EU-project EUMAC. J. Appl. Phycol. 11: 69–78.

Sieben, K., L. Ljunggren, U. Bergstrom and B.K. Eriksson. 2011a. A meso-predator release of stickleback promotes recruitment of macroalgae in the Baltic Sea. J. Exp. Mar. Biol. Ecol. 397: 79–84.

Sieben, K., A.D. Rippen and B.K. Eriksson. 2011b. Cascading effects from predator removal depend on resource availability in a benthic food web. Mar. Biol. 158: 391–400.

Smith, J.E., J.W. Runcie and C.M. Smith. 2005. Characterization of a large-scale ephemeral bloom of the green alga *Cladophora sericea* on the coral reefs of West Maui, Hawai'i. Mar. Ecol. Prog. Ser. 302: 77–91.

Sundbäck, K., B. Jönsson, P. Nilsson and I. Lindström. 1990. Impact of accumulating drifting macroalgae on a shallow-water sediment system: an experimental study. Mar. Ecol. Prog. Ser. 58: 261–274.

Sundbäck, K., A. Miles, S. Hulth, L. Pihl, P. Engstrom, E. Selander and A. Svenson. 2003. Importance of benthic nutrient regeneration during initiation of macroalgal blooms in shallow bays. Mar. Ecol. Prog. Ser. 246: 115–126.

Tagliapietra, D., M. Pavan and C. Wagner. 1998. Macrobenthic community changes related to eutrophication in Palude della Rosa (Venetian Lagoon, Italy). Estuar. Coast. Shelf Sci. 47: 217–226.

Teichberg, M., S.E. Fox, C. Aguila, Y.S. Olsen and I. Valiela. 2008. Macroalgal responses to experimental nutrient enrichment in shallow coastal Waters: growth, internal nutrient pools, and isotopic signatures. Mar. Ecol. Prog. Ser. 368: 117–126.

Teichberg, M., S.E. Fox, Y.S. Olsen, I. Valiela, P. Martinetto, O. Iribarne, E.Y. Muto, M.A.V. Petti, T.N. Corbisier, M. Soto-Jiménez, F. Páez-Osuna, P. Castro, H. Freitas, A. Zitelli, M. Cardinaletti and D. Tagliapietra. 2010. Eutrophication and macroalgal blooms in temperate and tropical coastal waters: nutrient enrichment experiments with *Ulva* spp. Glob. Change Biol. 16: 2624–2637.

Thamdrup, B., J.W. Hansen and B.B. Jsrgensen. 1998. Temperature dependence of oxygen respiration in a temperate coastal sediment. FEMS Microbiol. Ecol. 25: 189–200.

Thiel, M. and L. Watling. 1998. Effects of green algal mats on infaunal colonization of a New England mud flat–long-lasting but highly localized effects. Hydrobiologia 375/376: 177–189.

Thomsen, M.S. 2010. Experimental evidence for positive effects of invasive seaweed on native invertebrates via habitat-formation in a seagrass bed. Aquatic Invasions 5(4): 341–346.

Thomsen, M.S. and K. McGlathery. 2006. Effects of accumulations of sediments and drift algae on recruitment of sessile organisms associated with oyster reefs. J. Exp. Mar. Biol. Ecol. 328: 22–34.

Thomsen, M.S., T. de Bettignies, T. Wernberg, M. Holmer and B. Debeuf. 2012a. Harmful algae are not harmful to everyone. Harmful algae 16: 74–80.

Thomsen, M.S., T. Wernberg, A.H. Engelen, F. Tuya, M.A. Vanderklift, M. Holmer, K.J. McGlathery, F. Arenas, J. Kotta and B.R. Silliman. 2012b. A meta-analysis of seaweed impact on seagrasses: generalities and knowledge gaps. Plos One 7(1): 1–8.

Thornber, C.S., P. Dimilla, S.W. Nixon and R.A. McKinney. 2008. Natural and anthropogenic nitrogen uptake by bloom-forming macroalgae. Mar. Poll. Bull. 56: 261–269.

Tsai, C.-C., J.-S. Chang, F. Sheu, Y.-T. Shyu, A.Y.-C. Yua, S.-L. Wongd, C.-F. Daie and T.-M. Lee. 2005. Seasonal growth dynamics of *Laurencia papillosa* and *Gracilaria coronopifolia* from a highly eutrophic reef in southern Taiwan: temperature limitation and nutrient availability. J. Exp. Mar. Biol. Ecol. 315: 49–69.

Vahteri, P., A. Mäkinen, S. Salovius and I. Vuorinen. 2000. Are drifting algal mats conquering the bottom of the Archipelago Sea, SW Finland? Ambio 29: 338–343.

Valiela, I., J. McClelland, J. Hauxwell, P.J. Behr, D. Hersh and K. Foreman. 1997. Macroalgal blooms in shallow estuaries: controls and ecophysiological and ecosystem consequences. Limnol. Oceanogr. 42: 1105–1118.

Vermeij, M.J.A., M.L. Dailer and C.M. Smith. 2009. Nutrient enrichment promotes algal survival and dispersal of drifting fragments in an invasive tropical macroalga. Coral Reefs 28: 429–435.

Vetter, E.W. 1994. Hotspots of benthic production. Nature 372: 47.

Viaroli, P., M. Bartoli, R. Azzoni, G. Giordani, C. Mucchino, M. Naldi, D. Nizzoli and L. Tajé. 2005. Nutrient and iron limitation to *Ulva* blooms in a eutrophic coastal lagoon (Sacca di Goro, Italy). Hydrobiologia 550: 57–71.

Villano, N. and R.M. Warwick. 1995. Meiobenthic communities associated with the seasonal cycle growth of *Ulva rigida* Agardh in the Pallude della Rosa, Laggon of Venice. Estuar. Coast. Shelf Sci. 41: 181–194.

Villares, R., X. Puente and A. Carballeira. 1999. Nitrogen and phosphorous in *Ulva* sp. in the Galician Rias Bajas (northwest Spain): seasonal fluctuations and influence on growth. Bol. Inst. Esp. Oceanogr. 15: 337–341.

Virnstein, R.W. and P.A. Carbonara. 1985. Seasonal abundance and distribution of drift algae and seagrasses in the mid-Indian River lagoon, Florida. Aquat. Bot. 23: 67–82.

Vukadin, I. 1991. Impact of nutrient enrichment and their relation to the algal bloom in the Adriatic Sea. Mar. Poll. Bull. 23: 145–148.

Wallentinus, I. 1984. Comparisons of nutrient uptake rates for Baltic macroalgae with different thallus morphologies. Mar. Biol. 80: 215–225.

Wang, Y., Y. Wang, L. Zhu, B. Zhou and X. Tang. 2012. Comparative studies on the ecophysiological differences of two green tide macroalgae under controlled laboratory conditions. Plos One 7(8): 1–16.

Whalen, M.A., E. Duffy and J.B. Grace. 2013. Temporal shifts in top-down vs. bottom-up control of epiphytic algae in a seagrass ecosystem. Ecology 94(2): 510–520.

Widbom, B. and R. Elmgren. 1988. Response of benthic meiofauna to nutrient enrichment of experimental marine ecosystems. Mar. Ecol. Prog. Ser. 42: 257–268.

Worm, B. and H.K. Lotze. 2006. Effects of eutrophication, grazing, and algal blooms on rocky shores. Limnol. Oceanogr. 51: 569–579.

Worm, B., H.K. Lotze, C. Boström, R. Engkvist, V. Labanauskas and U. Sommer. 1999. Marine diversity shift linked to interactions among grazers, nutrients and propagule banks. Mar. Ecol. Prog. Ser. 185: 309–314.

Worm, B., H.K. Lotze and U. Sommer. 2000. Coastal food web structure, carbon storage, and nitrogen retention regulated by consumer pressure and nutrient loading. Limnol. Oceanogr. 45: 339–349.

Ye, N., X. Zhang, Y. Mao, C. Liang, D. Xu, J. Zou, Z. Zhuang and Q. Wang. 2011. "Green tides" are overwhelming the coastline of our blue planet: taking the world's largest example. Ecol. Res. 26: 477–485.

Zhang, X., H. Wang, Y. Mao, C. Liang, Z. Zhuang, Q. Wang and N. Ye. 2010. Somatic cells serve as a potential propagule bank of *Enteromorpha prolifera* forming a green tide in the Yellow Sea, China. J. Appl. Phycol. 22: 173–180.

II. Production of Macrophytes

6

Macrophyte Productivity and the Provisioning of Energy and Habitat to Nearshore Systems

Michael H. Graham,* Michael D. Fox and *Scott L. Hamilton*

Introduction

Foundation species are disproportionately important to the structure of their associated communities (Dayton 1972, Bruno and Bertness 2001) and are evident in both terrestrial and marine systems (e.g., mangroves, redwood trees, kelps, hermatypic corals). Through the direct provisioning of energy and habitat, foundation species facilitate the maintenance of community structure in different ways than keystone predators (e.g., sea otters, wolves) and ecosystem engineers (e.g., termites, beavers), although these terms are sometimes (and incorrectly) used interchangeably in the literature. Keystone predators actively regulate species interactions (Paine 1969a, Paine 1969b), while ecosystem engineers provide habitat architecture and modify the physical environment (Jones et al. 1994, Hastings et al. 2007). Like these critical species, however, the loss of foundation species from a system generally causes conspicuous declines in local biodiversity, productivity, and ecosystem functioning (Knowlton 2001, Graham 2004, Ellison et al. 2005). The utility of the foundation species concept is that variability in productivity and population dynamics of foundation species may be directly proportional to community characteristics and ecosystem function. Therefore, studies of the ecophysiology of foundation species are critically important for forecasting subsequent variability in associated communities and ecosystems. Because the role of foundation species may be context dependent (Hughes 2010), a

Moss Landing Marine Laboratories, 8272 Moss Landing Road, Moss Landing, CA, USA, 95039.
* Corresponding author: mgraham@mml.calstate.edu

detailed understanding of how macrophyte foundation species respond to external stressors and environmental conditions is paramount for developing successful management strategies, maintaining key ecosystem services, and promoting natural resilience in marine systems.

Macrophytes clearly play a foundational role in many marine ecosystems, though not all macrophytes can be considered foundation species. In any given system, most macroalgal and seagrass taxa are potential sources of fixed carbon and nutrients for primary consumers (Vadas 1977, Hay et al. 1994, Paine 2002, Sotka and Hay 2002, Amsler et al. 2005, Rasher et al. 2013). The role of marine macrophytes in habitat provisioning, however, is dependent on the size and distribution of the macrophyte relative to the scale of physical structure in the natural environment. The loss of a foliose red alga from offshore pinnacles, for example, may result in a decline in available food for herbivores, but the heterogeneous rocky substrates that remain will continue to serve as critical habitat for associated invertebrates and fishes. Therefore, in this system, the foliose red alga would not be considered a foundation species even though it is consumed as part of the food web. Conversely, large kelps growing on rocky coasts, or even short seagrasses growing on homogenous soft sediments, have a disproportionate effect on the provision of both energy and habitat to their respective systems. The functional importance of macrophytes as foundation species in marine systems is therefore dependent on the simultaneous provisioning of energy and critical habitat, and the magnitude of their relative contributions.

Here, we explore how understanding macrophyte productivity and morphology is useful for studying their role as foundation species in natural systems. Our goal is not to provide an exhaustive review of marine macrophyte productivity rates, nor the physiological and environmental processes that regulate such rates. Instead, our focus is to: (1) identify macrophyte taxa that are clearly disproportionate in their provisioning of energy and habitat to their associated communities; (2) describe some physiological features that facilitate the role of macrophytes as foundation species; and (3) discuss how variability in the productivity of marine macrophyte foundation species may impact their associated communities.

Macrophyte foundation species

The determination of whether a particular macroalgal or seagrass taxon is a foundation species is dependent on the context of the system within which the macrophyte is being studied. As such, it is not possible to provide a simple list of marine macrophyte foundation species, as the same species can function differently across systems. For example, *Laminaria hyperborea* is an important foundation species in rocky subtidal systems of Norway and the Northeast Atlantic, where much of the structure and energy flow of these systems is dependent on this single kelp species (Jupp and Drew 1974, Christie et al. 2003, Norderhaug et al. 2003, Abdullah and Fredriksen 2004, Bartsch et al. 2008). In contrast, there is little evidence for a foundational role of a closely related species *Laminaria farlowii* in California rocky subtidal systems, where high physical heterogeneity and the presence of large, perennial *Macrocystis* forests have been shown to be of overwhelming importance (Dayton 1985, Dayton et al. 1992).

So are kelps of the genus *Laminaria* foundation species or not? The answer depends on the characteristics of the system within which they are present and the scale at which they influence their associated communities. Ultimately, the only true test of a macrophyte's role as a foundation species is to study the structure of the community with and without the macrophyte in question. While an important source of energy and habitat, the loss of a stipitate *Laminaria* species from a California kelp forest is unlikely to fundamentally alter the structure and function of the system, whereas the loss of *Macrocystis* would drive fundamental changes in the community (Graham 2004, Arkema et al. 2009). It is therefore critical to understand the respective roles of macrophytes in different systems and how the role of a single species can change from location to location due to variations in environmental conditions and community structure. While the accurate description of a species as a foundation species can be challenging, it is important to understand which species have the potential to serve as foundation species within an ecological system and which do not.

To date, few studies have examined the effects of the presence/absence of a specific species on community structure to assess whether or not the "foundation species" label is appropriate. For example, Shelton (2010) nicely demonstrated that the surfgrass *Phyllospadix* serves as a foundation species in tidepools by provisioning habitat and modulating environmental conditions. Tidepools in which *Phyllospadix* was present were up to 10°C cooler and had more stable water temperatures, while tidepools that had *Phyllospadix* removed had warmer and more variable temperature, which drove substantial changes in community structure. In most cases, foundation species are identified through natural history observations that suggest particular species are critical to the provisioning of energy and habitat. Other studies have used stable isotopic analysis to quantify the disproportionate importance of a particular species to maintaining energy flow within a system (Duggins et al. 1989, Bustamante et al. 1995, Bustamante and Branch 1996, Miller et al. 2013). Generally, species-removal experiments are aimed at understanding the role of competition in regulating community structure and ecosystem function (Dayton 1975, Santelices and Ojeda 1984a, Clark et al. 2004) and are not always designed to directly address the role of a species as a foundation species. To get at the specific question of habitat provisioning, however, researchers have compared the diversity or community structure of a system with and without the putative foundation species, relying on both experimental and natural removal of key species (Duggins 1980, Connolly 1994, Attrill et al. 2000, Byrnes et al. 2011). Yet, we are unaware of a single study that has directly documented *both* the trophic and habitat consequences of the loss of a foundation species within a system. This is likely due to the large spatial scales over which macrophyte foundation species are distributed and the long timescales within which natural communities respond to physical disturbances or the loss of critical species.

Despite the dearth of experimental studies, it is clear that macrophyte foundation species exist in many systems (Figs. 1 and 2). Temperate coastlines are dominated by kelps and large fucoids that have exceptional biomass, high rates of primary productivity, and provide three dimensional habitat and structure (Steneck et al. 2002, Graham et al. 2007, Bartsch et al. 2008, Bolton 2010). Seagrasses generally grow on soft sediment substrates and provide a fundamental source of resources and shelter to diverse filter feeding invertebrate assemblages within the sediment, and to fish and

Figure 1. Marine macrophyte foundation species. (A) Subtidal forest of the giant kelp *Macrocystis pyrifera* off San Clemente Island, southern California, USA (photo credit: Enric Sala); (B) Shallow subtidal forest of the *Eisenia arborea* off San Clemente Island, southern California, USA (photo credit: Enric Sala); (C) Subtidal forest of the bull kelp *Nereocystis luetkeana*, central California, USA (photo credit: Scott Gabara); (D) Subtidal rhodolith bed of Santa Catalina Island, southern California, USA (photo credit: Scott Gabara); (E) Intertidal belt of the southern bull kelp *Durvillaea antarctica* off Bahia Mansa, southern Chile (photo credit: Michael Fox); (F) Floating mats of *Sargassum* in the Sargasso Sea, Atlantic Ocean (photo credit: Michael Fox); (G) Subtidal seagrass bed (photo credit: P. Rouja).

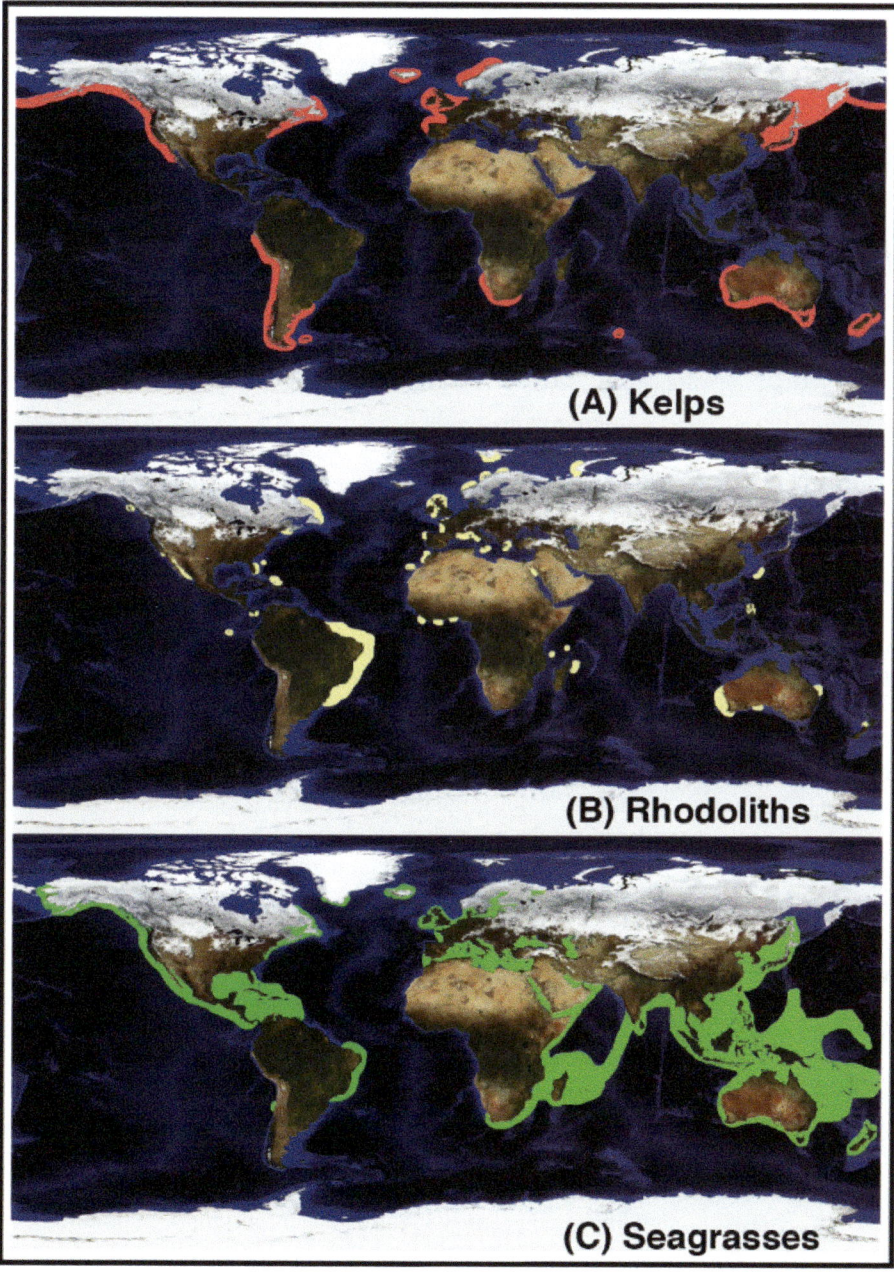

Figure 2. Global distribution of (A) kelps, (B) rhodoliths, and (3) seagrasses. Kelp distribution is not exhaustive for all kelp taxa, but represents areas where *Macrocystis*, *Nereocystis*, *Laminaria*, and *Ecklonia* serve a foundational role (modified from Steneck et al. 2002, Graham et al. 2007). Rhodolith distribution map is modified from Foster (2001). Seagrass distribution map is modified from Short et al. (2007) and represents currently accepted species in the families Hydrocharitaceae, Cymodoceaceae, Posidoniaceae, Zosteraceae, and Ruppiaceae.

invertebrates that live among the seagrass thalli in an otherwise featureless environment (Jackson et al. 2001, Hughes et al. 2008). The Sargasso Sea is a featureless, oligotrophic oceanic environment within which floating mats of *Sargassum natans* and *S. fluitans* provide intricate structure and presumably important energetic resources (Smith et al. 1973, Lapointe 1986). Despite little ecological research in this system, many species appear to have obligate associations with pelagic *Sargassum*, especially for critical early life history stages, which suggests that *Sargassum* likely functions as a foundation species in this system (Dooley 1972, Carr 1987, Moser et al. 1998). Similarly, rhodoliths, free living coralline algal nodules, provide well-documented physical structure to soft-sediment systems (Foster 2001, Steller et al. 2003, Hinojosa-Arango et al. 2014). Although coralline algal productivity rates are inherently low relative to other macroalgae, the tight association with cryptic invertebrates and the complex interstitial matrix of space within the rhodoliths suggests a strong role of rhodoliths as foundation species (Kamenos et al. 2004a,b,c, Steller and Cáceres-Martínez 2009, Hinojosa-Arango et al. 2014). Still, beyond general natural history observations and focused ecological studies, the joint provisioning of energy and habitat to natural systems by marine macrophytes remains poorly studied.

Here, we review the evidence for and identify marine macrophyte foundation species, limiting our discussion to instances in which enough data and information exist to make a strong argument for labeling a particular species using this concept. Many more macrophyte foundation species likely exist and it is important that we begin to document their presence with more experimental rigor than the typical natural history observations employed to date. With the onset of global climate change and the continued degradation of ecosystems through pollution and resource extraction, many macrophyte-based marine systems will face unprecedented challenges (Harley et al. 2012). Rigorous examination of the putative foundation species within a particular system will provide guidance for future ecological studies targeting ecosystem change and for management strategies designed to maintain and prevent future degradation of critical ecosystem services.

Mechanisms for high energy provision

Although direct tests of the disproportionate impact of foundation species on community structure through energy provisioning are not common, enough data exist to recognize the foundational role of certain macrophytes in their respective systems, allowing for a comparison of annual productivity rates among these taxa. We have compiled annual production (g C per m² per year) values for numerous marine macrophyte foundation species from published studies, focusing on those taxa (kelps, fucoids, seagrasses) that have been clearly identified as important to maintaining the structure of their associated communities (Table 1); this compilation is not exhaustive and relies primarily on species that have a clearly defined foundational role in their system. For comparative purposes, our list includes annual production values for non-marine foundation species as well as some kelp, fucoid, and seagrass taxa that are common, but have not been demonstrated to provide critical habitat and therefore do not meet our definition of foundation species.

Table 1.

Taxon	Location	Annual Production (g C m^{-2} yr^{-1})
Kelps		
Macrocystis laevis[1]	Marion Is., Prince Edward Is.	3500
Laminaria hyperborea[2]	Norway	3000
Saccharina latissima[3]	Nova Scotia, Canada	2000
Ecklonia radiata[4]	Western Australia	1600
Laminaria pallida[5]	South Africa	1330
Ecklonia radiata[6]	New South Whales, Eastern Australia	1100
Macrocystis pyrifera[7]	California, USA	800–1000
Ecklonia cava[8]	Japan	950
Macrocystis pyrifera[9, 10]	southern California, USA	120–680
Laminaria digitata[11]	Scotland, U.K.	405
Saccharina latissima[11]	Scotland, U.K.	135
Macrocystis integrifolia[12]	Barkley Sound, British Columbia, Canada	31
Saccharina latissima[13]	Young Sound, Greenland	0.1–2
Fucoids		
Sargassum platycarpum[14]	Curacao, Netherlands Antilles	2550
Cystoseira mediterranea[15]	Northwestern Mediterranean	900
Ascophyllum nodosum[16]	Cobscook Bay, Maine, USA	489–699
Ascophyllum nodosum[17]	Massachusetts, USA	353.7
Fucus vesiculosus[18]	Massachusetts, USA	127.8
Cystoseira spp.[19]	Eastern Canary Islands	1.5–10.5
Other macroalgae		
Cladophora gracilis[18]	Nauset Marsh, Massachusetts, USA	59–637
Himantothallus grandifolius[20]	Singy Island, Antarctica	16–56
Seagrasses		
Posidonia oceanica[21]	Gulf of Naples, Italy	1170
Cymodocea nodosa[22]	Urbinu Lagoon, Corsica, France	844
Zostera marina[23]	Netarts Bay, Oregon, USA	838
Zostera marina[7]	Denmark	800
Thalassia testudinum[7]	Caribbean	800
Zostera marina[24]	Puget Sound, Washington, USA	707

Table 1. contd....

Table 1. contd.

Taxon	Location	Annual Production (g C m^{-2} yr^{-1})
Halodule wrightii[25]	Mississippi Sound, Gulf of Mexico, USA	256
Thalassia testudinum[26]	Corpus Christi Bay, Texas, USA	253
Posidonia spp.[27]	Australia	140
Posidonia oceanica[28]	Gulf of Naples, Italy	60–184
Cymodocea nodosa[22]	Mar Menor Lagoon, Spain	47.8–48.9
Non-macrophytes		
Sequoia sempervirens[29]	northern California, USA	700–1000
Zooxanthellate corals[30]	various locations	744.6
Mangroves[31]	North Queensland, Australia	350–500

[1]Attwood et al. (1991); [2]Abdullah and Fredriksen (2004); [3]Mann (1972); [4]Kirkman (1984), as reported in Mann (2000); [5]Field et al. (1977), as reported in Mann (2000); [6]Larkum (1986), as reported in Mann (2000); [7]Mann (2000); [8]Reed et al. (2008); [9]Yokohama et al. (1987); [10]Jackson (1987); [11]Drew (1983); [12]Wheeler and Druehl (1986); [13]Borum et al. (2002); [14]Wanders (1976), as reported in Mann (1982); [15]Ballesteros (1989); [16]Vadas et al. (2004); [17]Roman et al. (1990); [18]Roman et al. (1991); [19]Johnston (1969), as reported in Mann (2000); [20]Drew and Hastings (1993); [21]Ott (1980); [22]Agostini et al. (2003); [23]Kentula and McIntire (1986); [24]Nelson and Waaland (1997); [25]Moncreiff et al. (1992); [26]Lee and Dunton (1996); [27]Pergent et al. (1994); [28]Busing and Fujimori (2005); [29]Muscatine (1990); [30]Mann (1982); [31]Miller (1972), as reported in Mann (1982).

The annual production of kelps (brown algae of the order Laminariales) is conspicuously higher than most marine foundation species, whether they be fucoids, seagrasses, or even mangroves and hermatypic corals (Table 1). Although annual production is highly variable among kelp taxa, species from at least four kelp genera (*Macrocystis, Laminaria, Saccharina,* and *Ecklonia*) have been reported to fix greater than 1000 g C per m^2 per year, with *Macrocystis* populations in the Prince Edward Islands having production rates as great as 3500 g C per m^2 per year (Attwood et al. 1991). These values should be interpreted with caution, however, as no studies have been conducted with the temporal sampling required to appropriately estimate annual rates of production. Therefore some values are likely to be overestimates of the actual annual productivity rates of these kelps. In each case, short-term production values were integrated over the year, obviating any seasonal or sub-annual trends in productivity, which is particularly important at higher latitudes because of the marked seasonal variation in solar irradiation (Deysher and Dean 1984, Jackson 1987, van Tussenbroek 1989). Although the methodologies for estimating primary production also varied among researchers, and thus among the taxa they studied, the among-taxa variability in estimated annual primary production is striking and suggests that marine macrophyte foundation species are "not created equally".

Macrocystis, Laminaria, and *Ecklonia* are globally distributed, and present in the Indian Ocean and both the northern and southern hemispheres of the Pacific and Atlantic Oceans. When production values of these taxa are integrated over the broad spatial scales across which the species are distributed, the contribution to local food webs is tremendous (Newell et al. 1982, Duggins et al. 1989, Graham 2004). With the exception of the prostrate kelp *Saccharina latissima* in Norway, the highest producing kelp taxa are also large in stature, extending either to the surface (e.g., canopy-forming *Macrocystis*) or well into the water column (e.g., the stipitate *Ecklonia radiata* or *Laminaria hyperborea*). Past research has shown that stature is important in determining the outcome of competitive interactions among kelp taxa and the structure of kelp assemblages (Dayton 1975, Dayton 1985, Foster and Schiel 1985), as well as the provisioning of habitat for associated species (Holbrook et al. 1990, Anderson 1994, Carr 1994). Kelps that reach further into the water column have superior access to light and can translocate internal resources and photosynthates to lower structures (juvenile fronds, sporophylls or holdfasts) via trumpet hyphae and/or sieve elements, thus maintaining a competitive advantage over low-lying taxa and increasing their overall productivity rates (Parker and Huber 1965, Schmitz and Lobban 1976, Lobban 1978, Schmitz and Srivastava 1979). Resource translocation also facilitates the higher productivity rates required to maintain biomass in such large kelp species, and the high per thallus productivity rates needed to support the production of their diverse associated foodwebs (Graham 2004).

The most productive kelp taxa tend to be found along mid-latitude temperate coastlines (Table 1, Fig. 2A), likely due to the combination of relatively low water temperatures and high nutrient concentrations driven by coastal upwelling, and greater integrated annual irradiance doses than higher latitude systems (Jackson 1987, Lüning 1990). Indeed, populations of *Macrocystis* in British Columbia and *Saccharina* in Greenland have production values 2-3 orders of magnitude less than their lower latitude counterparts (Table 1). Yet, although these values are low, they likely represent the consequences of light limitation associated with the effects of increasing latitude on primary production (Jackson 1987), rather than an inability of these taxa to function as foundation species within these systems (Graham et al. 2007). Critical experiments to test the role of kelps in the structure of nearshore communities in high latitude systems are rarely conducted, and when they are, the kelp taxa have been shown to play a critical role in enhancing biodiversity. This is exemplified by the high latitude kelp system in the Aleutian Islands, in which the annual canopy-forming kelp *Eularia fistulosa* clearly plays a foundational role (Estes and Palmisano 1974, Estes et al. 1978, Estes and Duggins 1995); unfortunately, the lack of primary production data and general ecological data for *Eularia* precludes comparisons to other foundation species. In the absence of additional field experimentation, it is not possible to determine whether the high variability in annual production among kelp taxa reflects subsequent variability in their role in the functioning of their associated communities.

Fucoids (rockweeds; brown algae of the order Fucales) lack trumpet hyphae or translocation mechanisms similar to those possessed by kelps (Moe and Silva 1981). Consequently, fucoids generally do not grow to the great lengths that many kelp taxa do, although they can form thick canopies in shallow temperate regions, especially in the absence of their kelp counterparts (Cousens 1984, Cousens 1985, Schiel and Foster 2006, Foster and Schiel 2010). Fucoid production values are also far less than those of the most productive kelps (Table 1), with the exception of the annual primary production values reported by Wanders (1976) for *Sargassum platycarpum* and Haxen and Grindley (1985) for *Durvillaea antarctica*. Interestingly, *Durvillaea antarctica*, which does possess hyphae similar to those in kelps (Nizamuddin 1968), has similar daily primary production values ($\sim$7.1 g C per m^2 per day; Haxen and Grindley (1985)) in the Prince Edward Islands as its kelp counterpart *Macrocystis* ($\sim$7–11.5 g C per m^2 per day; Attwood et al. 1991) in the same region. *Durvillaea* is the most massive of the fucoids (Fig. 1E), and when daily production values are integrated annually, they are comparable to the most productive kelp systems ($\sim$2500 g C per m^2 per year; Table 1). Despite the exceptional biomass and extensive distribution of *Durvillaea* in the Southern Hemisphere, very little is known about its ecological role as most studies to date have estimated productivity for economic means. The massive, complex morphology of *Durvillaea* suggests that its role as a foundation species is context-dependent, much like that seen with *E. menziesii* in California (Hughes 2010). As one of the most productive algal species, future studies should examine the foundational role of *Durvillaea* through the examination of the habitat is provides and the influence of its high rates of primary productivity on the trophic structure of the systems in which it occurs.

The literature is rampant with productivity values for macroalgae and phycologists have made comparisons among macroalgal taxa for decades. For example, Littler and Arnold (1982) compared primary productivity across 70 taxa from central California to Baja California, Mexico and found that hourly productivities (g C per g dry weight per hr) varied among taxa by 2-3 orders of magnitude. These data were interpreted to represent inherent differences in primary productivity driven by differences in thallus functional form (e.g., thin blades vs. crusts) and were useful for understanding the effect of environmental change on algal assemblages Littler and Littler (1980). The most productive taxa in the (Littler and Arnold 1982) dataset were thin *Ulva* species (9.5–11 g C per g dwt per hr) that had far greater productivity rates than the estimated values for fucoid and kelp species (0.2–1.1 g C per g dwt per hr). Species within the top 80% of hourly productivity values in Littler and Arnold (1982) study, however, have never been identified as foundation species, and in most field situations, they likely never will. Again, the reason is that such macroalgae are small and lack the conductive mechanisms to move photosynthate and support large thalli, and thus they do not provide habitat architecture, although they can provision energy to the associated community. Consequently, although kelps have relatively low dry weight-specific hourly productivities, when integrated across their large thalli they support extremely high production and play a more distinct foundational role within their respective systems because of their ability to maintain large stature and provide critical habitat for their associated communities.

Much of the annual productivity of kelps is ultimately consumed by herbivores and detritivores, fueling food webs from the bottom up. Kelp contributions as a source of energy for consumers can be detected observationally and by using natural isotopic tracers. For example, Duggins et al. (1989) used stable isotope analysis to show that secondary production (i.e., growth) of suspension feeding and detritivorous invertebrates, all the way up to predatory fishes, was enhanced on islands with kelp compared to islands lacking kelp. Similarly, Salomon et al. (2008) demonstrated that the flow of carbon to filter feeding oysters and mussels was strongly influenced by kelp contributions inside marine reserves characterized by high abundance of the kelp *Ecklonia radiata* and sea urchins, in contrast to fished areas outside of reserves where kelp densities were reduced by sea urchin grazing. Trophic subsidies from kelp-derived organic matter has also been shown to connect marine systems with terrestrial (Dugan et al. 2003), intertidal (Bustamante and Branch 1996), and submarine canyon ecosystems (Harrold et al. 1998, Vetter 1995) providing critical energetic subsidies that support higher than expected consumer biomass (Vanderklift and Wernberg 2008). Disturbances such as large storms can defoliate kelp beds and aid delivery to food-limited habitats outside the beds (Filbee-Dexter and Scheibling 2012), while the transport of kelp detritus over long distances can be aided by kelp morphology, such pneumatocysts and other float-like structures (Hobday 2000, Macaya et al. 2005, Hernandez-Carmona et al. 2006). Conversely, large-scale removal of kelp from a system can reduce food subsidies to local consumers and generate negative feed-back loops that can drive the system into an alternative stable state, such as the urchin barrens driven by El Niños in southern California (Ebeling et al. 1985, Tegner and Dayton 1991).

Seagrasses consistently exhibit high annual production values across a wide range of taxa (Table 1), likely due to the similar environmental conditions within which seagrasses grow. Seagrasses tend to flourish in more oligotrophic waters than kelps (Fig. 2C), preferring warm-temperate to tropical regions with high light (i.e., low turbidity), lacking a requirement for high ambient nutrient levels (Duarte 1995). Like kelps, however, high seagrass productivity rates are supported by internal structures (steles) that allow movement of photosynthates and gases through conducting cells (McRoy and McMillan 1997). Although few seagrasses attain large statures (Fig. 1G), especially relative to canopy-forming kelps, the seagrass translocation mechanisms allow for effective integration between the physiology of subsurface rhizomes (often growing in anoxic sediments) and photosynthetically active leaves (Marbà et al. 2006, Dean and Durako 2007). Thus, seagrasses are able to occupy relatively featureless, low productivity sediment-based systems and provide highly important resources and habitat structure to the organisms within these systems (Hemminga and Duarte 2000). The few studies that have directly addressed the functional role of seagrasses in their associated communities have clearly and consistently identified them as critical foundation species (Heck et al. 1995, Heck et al. 2003). Seagrasses provide energy for herbivorous fishes and invertebrates (Perkins-Visser et al. 1996, Beck et al. 2001, Heck Jr. et al. 2008), helping to fuel complex food webs, and they provide critical nursery habitat and shelter from predators for juvenile life history stages of many fishes and invertebrates (Heck et al. 1997).

In kelp forests, seagrass meadows, and rhodolith (maerl) beds, the foundational macrophytes also provide one of the dominant substrates for encrusting invertebrates and epiphytic algae. This coverage, particularly by epiphytic algae in seagrass meadows, constitutes a significant portion of the diets of associated grazers which in turn facilitates higher rates of primary production of the seagrasses (Moncreiff and Sullivan 2001, Duffy et al. 2003). As such, the indirect provisioning of resources should be considered when evaluating the foundational role of macrophytes. While secondary production is not the focus of this chapter, the influence of foundational macrophytes on this additional source of energy should not be overlooked. Given the relative importance of some of these epiphytic communities, a fundamental question to ask is how does this trophic subsidy change following the removal of a foundation species?

Clearly, additional macrophyte taxa likely support their associated communities in a foundational way, yet we lack the quantitative data necessary to classify them as true foundation species. *Himantothallus grandifolius* is a massive brown alga of the order Desmarestiales that can dominate nearshore regions of Antarctica (Moe and Silva 1977, Amsler et al. 1995). An estimated annual production of 16–56 g C per m^2 per year (Drew and Hastings 1993) for *H. grandifolius* exceeds that of high latitude kelps in the north Pacific and Atlantic (Table 1), yet the availability of this fixed carbon relative to background primary productivity in the region is unknown. Similarly, *Cladophora gracilis* can be locally abundant along northwest Atlantic shores with annual production values of 10–114 g C per m^2 per year (Roman et al. 1990), but *C. gracilis* would only be considered a foundation species if this production was disproportionate relative to other sources of fixed carbon available to this system, which remains unknown. Therefore, although some oligotrophic systems have such low background production potential that it is safe to classify abundant macrophytes as foundation species in the absence of direct studies (e.g., floating *Sargassum* mats in the mid-Atlantic (Carpenter and Cox 1974) or rhodoliths in the sub-tropics; Fig. 1D,F), researchers should be wary of making such classifications in more productive systems.

Habitat Provision

All marine macrophytes are inherently edible and at some non-trivial spatial scale provide habitat to organisms, whether they be microfauna, meiofaunal, or charismatic megafauna. Two criteria, however, are required to distinguish macrophyte foundation species from simple habitat-forming macrophytes. First, as previously discussed, macrophyte foundation species should *simultaneously* provide energy and habitat to their associated community. Second, energy and habitat provisioning by macrophyte foundation species should be *disproportionate* relative to other taxa in the system (Dayton 1972). Yet, unlike studies of energy flow through a system, it is not possible to use chemical tracers (i.e., stable isotopes) to study direct connections between fauna and the marine macrophytes with which they associate. Most researchers have focused on targeted removal of conspicuous macrophyte taxa to study the role of these putative foundation species in structuring nearshore systems (Dayton 1975, Konar 2000, Schiel 2006, Hughes 2010, Shelton 2010, Byrnes et al. 2011). It is difficult if not impossible, however, to disentangle the provision of energy from that of habitat using such simple experiments (Graham et al. 2008), especially for those faunal taxa that eat the macrophyte habitat within or upon which they live (e.g., limpets, amphipods, kelp crabs).

The role of habitat provisioning by marine macrophyte foundation species has not been well studied beyond simple removal experiments at relatively large spatial scales. It is clear in the ecological literature that species diversity can reflect the availability of complex habitats (Macarthur 1965, Heck et al. 1997, Friedlander and Parrish 1998, Tews et al. 2004). In addition to determining whether a particular foundation species does in fact disproportionately provide habitat to its associated system, it is important to study the distribution of biomass within and among the individual macrophyte thalli as this helps define the extent of habitat complexity. Marine macrophytes vary strikingly in their morphologies, even within species, and such morphological (and structural) variability is often linked to variability in physiological performance and productivity (Littler and Arnold 1982). Kelps with many fronds can be more productive than those with few fronds (Jackson 1977, Chapman and Lindley 1980, Zimmerman and Kremer 1986), and variability in frond shape can affect shading and mass transfer of nutrients (Gerard 1982, Carpenter 1990, Graham et al. 2007), both of which regulate physiology and subsequently ecosystem structure and function (Villegas et al. 2008). The two dominant canopy forming kelp species along the California coast, for example, provide fundamentally different habitat to their associated communities due to their morphological differences. *Nereocystis luetkeana* (bull kelp) is comprised of a single stipe that terminates in a large pneumatocyst and profusion of large blades at the surface (Fig. 1C); the surface canopy is the most structurally complex aspect of a *Nereocystis* bed, as the subsurface region simply consists of singular stalks. The subsurface region of a *Nereocystis* bed, therefore, provides a more featureless habitat than *Macrocystis* beds that have numerous bladed fronds extending throughout the water column (Fig. 1A). These structural differences in morphology can affect that role of kelps as a source of habitat for organisms in the community. Previous studies have suggested that fish abundance and diversity differ between adjacent *Nereocystis* and *Macrocystis* beds in central California (Bodkin 1986), although these results need to be supported by more rigorous studies.

Although relatively unstudied, morphological variability within and among marine macrophyte foundation species is likely to be a critical determinant of which taxa inhabit the macrophyte and how they utilize the habitat. Most kelps have holdfasts for attachment to substrate and many kelp holdfasts are formed from intertwined haptera that may or may not create interstitial spaces for organisms to recruit to and inhabit. Some kelps have large holdfasts with tightly bound haptera (e.g., *Egregia menziesii*), likely an adaptation for staying attached in wave exposed environments, that provide minimal habitat for associated taxa, whereas others have massive holdfasts with loosely bound haptera that can support 100s of taxa (e.g., *Macrocystis*; Andrews 1945, Ojeda and Santelices 1984). Furthermore, in the case of *Macrocystis*, the growth and biomass of the fronds on the thallus directly relates to the growth of haptera and the subsequent size of the holdfasts (Barilotti et al. 1985, McCleneghan and Houk 1985). It has also been shown that the density of fronds within a *Macrocystis* population is directly proportional to the abundance of fish that inhabit the fronds (Holbrook et al. 1990, Anderson 1994, Carr 1994), especially when the fronds reach the canopy. Again, the growth and biomass of the fronds on the thallus directly relates to the initiation and growth of new fronds (Lobban 1978, Fox 2013). Although fucoid holdfasts are generally discoid and featureless, the morphological variability among the finely dissected fronds common in most fucoids can provide substantial structural habitat. In general, natural systems founded by different marine macrophytes will likely be characterized by striking changes in the complexity of habitats provided by the macrophyte thalli.

The link between morphological variability in marine macrophyte foundation species and their associated fauna may be one of the most useful and interesting, though understudied, aspects of nearshore marine ecology. The population dynamics of kelps and fucoids is complex (Foster and Schiel 2010) and driven largely by interactions between the macroalgal life histories and variability in local climatic and physical process (i.e., waves, light, nutrients, sedimentation, etc.). For any given marine macrophyte foundation species, such interactions determine the distribution of individuals within a system, as well as recruitment, which regulates the time at which individuals of different ages and sizes appear, and thus the resulting complexity of the habitat. The only clonal kelp taxon that has been described as a foundation species is the vegetative form of *Macrocystis pyrifera* (previously known as *M. integrifolia*; Demes et al. 2009, Macaya and Zuccarello 2010) that inhabits shallow waters along the Pacific coasts of North and South America (Graham et al. 2007). Although frond physiology is likely similar between the clonal and aclonal *Macrocystis* morphs, the distribution of fronds within the thallus is strikingly different and thus has great implications for the structure of the system. In the aclonal form, fronds arise from the apex of single conical holdfasts and can range from a few to 100s of fronds per holdfast, often forming a tangled bundle that rises to the surface to form a canopy in waters as deep as 60 m (Graham et al. 2007). The clonal form, on the other hand, creeps along the substrate, in shallow nearshore waters, with a vegetatively elongating rhizome that sends fronds to the surface individually, rather than an entwined bundle. Therefore, taxa inhabiting shallow waters dominated by the clonal form of *Macrocystis* will encounter a much more uniform distribution of fronds, and thus habitat, than in a system dominated by the aclonal form. Such striking differences in the spatial distribution of fronds between these two *Macrocystis* morphs will likely have consequences to determining which taxa inhabit the system, and how they utilize the habitat.

Population cycling will create variability in the density, size, age structure, and distribution of individual thalli of aclonal marine macrophytes. Following a disturbance that creates open space, most macrophytes recruit in mass during episodes that can be either random or synchronized to environmental parameters (e.g., daylength). Regardless of how or why they do so, such recruitment episodes inherently result in the system being populated simultaneously by numerous small thalli (Dayton et al. 1992, Graham et al. 1997, Reed et al. 2009). Over time, these populations thin in density as individual thalli become larger and outcompete neighbors; this self-thinning process is characteristic of most large marine macrophyte populations (Cousens and Hutchings 1983, Dean et al. 1989, Dayton et al. 1992) and is likely due to a carrying capacity in overall biomass density (g/m^2) that can be supported by available resources (Jackson 1977, Jackson 1987, Tegner et al. 1997). As a consequence of self-thinning, the population shifts from being densely populated by small individuals to sparsely populated by very large ones. Using empirical data, North (1994) showed that regardless of thallus size (i.e., number of fronds per thallus), *Macrocystis* populations never exceeded frond densities of greater than 10 fronds/m^2. Therefore, plant density decreases with increasing thallus size, as individuals become larger. This same pattern can be observed in terrestrial forests and grasslands following the trajectory of a system post-disturbance towards its climax state (White and Harper 1970, Connell and Slatyer 1977, Hamilton et al. 1995).

Spatiotemporal shifts in the distribution and complexity of available habitat will likely occur for all marine macrophyte populations that experience self-thinning. As with most studies on productivity and habitat provisioning, the best data for studying size vs. density tradeoffs exist for the giant kelp, *Macrocystis pyrifera*. North (1994) originally studied southern California *Macrocystis* populations with plants up to 80 fronds each and densities from 0.01 to 1.6 plants per m^2, from which he derived his 10 fronds/m^2 carrying-capacity estimate. Later, van Tussenbroek (1989) used a similar approach to distinguish among various populations of *Macrocystis* in the Southern Hemisphere and, although she found the same self-thinning pattern that North (1994) did, she observed that different populations (e.g., Argentina vs. Chile vs. Falkland Islands) were in strikingly different locations on a size vs. density plot. Tegner et al. (1997) also showed that temporal variability in environmental conditions within a site (e.g., nutrient concentrations regulated by ENSO cycles) resulted in variability in carrying-capacity of *Macrocystis* populations (i.e., North's (1994) estimate of 10 fronds/m^2). We therefore compiled data from North (1994), van Tussenbroek (1989), and other globally distributed studies of *Macrocystis* size and density along with contemporary long-term monitoring studies in southern and central California to show the extreme variation that can exist in the distribution and complexity of available habitat within a single marine macrophyte foundation species (Fig. 3). Kelp forest taxa inhabiting a *Macrocystis* population can therefore move amongst 1-frond plants spaced every 0.25 m at the high density end of the spectrum, but experience shifts in space or time in which they may encounter 100-frond plants spaced every 10 m at the sparse end of the continuum. The consequences of such striking shifts in habitat-provisioning (driven by interactions between environmental parameters and kelp ecophysiology) on the diversity and productivity of nearshore kelp and fucoid systems have not been studied but are likely to be significant (Graham et al. 2008). The knowledge of such high variability in spatial habitat provisioning by certain species is fundamental to understanding how respective systems function and should be used to guide future research into the role of macrophytes as critical foundation species.

The variability of *Macrocystis* size vs. density on a global scale is exceptional (Fig. 3). It is important to note the geographic differences first observed by van Tussenbroek (1993) in the Southern Hemisphere appear to be consistent throughout the global range of this species. Of particular interest is the relative consistency with which particular regions occur on the plot. For example, many of the *Macrocystis* populations in Chile are dominated by the clonal '*integrifolia*-form' or by small statured annual populations of the aclonal '*pyrifera*-form' and therefore tend to structure unique high density, small sporophyte populations ranging from 1–9 sporophytes m^{-2} and averaging approximately 5.1 fronds per individual. Conversely, the *Macrocystis* populations along the California coast rarely exceed 0.5 sporophytes m^{-2} and average approximately 18.7 fronds per individual. The stark difference between these populations is most likely driven by changes in environmental parameters across regions. For example, the environmental conditions in Chile and Baja California are frequently unfavorable to high productivity rates and biomass production in *Macrocystis* due to varying combinations of temperature, nutrient, and salinity stress (Hernandez-Carmona et al. 2001, Buschmann et al. 2004, Edwards and Hernández-Carmona 2005, Buschmann

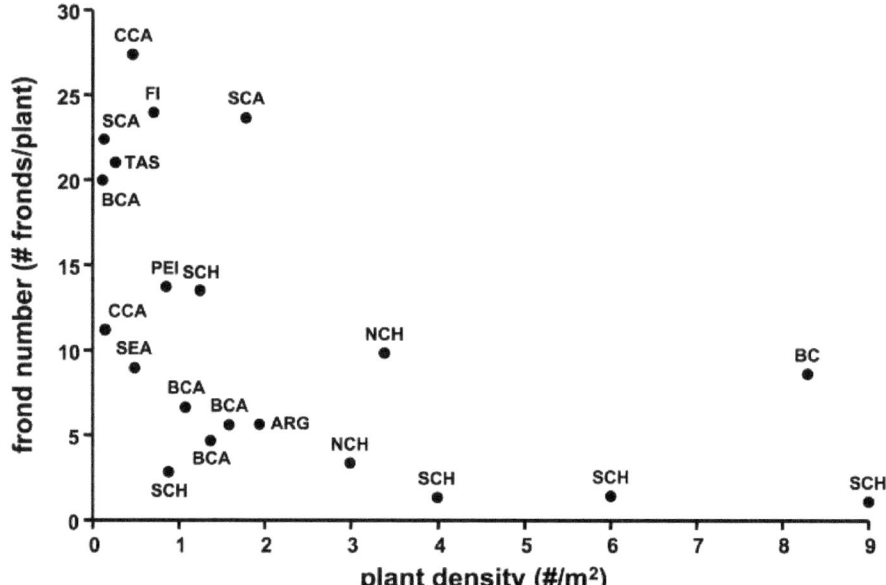

Figure 3. Maximum reported *Macrocystis* density vs. mean plant size globally. *Macrocystis* is the most globally distributed kelp species due to its large size, ability to float, and high degree of morphological and physiological plasticity. Throughout its distribution along the west coast of North America and the southern hemisphere, *Macrocystis* forests exhibit a wide range of size structure and density which can influence the resources it provides as a foundation species. The data shown in this figure were generated from an extensive literature and only represents studies that reported both plant density and size data from the same location. To illustrate the large scale application of this trend, the maximum reported densities (individuals · m^{-2}) were used for each site, and sites were chosen to be representative of the large geographic area they occur within. All size data were determined by counting the number of fronds > 1 m in height per sporophyte. Site abbreviations: ARG-Argentina; BC-British Columbia; BCA-Baja California, Mexico; CCA-Central California; FI-Falkland Islands; NCH-Northern Chile; PEI-Prince Edward Island; SCH-Southern Chile; SEA-Southeast Alaska; TAS-Tasmania.

et al. 2014). Similarly, in the higher latitude island populations and along California the *Macrocystis* populations inhabit stable, nutrient rich waters and the populations are primarily structured by wave disturbance, which favors the establishment of larger, less dense *Macrocystis* beds (Graham et al. 2007, Reed et al. 2008, Reed et al. 2011). This global perspective helps to illustrate the differential role of the same macrophyte species in structuring communities in regions defined by different environmental parameters. Thus, it is imperative that the high variability in the morphology and ecophysiology of marine macrophytes across regions and within species be considered when conducting studies on the role of macrophytes as foundation species.

The *Macrocystis* populations off California are arguably some of the well-studied populations of any marine macrophyte. As such, we compiled an extensive data set of plant size and density data from surveys ranging from Baja California to central California to investigate how regional variability in environmental conditions drives changes in habitat provisioning on a scale of 100s instead of 1000s of kilometers (Fig. 4). Perhaps, not surprisingly, we found that the carrying capacity first identified by North (1994) and Van Tussenbroek (1993) was confirmed by this new dataset.

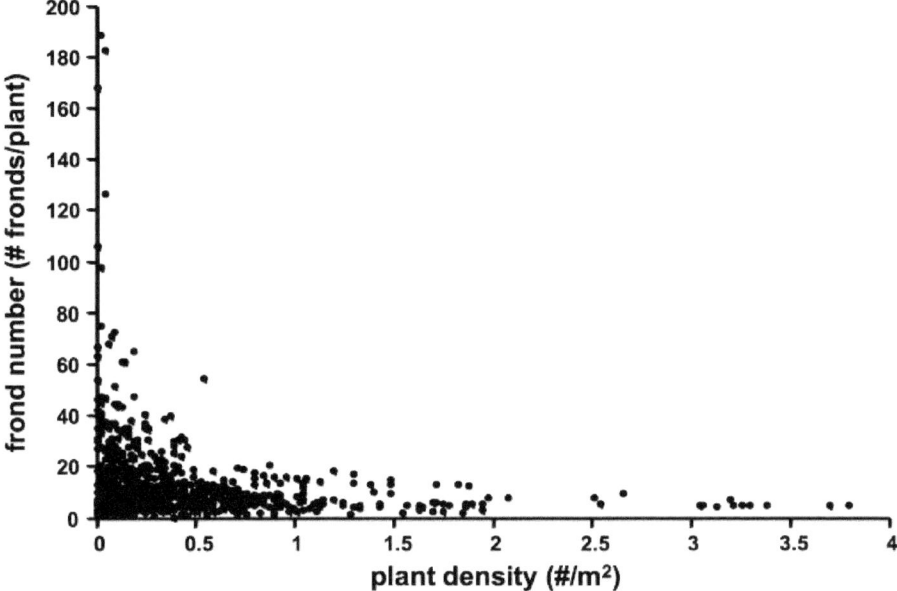

Figure 4. *Macrocystis* density vs. mean plant size for California. The density of *Macrocystis* sporophytes inversely influences the mean size of individuals within populations. High density *Macrocystis* forests are comprised of smaller individuals that turn over frequently, while low density populations are comprised mainly of larger, older sporophytes. Both biotic and abiotic forces such as competition for space, self-shading, and disturbance from wave action define the carrying capacity of *Macrocystis* forests in terms of plant density and size structure, which in turn alters the provisioning of resources by this species in different locations. Importantly, high regional variability in oceanographic conditions drives substantial changes in *Macrocystis* forest structure along the West Coast of North America and represents the range of forest structures shown globally in Fig. 3. The consideration of population-scale drivers on individual-scale processes is critically important for understanding how macrophyte foundation species provide resources and provision habitat differently in different systems and locations. The data for this figure were collected between 2006 and 2008 at 118 sites that range from central Baja California to Santa Cruz, California. The data were generously provided by Matt Edwards (Edwards 2004), the Partnership for the Interdisciplinary Studies of the Coastal Ocean (PISCO—UC Santa Barbara and UC Santa Cruz), the 2008 California Bight Survey (CRANE; Pondella et al. 2011). All data were collected using 60 m^2 swath surveys and compiled and reported here as individuals · m^{-2} and number of fronds > 1 m long per sporophyte.

Interestingly, the entire global range of densities and size structure can be found within one region, highlighting the extreme variability in plant size vs. density that occurs for this foundation species over small spatial scales. Examined regionally, Baja California *Macrocystis* populations typically occur at higher densities and smaller plant sizes relative to the populations of *Macrocystis* north of Point Conception, which have the largest plant sizes and lowest densities of all populations along the west coast of North America (Fig. 3). Again, these differences can be attributed to distinct differences in the oceanographic regimes that have been well defined in this region (Graham et al. 2008). The large variability in plant size and density, even just within California, has the potential to greatly influence the role that *Macrocystis* plays as a foundation species in this system. Surprisingly, the context-dependent role of *Macrocystis* has not been studied to date.

To our knowledge, processes regulating habitat provisioning in other macroalgal foundation species have rarely been studied. The Sargasso Sea ecosystem presents unique logistical challenges to studying the relationships between habitat provided by the *Sargassum* and the diversity and abundance of their associated fauna. Conceptually, however, this pelagic oceanic environment, the intricate morphology of the *Sargassum* mats, and the apparent adaptation of many fauna to this isolated environment, may make this system the most conceptually interesting to explore. In this system, temporal variability in the productivity and morphology of the *Sargassum* mats will likely regulate the quantity and quality, respectively, of the available habitat; whether or not the associated taxa will respond to such changes is unknown. In the more tractable benthic *Sargassum* beds in Bahia Concepcion, Mexico, *Sargassum* thalli are important to promoting local diversity, but the annual nature to *Sargassum* population dynamics in the region limits the magnitude of the impact (Hinojosa-Arango et al. 2014). Similarly, little work has been done to study variability in the quantity and quality of the available habitat provided by rhodoliths, although again, it is clear that the complexity of rhodolith habitats enhances faunal diversity relative to the surrounding benthos (Steller et al. 2003), and that rhodolith beds can be important to the population dynamics of commercially-important fauna (Hinojosa-Arango et al. 2014).

All seagrasses are rhizomatous, and thus inherently clonal. As such, it is difficult to study the relationship between plant density and size to determine whether self-thinning occurs within seagrass populations, as with large macroalgae. It is relatively straightforward, however, to estimate shoot densities (number of shoots per m^2) in the field to be compared to leaf biomass densities (g dry weight per m^2). Such studies are standard in the seagrass literature and yield similar magnitudes of variability as seen in the size vs. density data for kelps (Fig. 5C; Duarte and Sand-Jensen 1990, Horinouchi and Sano 1999, Cabaço et al. 2013). Again, spatiotemporal variability in seagrass habitat provisioning can be conspicuous. Olesen and Sand-Jensen (1994) observed striking differences in biomass vs. density cycles between Chesapeake Bay and Nova Scotia populations (Fig. 5A,B). Leaf biomass and shoot density in the warm temperate Chesapeake Bay varied seasonally by almost two orders-of-magnitude, whereas the cold temperate Nova Scotia population varied seasonally by more than order-of-magnitude in biomass but showed much less variability in shoot density. Such cycling will likely drive similar cycles in the availability of fixed carbon to direct consumers and detritivores, as well as habitat available for algal epiphytes that are known to be important to the functioning of these system. Yet, as with most systems structured by marine macrophytes, the effects on associated fauna remain unknown.

The future of ecosystems founded by marine macrophytes?

It is clear that large and abundant marine macrophytes have a foundational role in regulating the structure, diversity and productivity of nearshore ecosystems. It is also clear that the most important marine macrophyte foundation species are those that provide tremendous energy and habitat to their associated communities, relative to the background environment. If we remove these foundation species over non-trivial spatial scales, the systems will change dramatically (Graham 2004). It is unclear, however,

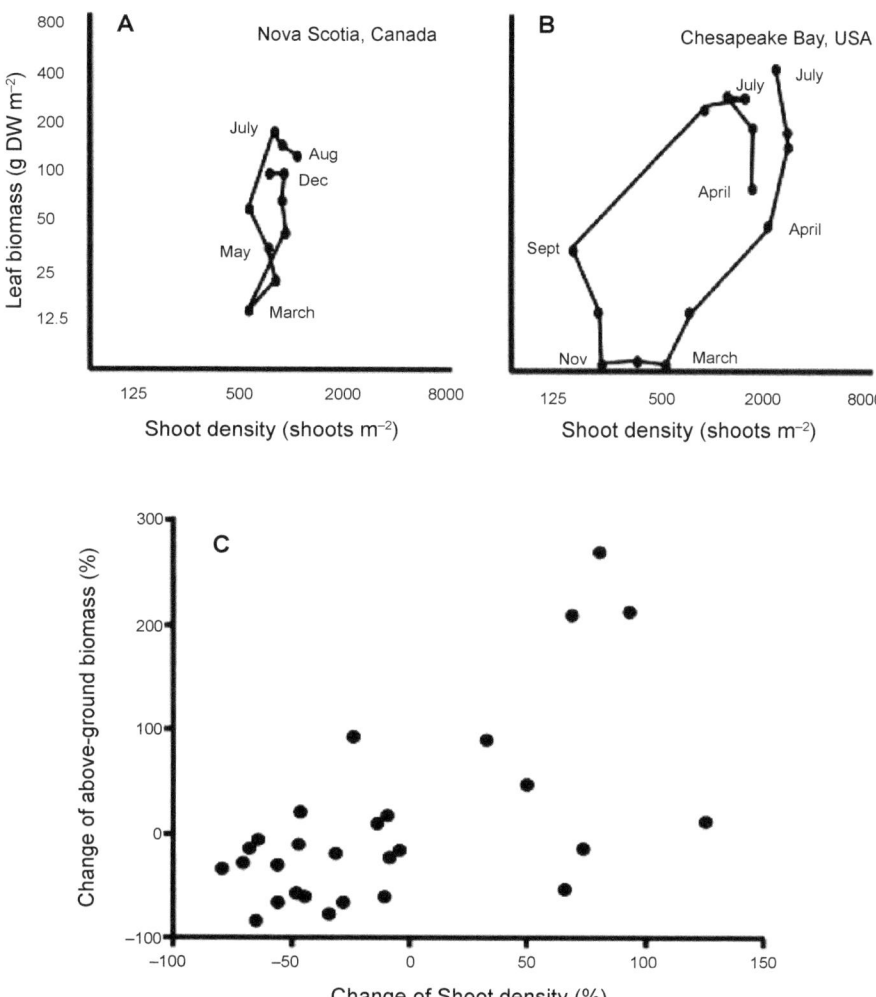

Figure 5. Changes in seagrass biomass vs. shoot density. For *Zostera marina*, populations in (A) Nova Scotia, Canada and (B) Chesapeake Bay, USA seasonally cycled through periods of high and low shoot density, with correlated changes in leaf biomass (Modified from Olesen and Sand-Jensen 1994). In a global survey, (C) above-ground biomass increases with shoot density for 14 seagrass species (Modified from Cabaco et al. 2013); original data were from experimental and descriptive studies of effects of nutrient addition of above-ground biomass and shoot density.

over what spatial and temporal scales such impacts will last, and whether sublethal perturbations to macrophyte ecophysiology will have similar impacts.

The tight linkage between environmental parameters and macrophyte ecophysiology suggests that changing climate will strongly impact systems founded by marine macrophytes (Harley et al. 2012). By definition, foundation species are disproportionate in their provisioning of energy to their associated system and changes in light, temperature, sedimentation or nutrients that increase or decrease macrophyte productivity will impact energy flow and habitat provisioning in the system. In some scenarios, it is straightforward to predict outcomes: seagrasses will decline with increasing turbidity (Moore et al. 1997, Orth et al. 2006); kelps will decline with decreasing nutrients (Jackson 1977, Zimmerman and Kremer 1986, Dayton et al. 1999); rhodoliths will decline with increasing ocean acidification (Jokiel et al. 2008); macroalgae and seagrasses may flourish due to increased availability of dissolved CO_2 (Johnson et al. 2012, Connell et al. 2013, Koch et al. 2013). These effects, however, will not be consistent across species (Porzio et al. 2011, Campbell and Fourqurean 2014). Still, it is less straightforward to predict outcomes when climate change or other human impacts result in multiple environmental parameters changing simultaneously (Harley et al. 2012). Changes in runoff to nearshore systems can result in enhanced nutrient concentrations that fuel growth of phytoplankton and/or microalgal epiphytes, both of which can decrease light levels and negatively impact seagrass populations. Similarly, population explosions of filamentous microalgae in areas with both increased temperature and nutrients can inhibit the recovery of kelp populations following disturbance (Connell and Russell 2010). So, although it is clear that systems founded by marine macrophytes will fluctuate to reflect changes in the distribution, abundance and productivity of these foundation species, predicting the direction of such changes as a function of climate is far less certain. This uncertainty is a fundamental hurdle in the development of management strategies and predictive models aimed at determining the effects of climate change and other anthropogenic stressors on ecosystems around the world. Using the concepts defined in this paper, we propose that the links between environmental conditions and macrophyte ecophysiology that drive changes in energy and habitat provisioning within and among foundational macrophyte taxa be used to guide future research into the role of marine macrophytes in regulating ecosystem structure and function.

Acknowledgements

We thank the Vantuna Research Group at Occidental College for collecting and providing the Bight '08 data; Dr. Matthew Edwards provided the raw data from his dissertation; PISCO data was provided by the Partnership for Interdisciplinary Studies of Coastal Oceans, a long-term ecological consortium funded primarily by the Gordon and Betty Moore Foundation and David and Lucile Packard Foundation.

References

Abdullah, M.I. and S. Fredriksen. 2004. Production, respiration and exudation of dissolved organic matter by the kelp *Laminaria hyperborea* along the west coast of Norway. Journal of the Marine Biological Association of the United Kingdom 84: 887–94.

Agostini, S., G. Pergent and B. Marchand. 2003. Growth and primary production of *Cymodocea nodosa* in a coastal lagoon. Aquatic Botany 76: 185–93.

Amsler, C.D., R.J. Rowley, D.R. Laur, L.B. Quetin and R.M. Ross. 1995. Vertical distribution of Antarctic peninsular macroalgae: cover, biomass and species composition. Phycologia 34: 424–30.

Amsler, C.D., K. Iken, J.B. McClintock, M.O. Amsler, K.J. Peters, J.M. Hubbard, F.B. Furrow and B.J. Baker. 2005. Comprehensive evaluation of the palatability and chemical defenses of subtidal macroalgae from the Antarctic Peninsula. Marine Ecology Progress Series 294: 141–59.

Anderson, T.W. 1994. Role of macroalgal structure in the distribution and abundance of a temperate reef fish. Marine Ecology Progress Series 113: 279–90.

Andrews, H.L. 1945. The kelp beds of the Monterey Region. Ecology 26: 24–37.

Arkema, K.K., D.C. Reed and S.C. Schroeter. 2009. Direct and indirect effects of giant kelp determine benthic community structure and dynamics. Ecology 90: 3126–37.

Attrill, M.J., J.A. Strong and A.A. Rowden. 2000. Are macroinvertebrate communities influenced by seagrass structural complexity? Ecography 23: 114–21.

Attwood, C., M.I. Lucas, T.A. Probyn, C.D. McQuaid and P.J. Fielding. 1991. Production and standing stocks of the kelp *Macrocystis laevis* Hay at the Prince Edward Islands, Subantarctic. Polar Biology 11: 129–33.

Ballesteros, E. 1989. Production of seaweeds in Northwestern Mediterranean marine communities: its relation with environmental factors. Scientia Marina 53: 357–364.

Barilloti, D.C., R.H. McPeak and P.K. Dayton. 1985. Experimental studies on the effects of commercial kelp harvesting in central and southern California *Macrocystis pyrifera* kelp beds. California Fish and Game 71: 4–20.

Bartsch, I., C. Wiencke, K. Bischof, C. Buchholz, B. Buck, A. Eggert, P. Feuerpfeil, D. Hanelt, S. Jacobsen, R. Karez, U. Karsten, M. Molis, M. Roleda, H. Schubert, R. Schumann, K. Valentin, F. Weinberger and J. Wiese. 2008. The genus *Laminaria sensu lato*: recent insights and developments. European Journal of Phycology 43: 1–86.

Beck, M.W., K.L. Heck, K.W. Able, D.L. Childers, D.B. Eggleston, B.M. Gillanders, B. Halpern, C.G. Hays, K. Hoshino and T.J. Minello. 2001. The identification, conservation, and management of estuarine and marine nurseries for fish and invertebrates. Bioscience 51: 633–41.

Bodkin, J. 1986. Fish assemblages in *Macrocystis* and *Nereocystis* kelp forests off central California. Fishery Bulletin 84: 799–808.

Bolton, J.J. 2010. The biogeography of kelps (Laminariales, Phaeophyceae): a global analysis with new insights from recent advances in molecular phylogenetics. Helgoland Marine Research 64: 263–79.

Borum, J.B., M.P. Pedersen, D.K.J. Krause-Jensen, P.C. Christensen and K.N. Nielsen. 2002. Biomass, photosynthesis and growth of *Laminaria saccharina* in a high-arctic fjord, NE Greenland. Marine Biology 141: 11–19.

Bruno, J. and M.D. Bertness. 2001. Positive interactions, facilitations, and foundation species. In: Bertness, M.D., S.D. Gaines and M. Hay (eds.). Marine Community Ecology. Sinauer Associates Sunderland, Massachusetts.

Buschmann, A.H., J.A. Vasquez, P. Osorio, E. Reyes, L. Filun, M.C. Hernandez-Gonzalez and A. Vega. 2004. The effect of water movement, temperature and salinity on abundance and reproductive patterns of *Macrocystis* spp. (Phaeophyta) at different latitudes in Chile. Marine Biology 145: 849–62.

Buschmann, A.H., S.V. Pereda, D.A. Varela, J. Rodríguez-Maulén, A. López, L. González-Carvajal, M. Schilling, E.A. Henríquez-Tejo and M.C. Hernández-González. 2014. Ecophysiological plasticity of annual populations of giant kelp (*Macrocystis pyrifera*) in a seasonally variable coastal environment in the Northern Patagonian Inner Seas of Southern Chile. Journal of Applied Phycology 26: 837–847.

Busing, R.T. and T. Fujimori. 2005. Biomass, production and woody detritus in an old coast redwood (*Sequoia sempervirens*) forest. Plant Ecology 177: 177–88.

Bustamante, R.H. and G.M. Branch. 1996. The dependence of intertidal consumers on kelp-derived organic matter on the west coast of South Africa. Journal of Experimental Marine Biology and Ecology 196: 1–28.

Bustamante, R.H., G.M. Branch and S. Eekhout. 1995. Maintenance of an exceptional intertidal grazer biomass in South Africa: subsidy by subtidal kelps. Ecology 76: 2314–29.

Byrnes, J.E., D.C. Reed, B.J. Cardinale, K.C. Cavanaugh, S.J. Holbrook and R.J. Schmitt. 2011. Climate-driven increases in storm frequency simplify kelp forest food webs. Global Change Biology 17: 2513–24.

Cabaço, S., E.T. Apostolaki, P. García-Marín, R. Gruber, I. Hernández, B. Martínez-Crego, O. Mascaró, M. Pérez, A. Prathep, C. Robinson, J. Romero, A.L. Schmidt, F.T. Short, B.I. van Tussenbroek and R. Santos. 2013. Effects of nutrient enrichment on seagrass population dynamics: evidence and synthesis from the biomass–density relationships. Journal of Ecology 101: 1552–62.

Campbell, J.E. and J.W. Fourqurean. 2014. Ocean acidification outweighs nutrient effects in structuring seagrass epiphyte communities. Journal of Ecology 102: 730–737.

Carpenter, E.J. and J.L. Cox. 1974. Production of pelagic *Sargassum* and a blue-green epiphyte in the western Sargasso Sea. Limnology and Oceanography 19: 429–36.

Carpenter, R.C. 1990. Competition among marine macroalgae: a physiological perspective. Journal of Phycology 26: 6–12.

Carr, A. 1987. New perspectives on the pelagic stage of sea turtle development. Conservation Biology 1: 103–21.

Carr, M.H. 1994. Effects of macroalgal dynamics on recruitment of a temperate reef fish. Ecology 75: 1320–33.

Chapman, A.R.O. and J.E. Lindley. 1980. Seasonal growth of *Laminaria solidungula* in the Canadian high Arctic in relation to irradiance and dissolved nutrient concentrations. Marine Biology 57: 1–5.

Christie, H., N.M. Jørgensen, K.M. Norderhaug and E. Waage-Nielsen. 2003. Species distribution and habitat exploitation of fauna associated with kelp (*Laminaria hyperborea*) along the Norwegian Coast. Journal of the Marine Biological Association of the United Kingdom 83: 687–99.

Clark, R.P., M.S. Edwards and M.S. Foster. 2004. Effects of shade from multiple kelp canopies on an understory algal assemblage. Marine Ecology Progress Series 267: 107–19.

Connell, J.H. and R.O. Slatyer. 1977. Mechanisms of succession in natural communities and their role in community stability and organization. The American Naturalist 111: 1119–44.

Connell, S.D. and B.D. Russell. 2010. The direct effects of increasing CO_2 and temperature on non-calcifying organisms: increasing the potential for phase shifts in kelp forests. Proceedings of the Royal Society B: Biological Sciences 277: 1409–15.

Connell, S.D., K.J. Kroeker, K.E. Fabricius, D.I. Kline and B.D. Russell. 2013. The other ocean acidification problem: CO_2 as a resource among competitors for ecosystem dominance. Philosophical Transactions of the Royal Society B: Biological Sciences 368.

Connolly, R.M. 1994. Removal of seagrass canopy: effects on small fish and their prey. Journal of Experimental Marine Biology and Ecology 184: 99–110.

Cousens, R. 1984. Estimation of annual production by the intertidal brown alga *Ascophyllum nodosum* (L.) Le Jolis. Botanica Marina 27: 217–227.

Cousens, R. 1985. Frond size distributions and the effects of the algal canopy on the behaviour of *Ascophyllum nodosum* (L.) Le Jolis. Journal of Experimental Marine Biology and Ecology 92: 231–49.

Cousens, R. and M.J. Hutchings. 1983. The relationship between density and mean frond weight in monospecific seaweed stands. Nature 301: 240–41.

Dayton, P.K. 1972. Toward an understanding of community resilience and the potential effects of enrichments to the benthos at McMurdo Sound, Antarctica. pp. 81–96. *In*: Proceedings of the Colloquium on Conservation Problems in Antarctica. Allen Press, Lawrence, KS.

Dayton, P.K. 1975. Experimental studies of algal canopy interactions in a sea otter-dominated kelp community at Amchitka Island, Alaska. Fishery Bulletin 78: 230–37.

Dayton, P.K. 1985. Ecology of kelp communities. Annual Review of Ecology and Systematics 16: 215–45.

Dayton, P.K., M.J. Tegner, P.E. Parnell and P.B. Edwards. 1992. Temporal and spatial patterns of disturbance and recovery in a kelp forest community. Ecological Monographs 62: 421–45.

Dayton, P.K., M.J. Tegner, P.B. Edwards and K.L. Riser. 1999. Temporal and spatial scales of kelp demography: the role of oceanographic climate. Ecological Monographs 69: 219–50.

Dean, R.J. and M.J. Durako. 2007. Carbon sharing through physiological integration in the threatened seagrass *Halophila johnsonii*. Bulletin of Marine Science 81: 21–35.

Dean, T.A., K. Thies and S.L. Lagos. 1989. Survival of juvenile giant kelp: the effects of demographic factors, competitors, and grazers. Ecology 70: 483–95.

Demes, K.W., M.H. Graham and T.S. Suskiewicz. 2009. Phenotypic plasticity reconciles incongruous molecular and morphological taxonomies: the giant kelp, *Macrocystis* (Laminariales, Phaeophyceae), is a monospecific genus. Journal of Phycology 45: 1266–69.

Deysher, L.E. and T.A. Dean. 1984. Critical irradiance levels and the interactive effects of quantum irradiance and does on the gametogenesis in the giant kelp, *Macrocystis pyrifera*. Journal of Phycology 20: 520–24.

Dooley, J.K. 1972. Fishes associated with pelagic *Sargassum* complex, with a discussion of the *Sargassum* community. Contributions in Marine Science 16: 1–32.

Drew, E. 1983. Physiology of *Laminaria*. Marine Ecology 4: 227–50.

Drew, E. and R. Hastings. 1992. A year-round ecophysiological study of *Himantothallus grandifolius* (Desmarestiales, Phaeophyta) at Signy Island, Antarctica. Phycologia 31: 262–77.

Duarte, C.M. 1995. Submerged aquatic vegetation in relation to different nutrient regimes. Ophelia 41: 87–112.

Duarte, C.M. and K. Sand-Jensen. 1990. Seagrass colonization: biomass development and shoot demography in *Cymodocea nodosa* patches. Marine Ecology Progress Series 67: 93–103.

Duffy, J.E., J. Paul Richardson and E.A. Canuel. 2003. Grazer diversity effects on ecosystem functioning in seagrass beds. Ecology letters 6: 637–45.

Dugan, J.E., D.M. Hubbard, M.D. McCrary and M.O. Pierson. 2003. The response of macrofauna communities and shorebirds to macrophyte wrack subsidies on exposed sandy beaches of southern California. Estuarine, Coastal and Shelf Science 58: 25–40.

Duggins, D.O. 1980. Kelp beds and sea otters: an experimental approach. Ecology 61: 447–53.

Duggins, D.O., C.A. Simenstad and J.A. Estes. 1989. Magnification of secondary production by kelp detritus in coastal marine ecosystems. Science 245: 170–73.

Ebeling, A.W., D.R. Laur and R.J. Rowley. 1985. Severe storm disturbances and reversal of community structure in a southern California kelp forest. Marine Biology 84: 287–294.

Edwards, M.S. 2004. Estimating scale-dependency in disturbance impacts: El Niños and giant kelp forests in the Northeast Pacific. Oecologia 138: 436–47.

Edwards, M.S. and G. Hernández-Carmona. 2005. Delayed recovery of giant kelp near its southern range limit in the North Pacific following El Niño. Marine Biology 147: 273–79.

Ellison, A.M., M.S. Bank, B.D. Clinton, E.A. Colburn, K. Elliott, C.R. Ford, D.R. Foster, B.D. Kloeppel, J.D. Knoepp, G.M. Lovett, J. Mohan, D.A. Orwig, N.L. Rodenhouse, W.V. Sobczak, K.A. Stinson, J.K. Stone, C.M. Swan, J. Thompson, B. Von Holle and J.R. Webster. 2005. Loss of foundation species: consequences for the structure and dynamics of forested ecosystems. Frontiers in Ecology and the Environment 3: 479–86.

Estes, J.A. and J.F. Palmisano. 1974. Sea otters: their role in structuring nearshore communities. Science 185: 1058–60.

Estes, J.A. and D.O. Duggins. 1995. Sea otters and kelp forests in Alaska: generality and variation in a community ecological paradigm. Ecological Monographs 65: 75–100.

Estes, J.E., N.S. Smith and J.F. Palmisano. 1978. Sea otter predation and community organization in the western Aleutian Islands, Alaska. Ecology 59: 822–33.

Filbee-Dexter, K. and R.E. Scheibling. 2012. Hurricane-mediated defoliation of kelp beds and pulsed delivery of kelp detritus to offshore sedimentary habitats. Marine Ecology Progress Series 455: 51–64.

Foster, M.S. 2001. Rhodoliths: between rocks and soft places. Journal of Phycology 37: 659–67.

Foster, M.S. and D.R. Schiel. 1985. The ecology of giant kelp forests in California: a community profile. U.S. Fish and Wildlife Service 87: 1–152.

Foster, M.S. and D.R. Schiel. 2010. Loss of predators and the collapse of southern California kelp forests(?): Alternatives, explanations and generalizations. Journal of Experimental Marine Biology and Ecology 393: 59–70.

Fox, M.D. 2013. Resource translocation drives δ^{13}C fractionation during recovery from disturbance in giant kelp, *Macrocystis pyrifera*. Journal of Phycology 49: 811–15.

Friedlander, A.M. and J.D. Parrish. 1998. Habitat characteristics affecting fish assemblages on a Hawaiian coral reef. Journal of Experimental Marine Biology and Ecology 224: 1–30.

Gerard, V.A. 1982. *In situ* rates of nitrate uptake by giant kelp, *Macrocystis pyrifera* (L.) C. Agardh: tissue differences, environmental effects, and predictions of nitrogen-limited growth. Journal of Experimental Marine Biology and Ecology 62: 211–24.

Graham, M.H. 2004. Effects of local deforestation on the diversity and structure of southern California giant kelp forest food webs. Ecosystems 7: 341–57.

Graham, M.H., C. Harrold, S. Lisin, K. Light, J.M. Watanabe and M. Foster. 1997. Population dynamics of giant kelp *Macrocystis pyrifera* along a wave exposure gradient. Marine Ecology Progress Series 148: 269–79.

Graham, M.H., J.A. Vasquez and A.H. Buschmann. 2007. Global ecology of the giant kelp *Macrocystis*: from ecotypes to ecosystems. Oceanography and Marine Biology: An Annual Review 45: 39–88.

Graham, M.H., B.S. Halpern and M.H. Carr. 2008. Diversity and dynamics of California subtidal kelp forests. pp. 103–34. *In*: Branch, G.M. and T.M. McClanahan (eds.). Food Webs and the Dynamics of Marine Benthic Ecosystems. Oxford University Press, Oxford.

Hamilton, N.S., C. Matthew and G. Lemaire. 1995. In defense of the -3/2 boundary rule: a re-evaluation of self-thinning concepts and status. Annals of Botany 76: 569–77.

Harley, C.D., K.M. Anderson, K.W. Demes, J.P. Jorve, R.L. Kordas, T.A. Coyle and M.H. Graham. 2012. Effects of climate change on global seaweed communities. Journal of Phycology 48: 1064–78.

Harrold, C., K. Light and S. Lisin. 1998. Organic enrichment of submarine-canyon and continental-shelf benthic communities by macroalgal drift imported from nearshore kelp forests. Limnology and Oceanography 43: 669–678.

Hastings, A., J.E. Byers, J.A. Crooks, K. Cuddington, C.G. Jones, J.G. Lambrinos, T.S. Talley and W.G. Wilson. 2007. Ecosystem engineering in space and time. Ecology letters 10: 153–64.

Haxen, P. and J. Grindley. 1985. *Durvillaea antarctica* production in relation to nutrient cycling at Marion Island. pp. 637–640. *In*: Siegfried W.R., P.R. Condy and R.M. Laws (eds.). Antarctic Nutrient Cycles and Food Webs. Springer, Berlin Heidelberg.

Hay, M.E., Q.E. Kappel and W. Fenical. 1994. Synergisms in plant defenses against herbivores: Interactions of chemistry, calcification, and plant quality. Ecology 75: 1714–26.

Heck, K.L., K. Able, C. Roman and M. Fahay. 1995. Composition, abundance, biomass, and production of macrofauna in a New England estuary: comparisons among eelgrass meadows and other nursery habitats. Estuaries 18: 379–89.

Heck, K.L., D. Nadeau and R. Thomas. 1997. The nursery role of seagrass beds. Gulf of Mexico Science 15: 50–54.

Heck, K.L., G. Hays and R. Orth. 2003. Critical evaluation of the nursery role hypothesis for seagrass meadows. Marine Ecology Progress Series 253: 123–36.

Heck, K.L., T.J. Carruthers, C.M. Duarte, A.R. Hughes, G. Kendrick, R.J. Orth and S.W. Williams. 2008. Trophic transfers from seagrass meadows subsidize diverse marine and terrestrial consumers. Ecosystems 11: 1198–210.

Hemminga, M.A. and C.M. Duarte. 2000. Seagrass Ecology. Cambridge University Press.

Hernandez-Carmona, G., D. Robledo and E. Serviere-Zaragoza. 2001. Effect of nutrient availability on *Macrocystis pyrifera* recruitment and survival near its southern limit off Baja California. Botanica Marina 44: 221–29.

Hernández-Carmona, G., B. Hughes and M.H. Graham. 2006. Reproductive longevity of drifting kelp *Macrocystis pyrifera* (Phaeophyceae) in Monterey Bay, USA. Journal of Phycology 42: 1199–1207.

Hinojosa-Arango, G., R. Rioja-Nieto, Á.N. Suárez-Castillo and R. Riosmena-Rodríguez. 2014. Using GIS methods to evaluate rhodolith and *Sargassum* beds as critical habitats for commercially important marine species in Bahía Concepción, BCS, México. Cryptogamie Algologie 35: 49–65.

Hobday, A.J. 2000. Abundance and dispersal of drifting kelp *Macrocystis pyrifera* rafts in the Southern California Bight. Marine Ecology Progress Series 195: 101–116.

Holbrook, S.J., M.H. Carr, R.J. Schmitt and J.A. Coyer. 1990. Effect of giant kelp on local abundance of reef fishes: the importance of ontogenetic resource requirements. Bulletin of Marine Science 47: 104–14.

Horinouchi, M. and M. Sano. 1999. Effects of changes in seagrass shoot density and leaf height on abundances and distribution patterns of juveniles of three gobiid fishes in a *Zostera marina* bed. Marine Ecology Progress Series 183: 87–94.

Hughes, A.R., S.L. Williams, C.M. Duarte, K.L. Heck and M. Waycott. 2008. Associations of concern: declining seagrasses and threatened dependent species. Frontiers in Ecology and the Environment 7: 242–46.

Hughes, B.B. 2010. Variable effects of a kelp foundation species on rocky intertidal diversity and species interactions in central California. Journal of Experimental Marine Biology and Ecology 393: 90–99.

Jackson, E.L., A.A. Rowden, M.J. Attrill, S. Bossey and M. Jones. 2001. The importance of seagrass beds as a habitat for fishery species. Oceanography and Marine Biology: An Annual Review 39: 269–304.

Jackson, G.A. 1977. Nutrients and production of giant kelp, *Macrocystis pyrifera*, off southern California. Limnology and Oceanography 22: 979–95.

Jackson, G.A. 1987. Modeling the growth and harvest yield of the giant kelp *Macrocystis pyrifera*. Marine Biology 95: 611–624.

Johnson, V.R., B.D. Russell, K.E. Fabricius, C. Brownlee and J.M. Hall-Spencer. 2012. Temperate and tropical brown macroalgae thrive, despite decalcification, along natural CO_2 gradients. Global Change Biology 18: 2792–803.

Johnston, C. 1969. The ecological distribution and primary production of macrophytic marine algae in the eastern Canaries. Internationale Revue der gesamten Hydrobiologie und Hydrographie 54: 473–90.

Jokiel, P., K. Rodgers, I. Kuffner, A. Andersson, E. Cox and F. Mackenzie. 2008. Ocean acidification and calcifying reef organisms: a mesocosm investigation. Coral Reefs 27: 473–83.

Jones, C.G., J.H. Lawton and M. Shachak. 1994. Organisms as ecosystem engineers. Oikos 69: 373–86.

Jupp, B.P. and E.A. Drew. 1974. Studies on the growth of *Laminaria hyperborea* (Gunn.) Fosl. I. Biomass and productivity. Journal of Experimental Marine Biology and Ecology 15: 185–96.

Kamenos, N.A., P.G. Moore and J.M. Hall-Spencer. 2004a. Nursery-area function of maerl grounds for juvenile queen scallops *Aequipecten opercularis* and other invertebrates. Marine Ecology Progress Series 274: 183–189.

Kamenos, N.A., P.G. Moore and J.M. Hall-Spencer. 2004b. Small-scale distribution of juvenile gadoids in shallow inshore waters; what role does maerl play? ICES Journal of Marine Science: Journal du Conseil 61: 422–429.

Kamenos, N.A., P.G. Moore and J.M. Hall-Spencer. 2004c. Maerl grounds provide both refuge and high growth potential for juvenile queen scallops (*Aequipecten opercularis* L.). Journal of Experimental Marine Biology and Ecology 313: 241–254.

Kentula, M.E. and C.D. McIntire. 1986. The autecology and production dynamics of eelgrass (*Zostera marina* L.) in Netarts Bay, Oregon. Estuaries 9: 188–99.

Kirkman, H. 1984. Standing stock and production of *Ecklonia radiata* (C. Ag.) J. Agardh. Journal of Experimental Marine Biology and Ecology 76: 119–30.

Knowlton, N. 2001. The future of coral reefs. Proceedings of the National Academy of Sciences of the USA 98: 5419–25.

Koch, M., G. Bowes, C. Ross and X.H. Zhang. 2013. Climate change and ocean acidification effects on seagrasses and marine macroalgae. Global Change Biology 19: 103–32.

Konar, B. 2000. Seasonal inhibitory effects of marine plants on sea urchins: structuring communities the algal way. Oecologia 125: 208–17.

Lapointe, B.E. 1986. Phosphorus-limited photosynthesis and growth of *Sargassum natans* and *Sargassum fluitans* (Phaeophyceae) in the western North Atlantic. Deep Sea Research Part A. Oceanographic Research Papers 33: 391–99.

Larkum, A.W.D. 1986. A study of growth and primary production in *Ecklonia radiata* (C. Ag.) J. Agardh (Laminariales) at a sheltered site in Port Jackson, New South Wales. Journal of Experimental Marine Biology and Ecology 96: 177–90.

Lee, K.S. and K.H. Dunton. 1996. Production and carbon reserve dynamics of the seagrass *Thalassia testudinum* in Corpus Christi Bay, Texas, USA. Marine Ecology Progress Series 143: 201–210.

Littler, M.M. and D.S. Littler. 1980. The evolution of thallus form and survival strategies in benthic marine macroalgae: field and laboratory tests of a functional form model. American Naturalist 116: 25–44.

Littler, M.M. and K.E. Arnold. 1982. Primary productivity of marine macroalgal functional form groups from southwestern North America. Journal of Phycology 18: 307–11.

Lobban, C.S. 1978. Translocation of ^{14}C in *Macrocystis pyrifera* (giant kelp). Plant Physiology 61: 585–89.

Lüning, K. 1990. Seaweeds: Their Environment, Biogeography, and Ecophysiology. Wiley, New York.

Macarthur, R.H. 1965. Patterns of species diversity. Biological Reviews 40: 510–33.

Macaya, E.C. and G.C. Zuccarello. 2010. DNA barcoding and genetic divergence in the giant kelp *Macrocystis* (Laminariales). Journal of Phycology 46: 736–42.

Macaya, E.C., S. Boltana, I.A. Hinojosa, J.E. Macchiavello, N.A. Valdivia, N.R. Vasquez and M. Thiel. 2005. Presence of sporophylls in floating kelp rafts of *Macrocystis* spp. (Phaeophyceae) along the Chilean Pacific Coast. Journal of Phycology 41: 913–922.

Mann, K.H. 1972. Ecological energetics of the seaweed zone in a marine bay on the Atlantic coast of Canada. II. Productivity of the seaweeds. Marine Biology 14: 199–209.

Mann, K.H. 1982. Ecology of Coastal Waters. A Systems Approach. University of California Press, Berkeley, 322 pp.

Mann, K.H. 2000. Ecology of Coastal Waters with Implications for Management. Blackwell Science Massachusetts, USA, 406 pp.

Marbà, N., M.A. Hemminga and C.M. Duarte. 2006. Resource translocation within seagrass clones: allometric scaling to plant size and productivity. Oecologia 150: 362–72.

McCleneghan, K. and J.L. Houk. 1985. The effects of canopy removal on holdfast growth in *Macrocystis pyrifera* (Phaeophyta; Laminariales). California Fish and Game 71: 21–27.

McRoy, C.P. and C. McMillan. 1977. Production ecology and physiology of seagrasses. pp. 53–87. *In*: McRoy, C.P. and C. Helfferich (eds.). Seagrass Ecosystems. A Scientific Perspective. Marcel Dekker, Inc., New York.

Miller, P.C. 1972. Bioclimate, leaf temperature, and primary production in red mangrove canopies in south Florida. Ecology 53: 22–45.

Miller, R.J., H.M. Page and M.A. Brzezinski. 2013. $\delta^{13}C$ and $\delta^{15}N$ of particulate organic matter in the Santa Barbara Channel: drivers and implications for trophic inference. Marine Ecology Progress Series 474: 53–66.

Moe, R.L. and P.C. Silva. 1977. Antarctic marine flora: uniquely devoid of kelps. Science 196: 1206–08.

Moe, R.L. and P.C. Silva. 1981. Morphology and taxonomy of *Himantothallus* (including *Phaeoglossum* and *Phyllogigas*), an Antarctic member of the Desmarestiales (Phaeophyceae). Journal of Phycology 17: 15–29.

Moncreiff, C.A. and M.J. Sullivan. 2001. Trophic importance of epiphytic algae in subtropical seagrass beds: evidence from multiple stable isotope analyses. Marine Ecology Progress Series 215: 93–106.

Moore, K.A., R.L. Wetzel and R.J. Orth. 1997. Seasonal pulses of turbidity and their relations to eelgrass (*Zostera marina* L.) survival in an estuary. Journal of Experimental Marine Biology and Ecology 215: 115–34.

Moser, M., P. Auster and J. Bichy. 1998. Effects of mat morphology on large *Sargassum*-associated fishes: observations from a remotely operated vehicle (ROV) and free-floating video camcorders. Environmental Biology of Fishes 51: 391–98.

Muscatine, L. 1990. The role of symbiotic algae in carbon and energy flux in reef corals. pp. 75–84. *In*: Dubinsky, Z. (ed.). Ecosystems of the World 25: Coral Reefs. Elsevier, New York.

Nelson, T.A. and J.R. Waaland. 1997. Seasonality of eelgrass, epiphyte, and grazer biomass and productivity in subtidal eelgrass meadows subjected to moderate tidal amplitude. Aquatic Botany 56: 51–74.

Newell, R., J. Field and C. Griffiths. 1982. Energy balance and significance of microorganisms in a kelp bed community. Marine Ecology Progress Series 8: 103–13.

Nizamuddin, M. 1968. Observations on the order Durvilleales J. Petrov, 1965. Botanica Marina 11: 115–17.

Norderhaug, K.M., S. Fredriksen and K. Nygaard. 2003. Trophic importance of *Laminaria hyperborea* to kelp forest consumers and the importance of bacterial degradation to food quality. Marine Ecology Progress Series 255: 135–44.

North, W.J. 1994. Review of Macrocystis biology. pp. 447–527. Biology of Economic Algae. SPB Academic Publishing, The Hague, The Netherlands.

Ojeda, F. and B. Santelices. 1984. Invertebrate communities in holdfasts of the kelp *Macrocystis pyrifera* from southern Chile. Marine Ecology Progress Series 16: 65–73.

Olesen, B. and K. Sand-Jensen. 1994. Biomass-density patterns in the temperate seagrass *Zostera marina*. Marine Ecology Progress Series 109: 283–83.

Orth, R.J., T.J. Carruthers, W.C. Dennison, C.M. Duarte, J.W. Fourqurean, K.L. Heck, A.R. Hughes, G.A. Kendrick, W.J. Kenworthy and S. Olyarnik. 2006. A global crisis for seagrass ecosystems. Bioscience 56: 987–96.

Ott, J.A. 1980. Growth and production in *Posidonia oceanica* (L.) Delile. Marine Ecology 1: 47–64.

Paine, R.T. 1969a. A note on trophic complexity and community stability. The American Naturalist 103: 91–93.

Paine, R.T. 1969b. The *Pisaster-Tegula* interaction: prey patches, predator food preference, and intertidal community structure. Ecology 50: 950–61.

Paine, R.T. 2002. Trophic control of production in a rocky intertidal community. Science 296: 736–739.

Parker, B.C. and J. Huber. 1965. Translocation in *Macrocystis*. II. Fine structure of the sieve tubes. Journal of Phycology 1: 172–79.

Pergent, G., J. Romero, C. Pergent-Martini, M.A. Mateo and C.F. Boudouresque. 1994. Primary production, stocks and fluxes in the Mediterranean seagrass *Posidonia oceanica*. Marine Ecology Progress Series 106: 139–39.

Perkins-Visser, E., T.G. Wolcott and D.L. Wolcott. 1996. Nursery role of seagrass beds: enhanced growth of juvenile blue crabs (*Callinectes sapidus* Rathbun). Journal of Experimental Marine Biology and Ecology 198: 155–73.

Pondella, D., J. Williams, J. Claisse, R. Schaffner, K. Ritter and K. Schiff. 2011. Southern California Bight 2008 Regional Monitoring Program: Volume V. Rocky Reefs. Southern California Coastal Water Research Project, Costa Mesa, CA. 92 p.

Porzio, L., M.C. Buia and J.M. Hall-Spencer. 2011. Effects of ocean acidification on macroalgal communities. Journal of Experimental Marine Biology and Ecology 400: 278–87.

Rasher, D.B., A.S. Hoey and M.E. Hay. 2013. Consumer diversity interacts with prey defenses to drive ecosystem function. Ecology 94: 1347–58.

Reed, D.C., A. Rassweiler and K.K. Arkema. 2008. Biomass rather than growth rate determines variation in net primary production by giant kelp. Ecology 89: 2493–505.

Reed, D.C., A. Rassweiler and K. Arkema. 2009. Density derived estimates of standing crop and net primary production in the giant kelp *Macrocystis pyrifera*. Marine Biology 156: 2077–83.

Reed, D.C., A. Rassweiler, M.H. Carr, K.C. Cavanaugh, D.P. Malone and D.A. Siegel. 2011. Wave disturbance overwhelms top-down and bottom-up control of primary production in California kelp forests. Ecology 92: 2108–16.

Roman, C., K. Able, M. Lazzari and K. Heck. 1990. Primary productivity of angiosperm and macroalgae dominated habitats in a New England salt marsh: a comparative analysis. Estuarine, Coastal and Shelf Science 30: 35–45.

Salomon, A.K., N.T. Shears, T.J. Langlois and R.C. Babcock. 2008. Cascading effects of fishing can alter carbon flow through a temperate coastal ecosystem. Ecological Applications 18: 1874–1887.

Santelices, B. and F. Ojeda. 1984a. Effects of canopy removal on the understory algal community structure of coastal forests of *Macrocystis pyrifera* from southern South America. Marine Ecology Progress Series 14: 165–73.

Santelices, B. and F. Ojeda. 1984b. Population dynamics of coastal forests *Macrocystis pyrifera* in Puerto Toro, Isla Navarino, Southern Chile. Marine Ecology Progress Series 14: 175–83.

Schiel, D.R. 2006. Rivets or bolts? When single species count in the function of temperate rocky reef communities. Journal of Experimental Marine Biology and Ecology 338: 233–52.

Schiel, D.R. and M.S. Foster. 2006. The population biology of large brown seaweeds: ecological consequences of multiphase life histories in dynamic coastal environments. Annual Review of Ecology, Evolution, and Systematics 37: 343–72.

Schmitz, K. and C. Lobban. 1976. A survey of translocation in Laminariales (Phaeophyceae). Marine Biology 36: 207–16.

Schmitz, K. and L. Srivastava. 1979. Long distance transport in *Macrocystis integrifolia*: I. Translocation of ^{14}C-labeled assimilates. Plant Physiology 63: 995–1002.

Shelton, A.O. 2010. Temperature and community consequences of the loss of foundation species: surfgrass (*Phyllospadix* spp., Hooker) in tidepools. Journal of Experimental Marine Biology and Ecology 391: 35–42.

Short, F., T. Carruthers, W. Dennison and M. Waycott. 2007. Global seagrass distribution and diversity. Journal of Experimental Marine Biology and Ecology 350: 3–20.

Smith, K.L., K.A. Burns and E.J. Carpenter. 1973. Respiration of the pelagic *Sargassum* community. Deep Sea Research and Oceanographic Abstracts 20: 213–17.

Sotka, E.E. and M.E. Hay. 2002. Geographic variation among herbivore populations in tolerance for a chemically-rich seaweed. Ecology 83: 2721–35.

Steller, D.L. and C. Cáceres-Martínez. 2009. Coralline algal rhodoliths enhance larval settlement and early growth of the Pacific calico scallop *Argopecten ventricosus*. Marine Ecology, Progress Series 396: 49–60.

Steller, D.L., R. Riosmena-Rodríguez, M.S. Foster and C.A. Roberts. 2003. Rhodolith bed diversity in the Gulf of California: the importance of rhodolith structure and consequences of disturbance. Aquatic Conservation: Marine and Freshwater Ecosystems 13: S5–S20.

Steneck, R.S., M.H. Graham, B.J. Bourque, D. Corbett, J.M. Erlandson, J.A. Estes and M.J. Tegner. 2002. Kelp forest ecosystems: biodiversity, stability, resilience and future. Environmental Conservation 29: 436–59.

Tegner, M. and P. Dayton. 1991. Sea urchins, El Niños, and the long-term stability of southern California kelp forest communities. Marine Ecology Progress Series 77: 49–63.

Tegner, M.J., P.K. Dayton, P.B. Edwards and K.L. Riser. 1997. Large-scale, low-frequency oceanographic effects on kelp forest succession: a tale of two cohorts. Marine Ecology Progress Series 146: 117–34.

Tews, J., U. Brose, V. Grimm, K. Tielbörger, M.C. Wichmann, M. Schwager and F. Jeltsch. 2004. Animal species diversity driven by habitat heterogeneity/diversity: the importance of keystone structures. Journal of Biogeography 31: 79–92.

Vadas, R.L. 1977. Preferential feeding: an optimization strategy in sea urchins. Ecological Monographs 47: 337–71.

Vadas, S., L. Robert, W.A. Wright and B.F. Beal. 2004. Biomass and productivity of intertidal rockweeds (*Ascophyllum nodosum* LeJolis) in Cobscook Bay. Northeastern Naturalist 11: 123–42.

van Tussenbroek, B.I. 1989. Seasonal growth and composition of fronds of *Macrocystis pyrifera* in the Falkland Islands. Marine Biology 100: 419–30.

Van Tussenbroek, B.I. 1993. Plant and frond dynamics of the giant kelp, *Macrocystis pyrifera*, forming a fringing zone in the Falkland Islands European Journal of Phycology 28: 161–65.

Vanderklift, M.A. and T. Wernberg. 2008. Detached kelps from distant sources are a food subsidy for sea urchins. Oecologia 157: 327–335.

Vetter, E.W. 1995. Detritus-based patches of high secondary production in the nearshore benthos. Marine Ecology Progress Series 120: 251–262.

Villegas, M.J., J. Laudien, W. Sielfeld and W.E. Arntz. 2008. *Macrocystis integrifolia* and *Lessonia trabeculata* (Laminariales; Phaeophyceae) kelp habitat structures and associated macrobenthic community off northern Chile. Helgoland Marine Research 62: 33–43.

Wanders, J. 1976. The role of benthic algae in the shallow reef of Curaçao (Netherlands Antilles) II: Primary productivity of the *Sargassum* beds on the northeast coast submarine plateau. Aquatic Botany 2: 327–35.

Wheeler, W. and L. Druehl. 1986. Seasonal growth and productivity of *Macrocystis integrifolia* in British Columbia, Canada. Marine Biology 90: 181–86.

White, J. and J.L. Harper. 1970. Correlated changes in plant size and number in plant populations. Journal of Ecology 58: 467–85.

Yokohama, Y., J. Tanaka and M. Chihara. 1987. Productivity of the *Ecklonia cava* community in a bay of Izu Peninsula on the Pacific coast of Japan. Journal of Plankton Research 100: 129–41.

Zimmerman, R.C. and J.N. Kremer. 1986. *In situ* growth and chemical composition of the giant kelp, *Macrocystis pyrifera*: response to temporal changes in ambient nutrient availability. Marine Ecology Progress Series 27: 277–85.

Secondary Production

Hartvig Christie and *Kjell Magnus Norderhaug*

Introduction

The primary production of macrophyte foundation species and their epiphytic algae constitutes the nutrient pool for the next ecosystem level, the secondary production. The secondary production within a macrophyte habitat depends on the amount of the primary production, how much of the primary production that is available to consumers (not exported from the system), and how efficient the available primary production can be utilised as food. Utilisation efficiency depends on factors like the nutrient value (e.g., content of proteins), algal content of supportive molecules (e.g., little digestable lignin) and content of deterring metabolites (antigrazing molecules). The content of supportive material and deterring metabolites may be high and nutrient value low in many perennial macrophytes compared to pelagic primary producers (phytoplankton), and the transfer efficiency from primary to secondary producers are consequently expected to be lower in macrophyte beds compared to pelagic communities. More general assumptions from pelagic communities of up to 20% transfer rates of primary production in marine ecosystems (e.g., Skjoldal et al. 2004, but see Pauly and Christensen 1995) will probably not be relevant in macrophyte communities. Few attempts have been made to calculate the total secondary production in seaweed and seagrass systems (see later).

In an evolutionary perspective, a sustainable balance has developed between the primary producers (the macrophytes) and the grazers/herbivores. As the macrophytes do not only serve as an important nutrition source, but also serve as a three dimensional habitat for most associated mobile and sessile fauna (Schultze et al. 1990, Christie et al. 2009), these systems depend on persistence of the perennial macrophytes. Thus

Norwegian Institute for Water Research (NIVA).
* Corresponding author: hartvig.christie@niva.no

the macrophyte ecosystem persistence and resilience depend on that the secondary producers do not graze and remove more than the growth of the foundation habitat species. Most marine systems from the littoral zone to the deep water depend on the energy produced by the primary production in the euphotic zone. Then, the secondary production within a macrophyte bed should normally be less than exploiting all primary production in the bed so that a part of the energy/carbon produced may be available as decaying organic matter for fauna communities outside the bed (Duggins et al. 1989, Krumhansl and Scheibling 2012) or to deeper systems (Vetter 1995, 1996). In some cases, the energy transfer does not follow pathways of sustainable persistence, and therefore macrophytes are declining globally (Nelleman et al. 2009). These declines are often a result of alterations among secondary producers initiated by bottom up (e.g., eutrophication) or top down (overgrazing) processes.

Energy transport and energy budgets are difficult to calculate in ecosystems with high structural and functional diversity, and studies of food webs from the macrophytes and up the food chain is complex. The secondary producers include microorganisms (Armstrong et al. 2000, Bengtsson et al. 2010, 2012), meiofauna (including, e.g., nematods, insects, oligochates, ostracods, harpacticoid copepods, Hagerman 1966, De Troch et al. 2003, Arroyo et al. 2004, Steinarsdottir et al. 2010), macrofauna including sessile and mobile animals (Hagerman 1966, Schultze et al. 1990, Fredriksen et al. 2005, Bartsch et al. 2008, Christie et al. 2009), and megaherbivores including herbivore fish (e.g., Thibaut et al. 2012), turtles (Christianen et al. 2012) and mammals (dugongs and manatees, Carrol 1988). The feeding strategies and roles in energy transport may in many cases be difficult to determine. The consumers may graze directly on the perennial macrophyte host (Mann 1982, Toth and Pavia 2002), on algal epiphytes on the host (Duffy and Hay 2000), on exudates and microflora connected to the host's surface (gastropod scraping by radula), by direct feeding on DOC (dissolved organic carbon) and POC (particulate organic carbon) or filtering of POC of benthic or pelagic origin (Nair and Anger 1979a, Seiderer and Newell 1985, Duggins and Eckman 1997, Norderhaug et al. 2003). To compensate for a low algal food value, omnivory is common, especially among mobile macrofauna, and some species change diet and become increasingly predatory with age (Nair and Anger 1979b, Cruz-Rivera and Hay 2000a,b). Some of these pathways have been identified (Newell et al. 1982, Newell and Field 1983, Norderhaug et al. 2003, Krumhansl and Scheibling 2012), but because of the above mentioned variation in feeding habits it is difficult to quantify energy transfer rates from primary to secondary producers in seaweeds and seagrasses.

Feeding and energy transfer pathways

There are mainly two ways of transferring energy from primary to secondary producers, directly via grazing or indirectly via fragmentation and decaying detritus through microbial loops. Pathways for energy transport are described below.

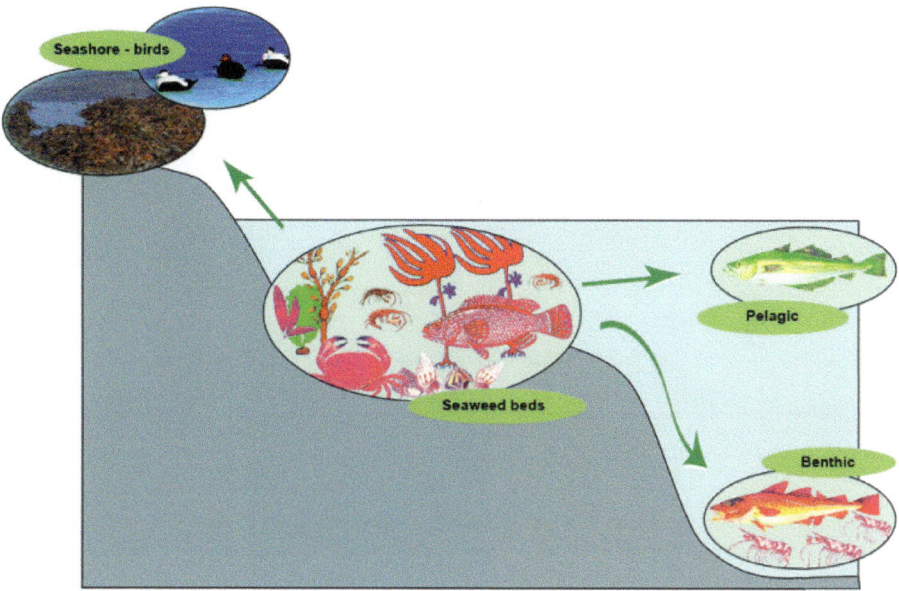

Figure 1. An example on seaweed system with energy pathways. While the primary production is within the macrophyte system, different fauna components are exploiting the energy (biomass) produced which is available both inside the system and by transport to different systems outside the seaweed as the arrows show.

Mesograzers and larger herbivores feeding directly on macrophytes

Ecological strategy among macrophytes differs according to life history traits and size. Perennials have various types of grazer defence, mainly chemicals including phlorotannins, halogen or sulphur compounds, and supportive molecules (mainly lignins) that dilute the nutrient content and give them a rigid structure which is difficult to graze (Padilla 1985, Steinberg 1992, Paul et al. 2001, Van Alstyne 2001). Many ephemeral algae cannot invest large resources in chemical and supportive defences and are more delicate and susceptible to grazers (palatable). Ephemeral survival depends on opportunistic traits like rapid growth and rapid and high reproductive effort. Another strategy to withstand grazing pressure is to be of low preference or nutrient value by low concentrations of nitrogen rich moleclues/amino acids (Cruz-Rivera and Hay 2000b, see also Fig. 2 in Krumhansl and Scheibling 2012). Then the consumers must compensate for this lack of nutrients, by using different food sources to have a balanced diet, and omnivory (including grazing, filtration, detrivory and predation) is found among secondary producers (Nair and Anger 1979a,b, Shillaker and Moore 1987, Cruz-Rivera and Hay 2000a,b, Toth and Pavia 2002, Norderhaug et al. 2003). Low nutritional value in algae also increases the effect from chemical

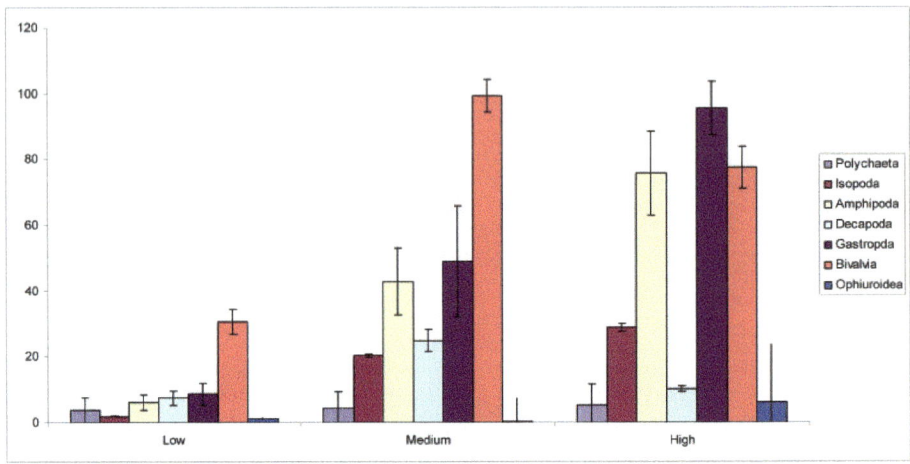

Figure 2. Example of secondary production by important taxa associated to NE Atlantic kelps in an wave exposure gradient. The average total annual secondary production is given as g D.W. per m² per yr (±SE). From Norderhaug and Christie (2011).

defence molecules (Paul et al. 2001) and grazing may induce chemical defences in algae (Amsler 2001). Compensatory feeding leading to excess grazing have been shown (Cruz-Rivera and Hay 2000b). Some grazers (amphipods) are very tolerant to chemical defence molecules, and use chemically defended algae both as food and shelter against less tolerant omnivore fish grazing on epiphytes and preying on small grazers (Hay et al. 1988).

Despite being not attractive to grazers, most macrophytes are directly grazed to some extent and grazers have to be controlled by predators in order to maintain the system (see later). Common mesograzers as gastropods, amphipods and isopods may have significant effect on both germlings and adult macrophytes, which is well described from the Baltic (e.g., Schramm and Nienhuis 1996). Grazers including gastropods, amphipods and isopods have been shown to graze their host seaweed or seagrass destructive (e.g., Hay and Duffy 2000, Fredriksen et al. 2004, Worm and Lotze 2006), and must be controlled by predators to maintain a stable macrophyte community (see Baden et al. 2012). Another factor reducing the risk of overgrazing is functional redundancy (Duffy et al. 2001) where a diversity of feeding strategies among the grazers contributes to ensure a diversity of plants.

One particularly ecological important grazer is sea urchins, which may form fronts and perform destructive grazing of macrophytes on a large scale (thousands of km²). Destructive sea urchin grazing has been reported globally; in the NE Pacific (Dayton et al. 1992, Peterson and Estes 2001) NW Atlantic (Scheibling and Hennigar 1997, Steneck et al. 2002), NE Atlantic (Norderhaug and Christie 2009), Australian waters (Ling 2008), and productive macrophyte communities have been turned into vast marine deserts, called barren grounds. These events with dramatic reduction in community diversity and production demonstrate the importance of macrophytes as habitats. This have received much attention and been debated around the world since the 1960s because of the large and dramatic consequences (Lawrence 1975), and different reasons including parasites, diseases in combination with climatic variation and overfishing have been suggested as causes behind the community changes (Dayton et al. 1992, Peterson and Estes 2001, Scheibling and Hennigar 1997, Steneck et al. 2002, Norderhaug and Christie 2009). So far evidences suggest that anthropogenic factors have caused ecosystem imbalance and resulted in such large scale events that probably in a historically perspective are new events, at least in temperate macrophyte beds (Steneck et al. 2004).

There are even larger secondary producers among herbivore vertebrates, mainly in tropical and subtropical waters. Herbivory may be too energetically inefficient for large, active animals in colder water at higher latitudes. The largest herbivores are the sea-cows (sirenian marine mammals, dugongs and manatees) that mainly eat seagrasses up to 15 kg a day (Carroll 1988). Also turtles are among the herbivores with an ecological role in structuring and stabilizing macrophyte beds (Christianen et al. 2012) but these megaherbivores are declining globally and their ecological role are probably local. The high diversity of herbivore fish in tropical waters (e.g., Thibaut et al. 2012) are not using the macroalgae as habitat, but are rather important by their grazing of algae to prevent overgrowing and loss of coral reefs (Norstrom et al. 2009, Nystrom et al. 2012).

Mesograzers feeding on epiphytes

Perennial macrophytes, both seaweeds and seagrasses, are substrates for epiphytic organisms, including red, green and brown macroalgae, diatoms and sessile animals (including sponges, bryozoans, hydroids, colonial and solitary ascidians, tube-building polychaets and bivalves). The epiphytes, or more loose associated filamentous algae, are opportunists that are competitors for light, nutrients and other resources and thus can harm the habitat building perennials (e.g., Baden et al. 2012). As epiphytes mainly are more palatable than their host the secondary producers will normally prefer the epiphytes to the perennials (Jepson et al. 2008) and a side effect from this is to clean the perennials and enhance their fitness. Thus, grazers benefit the perennials and increase the stability of macrophyte bed structure, and simultaneously serve their function as important secondary producers in the system. Such ecological role of mobile macrofauna has received considerable attention (e.g., Moksnes et al. 2008, Eriksson et al. 2011). A less studied, but significant part of the community is meiofauna which also may contribute to grazing of different epiphytic algae (De Troch et al. 2003, Arroyo et al. 2004, Steinarsdottir et al. 2010).

Mesograzers feeding on POC and DOC from macrophytes

The nutrient content of larger macrophytes is commonly low and the content of supportive material makes many macrophytes stiff and difficult to graze. Normally, foundation macrophyte species are therefore not grazed directly but enter the food web as Particulate and Dissolved Organic Carbon (POC and DOC). A key process in making POC and DOC more available and favourable to consumers are bacterial degradation including nitrogen enrichment. In many macrophyte beds there are a high number of species and particular high number of individuals of small mobile grazers. A combination of feeding experiments and stabile isotope studies (Dunton 2001, Norderhaug et al. 2003, 2006, Fredriksen 2003) suggest that these mesograzers are not grazing on the macrophytes but feed on particulate matter from the macrophytes. As the macrophytes grow, smaller and larger particles tear off, and these are colonised by microorganisms. The decaying process increase the nitrogen content of the combined POC and microorganism material and increase the nutrition value to the consumer (Norderhaug et al. 2003). At the same time, this microbial loop reduces the transfer efficiency from primary producer to secondary consumer. But because macrophyte systems in general have such a high primary production this food source will still be able to support a high secondary production. At the same time, dominance of detritus feeding to grazing represents an effective latch to maintain the vital macrophyte role as a foundation species and prevent overgrazing.

Similar to the mobile secondary producers, there are sessile animals in macrophyte beds, both as epiphytes and on the bottom between the plants. Those animals are filter feeders, and according to (Newell and Field 1983, Duggins et al. 1989) most of their carbon supply comes from macrophyte origin.

DOC is organic molecules exuded by the macrophytes. These molecules are dissolved in the water masses, and probably enriched by microorganisms as described for POC. The organic exudates from macrophytes may be as high as 20–30% of the total primary production (Barron et al. 2003, Abdullah and Fredriksen 2004). Similar exuded molecules may also cover and protect the surfaces of perennial macrophytes. There they are attracted by microorganisms (see Bengtsson et al. 2010, 2012) which also increase the nutrition value (increasing N content) and then grazed by mesograzers scraping this surface slurry on perennials.

Complex and multitrophic level energy transfer among secondary producers

As mentioned, many macrophytes like larger perennial algae and seagrasses are not directly grazed as their carbon production is entering the food chain as decaying organic material (Newell and Field 1983, Norderhaug et al. 2003, Krumhansl and Scheibling 2012). Then the produced energy are to some extent metabolised through microbial or microorganism respiration before reaching the secondary producers, and bacterial degradation may increase the food value of detritus (Duggins and Eckmann 1997). As many of the secondary producers, both filtering and active particle feeding, utilize the organic material available, these "secondary producers" can really inhabit different levels in the food web dependent of food origin. The energy transfer will also be less efficient because carbon is mineralized and energy is burned through microbial loops and other complex food web pathways.

Examples of animals that have multitrophic traits are crustaceans and other fauna associated to macrophytes that change between detritivores, omnivorous or predatory traits (Nair and Anger 1979a,b, Shillaker and Moore 1987, Cruz-Rivera and Hay 2000a,b, Toth and Pavia 2002, Norderhaug et al. 2003). These animals with roles on different levels in the food chain may be herbivore as juvenile and develop to omnivore, carnivore/cannibalistic as adults (see Christie and Kraufvelin 2004). Then, a set of alternative food webs based on macrophyte produced carbon is possible even within single species. There are high production and diversity among microorganisms (Bengtsson et al. 2010, 2012), protozoa (e.g., Gismervik 2004) and meiofauna at different trophic levels (Steinarsdottir et al. 2009, 2010). Even if stable carbon isotopes analysis indicates that the ultimate carbon source is from the macrophytes (Fredriksen 2003, Jepsen et al. 2008, Steinarsdottir et al. 2009) there may in fact be multitrophic levels and less efficient energy transfers within the fauna that are labelled as secondary producers among diverse species and individual sizes.

Mesograzer control (mesograzer to macrophyte, predator to mesograzer)

Grazing may remove the perennials but is at the same time important to control growth of epiphytic competitors. Consequently, grazing is important for the ecosystem stability and a sustainable macrophyte ecosystem depend on intermediate grazing pressure. If mesograzers are too abundant the ecological stability decreases and macrophytes are at risk of being overgrazed (Fredriksen et al. 2004, Duffy and Hay 2000), and if secondary producers are too few there is a risk of short lived epiphytes to overgrow the foundation species and change the system from a stable perennial habitat to a more fluctuating seasonal habitat (Duarte 1995, Baden et al. 2012).

Overfishing of larger predatory fish may lead to cascading perturbations down the food web. A shift to higher abundance of mesopredators like small fish caused by removal of large predatory fish (mesopredator release, Eriksson et al. 2011) has led to increasing predation and reduced secondary production. This in turn, may result in filamentous algal overgrowing seagrasses (Moksnes et al. 2008, Baden et al. 2012) and seaweeds (Eriksson et al. 2009). Contrary, mesopredators like crabs may reduce sea urchins and in this case increasing mesopredators lead to reduced grazing of perennials and thus benefit the macrophytes (Steneck et al. 2013) and secondary producers.

Like grazers can structure macrophyte communities, the composition of macrophyte species can influence on composition and magnitude of the secondary producers. Large perennial macrophytes may be suitable habitat for sessile fauna as well as for mobile secondary producers (Schultze et al. 1990), while all macrophytes are habitat for mobile meiofauna and mesograzers (Arroyo et al. 2004, Christie et al. 2009). The size and composition of macrophytes is important for presence and abundance of associated fauna. Increasing habitat volume and more complex algal structures have been found to support higher diversity and higher abundances of mesograzers (Hacker and Steneck 1990, Christie et al. 2007, Norderhaug et al. 2007).

Secondary production based on macrophyte carbon outside the macrophyte bed

A large part of the primary production is exported and consumed elsewhere (Duggins et al. 1989, Krumhansl and Scheibling 2012). The deposition of kelp material in deeper trenches described by Vetter (1995, 1996) give energy to a high secondary production and one of the highest abundances of small crustacean herbivores (leptostacans and amphipods) observed. These are among the highest productive deep water communities ever identified. POC and DOC from the macrophyte beds may be consumed and form important food resources for filtering and detritus feeding animals at different distances from the bed. Macrophyte materials are deposited on the shores and are there utilised by amphipods, oligochaets and different insects (here you often have a lot of harpacticoid copepods). Then significant parts of the macrophyte productions fuel secondary production in supalittoral, intertidal and subtidal communities on coastal waters (Vetter 1995, Bustamente and Branch 1996, Krumhansl and Scheibling 2012, see also Fig. 1).

Secondary producers inside macrophyte beds not using macrophyte carbon

In the seaweed and seagrass beds there are various forms of filtering or detritus feeding animals that may have an extent of their carbon from pelagic (phytoplankton origin) or allocthonous material. More detailed knowledge on these secondary producers' food sources and food web energy transport inside macrophyte beds are important tasks for future scientific studies.

Estimating secondary production

According to Brey (1990) there is strong positive correlation between P (production) and B (biomass) for invertebrates generally, but not very strong between P and W (individual weight). Negative correlation between P/B and W means that smaller animals have higher P/B ratio. P/B ratio varied between 0.05 and 15.11 for the animals analysed by Brey (1990). This large variations combined with complex trophic structures makes it difficult to estimate secondary production for the systems or whole community composed by a diversity of species (see Asmus 1987, Fredette et al. 1990, Edgar 1990, Cartes et al. 2001, Chrisite et al. 2003).

Methods for secondary production estimates

Different methods can be applied to calculate secondary production. Most are based on field sampling of the species of interest through time. Some early freshwater studies were based on biomass growth and mortality of cohorts which could be followed through time (e.g., Waters 1977, Benke 1984).

A widely used method from marine systems is the size-frequency method developed by Hynes and Coleman (1968), later modified by Hamilton (1969), to calculate the production on species level. It is based on the loss of biomass by size-frequency classes and is stated as:

$$P = i\sum_{j=i}^{i}(N_j - N_{j+1}) \times \sqrt{W_j \times W_{j+1}}$$

Where P is the estimated production, I is the number of size classes used in the calculation, N_j is the mean number of individuals in size class j averaged over all the sampling events and W_j the mean dry weight of individuals belonging to size class j. One weakness by using this method is that many of the species are omnivores or change diet with age (Nair and Anger 1979a,b, Shillaker and Moore 1987, Cruz-Rivera and Hay 2000a,b, Toth and Pavia 2002, Norderhaug et al. 2003, see also 2.4), so in some respects species on different food web levels are compared (Benke et al. 2001).

Other authors have developed models to take into account other factors, e.g., temperature and depth (Tumbiolo and Downing 1994, especially relevant for soft bottom communities across depth gradients). Most invertebrates associated to macrophyte communities have a life cycle with one main reproduction event per year. However, some species reproduce irregularly or continuously through the year and this violates the assumptions of the size-frequency calculations and reduce the accuracy of the results. If the time from hatching to attaining the largest size class (Cohort Production Interval, CPI) is known, it can be accounted for by multiplying the yearly production by 1/CPI (Vetter 1996). Cartes et al. (2001) reviewed use of different models.

The main scientific interest when quantifying secondary production is typically to calculate the total secondary production per year, and compare the secondary production to the production at lower and higher ecosystem levels. If the size-frequency method is used, the production of all the species has to be calculated individually. A majority of the published literature on secondary production has focused on single or few species (e.g., Asmus 1987, Edgar 1993, Vetter 1995, 1996, Sejr et al. 2002, Jeong et al. 2006), but a few studies have taken into account a large part of the secondary production (see Table 1). Fredette et al. (1990) calculated the production of 13 species and 20% of the identified fauna in an eelgrass community, and Norderhaug and Christie (2011) calculated the secondary production from 30 species accounting for 96% of the total mobile fauna identified in a kelp forest community.

Other indirect method for estimation of secondary production is to use the P/B ratios given by Brey (1990) for different species group, then estimates is only based on a biomass measurement. P/B ratios varied strongly between different taxa, but a ratio on 10 and even higher indicate high production and turnover among many small invertebrates. In large macroalgae community mesocosm experiments, Christie and Kraufvelin (2004) followed export of mobile mesograzers throughout two years, and thus a possible method for indicating secondary production that is impossible in natural ecosystems. Daily exports of 1-2% of the pt. standing stock of these animals also indicated very high turnover and high secondary production. However, such high mobility and thus daily export and import of many secondary producers (see Jørgensen and Christie 2003) are factors complicating exact estimates of secondary production.

By combining the results from a number of studies, meta-analysis can provide more general and robust production estimates (Kaiser et al. 2011). Cusson and Bourget (2005) calculated the annual secondary production of benthic invertebrates globally using 547 datasets. Modelling techniques developed the last decades provide new tools for more effective calculations of secondary production (see Cartes et al. 2001 and references therein). Artificial neural networks are flexible tools suitable for community

Table 1. Some examples of secondary production measurements given as production in g dry weight (DW) or ash-free dry weight (AFDW) per m² per year at different benthic communities. The numbers referring to Brey (1990) are estimates based on other's research data. The concerns related to comparisons of the numbers given are discussed in the text.

Habitat	g DW or g AFDW m⁻² yr⁻¹	Reference
Kelp, exposed	308 (DW)	Norderhaug and Christie 2011
Kelp, medium exposed	250 (DW)	Norderhaug and Christie 2011
Kelp, low exposure	68 (DW)	Norderhaug and Christie 2011
seagrass	200 (DW)	Fredette et al. 1990
Nebalia on kelp detritus	2400 (DW)	Vetter 1995
Seagrass	47.1 (AFDW)	Edgar 1990b
Unvegetated	8.8 (AFDW)	Edgar 1990b
Seagrass epifauna	17.2 (AFDW)	Edgar et al. 1994
Epifauna (unvegetated)	3.3 (AFDW)	Edgar et al. 1994
Infauna seagrass	62 (AFDW)	Edgar et al. 1994
Infauna unvegetated	54 (AFDW)	Edgar et al. 1994
Infauna, soft bottom	2.08 (AFDW)	Brey 1990
Infauna, soft bottom	13.17 (AFDW)	Brey 1990
Infauna, soft bottom	21.42 (AFDW)	Brey 1990
Infauna, soft bottom	12.35 (AFDW)	Brey 1990
Mussel and seagrass bed	428 (AFDW)	Valentine and Heck 1993
Seagrass	245 (AFDW)	Valentine and Heck 1993
Seagrass	180 (AFDW)	Valentine and Heck 1993
Unvegetated	42 (AFDW)	Valentine and Heck 1993

production calculations (Brey et al. 1996), and may hence be useful to meet the need for total ecosystem level production calculations.

Quantitative studies of secondary production

Macrophyte communities are among the highest productive ecosystems on the planet (Nellemann et al. 2009) and the secondary production is also found high compared to other benthic communities (see Cusson and Bourget 2005). The few existing studies indicate low transfer rate from primary to secondary producers. Norderhaug and Christie (2011) reported a secondary production depending on the wave exposure level; between 68 to 308 g D.W. per m² per year in *Laminaria* kelp forest on the Norwegian coast. Highest production was found at highest wave exposure level (see Table 1). The estimated secondary production accounted for 3 to 8% of the kelp primary production. Sessile fauna were not taken into account in the study. However, large export of kelp detritus has been observed from this system (own observation, see also Krumhansl and Scheibling 2012). The production numbers are in accordance with a few other studies focusing on the total secondary production in macrophyte beds. Fredette et al. (1990) estimated a secondary production of approximately 200 g D.W. m^{-2} year^{-1} in seagrass beds in Virginia, USA. Jeong et al. (2006) report an annual production on 20.07 g D.W. per m^2 for the amphipod *Jassa slatteryi*, indicating a high total community secondary production.

Available secondary production calculations are in accordance with the assumption of macrophyte communities being excess export systems where only a limited part of the primary production is used within the system (Duarte and Cebrián 1996). The highest secondary production ever measured (as far as we can see, see also Table 1) is on exported macroalgae detritus beds (Vetter 1995) where leptostracans and amphipods occurred in very high densities. With exception of this detritus deep water community, literature reviews of P/B ratios and secondary productions have reported highest secondary production in shallow macrophyte vegetated communities, and particularly shallow rocky habitats with macroalgae (Ricciardi and Bourget 1999, Cusson and Bourget 2005, Cowles et al. 2009, Wong et al. 2011). In Table 1 there are listed some secondary production measurements that indicate that vegetated habitats show higher secondary production than unvegetated areas. Both macroalgae and seagrass communities have high secondary production in addition to the production of exported POM/detritus from the systems that fuel a secondary production outside the macrophyte and that may be far higher than the secondary production within the macrophytes (see the numbers from Vetter 1995 from kelp detritus on deep water in Table 1). It is most reliable to compare results from the same study if available, while comparisons between studies are more difficult due to different sampling and calculation methods used, different species (soft or hard animals, herbivores or detritivores), different size groups used, how data are presented (as DW, AFDW, per year or per day, etc.), or how much of the organic material are grazed within the community studied. The large differences between the secondary productions in kelp beds in the three levels of wave exposure reported by Norderhaug and Christie (2011, see Table 1) may reflect that kelps in exposed waters provide more space and habitats for secondary producers (Christie et al. 2003, Norderhaug et al. 2007) in addition to the differences in primary production.

Concluding remarks

Although limited data of quantitative measurements of total secondary production in macrophyte ecosystems have been published, available data suggest that the large amount of energy produced per unit area by macrophytes give rise to a high within-system secondary production and an even higher part is exported from the system to adjacent unvegetated areas. Many studies indicate a significant export of seaweed and seagrass carbon to secondary producers in ecosystems outside the macrophyte system itself. There are many studies quantifying primary production of macrophytes, but there is a challenge and a need for developing better methods for quantitative determination of secondary production. This will be important for understanding energy transport from the plants and up the food chain, for carbon and nutrient cycling and also for eventually amount of CO_2 fixation and capture.

In this section, possible pathways of energy transfer in the systems have been described. As the organisms that are expected to utilize the primary production (as secondary producers) are in fact composed by a high diversity of taxa of different sizes from bacteria to sea-cows and with a variety of ecological traits, there is a real challenge to make generalizations about the energy transfer efficiency in these systems. The high amount of degrading microorganisms, herbivore meiofauna and macrofauna, detritivores, omnivores and probable partly micropredator meio- and macrofauna belonging to this group, all indicate a variety of food chains and more or less efficient transfer of the macrophyte energy release. Focus on these smaller organisms combined with molecular and stable isotope studies may be one way of approaching further insight in these challenges.

The macrophytes are both food and habitat for variety of organisms belonging to this diverse group called secondary producers. For the persistence of a system with the risk of consumers that consume too much of their habitat, a number of internal and external regulating and feedback mechanisms have evolved (e.g., mainly eating particulate matter from the macrophytes, dispersal of excess faunal production, predator control and different trophic interactions). Reduced nutrient value of the most important foundation species among macrophytes and a variety of energy pathways and redundancy of organisms with different feeding traits contribute to high system diversity and stability.

References

Abdullah, M.I. and S. Fredriksen. 2004. Production, respiration and exudation of dissolve organic matter by the kelp *Laminaria hyperborea* (Gunnerus) Foslie along the west coast of Norway. J. Mar. Biol. Ass. U.K. 84: 887–894.

Amsler, C.D. 2001. Induced defences in macroalgae: the herbivore makes a difference. J. Phycol. 37: 353–356.

Armstrong, E., A. Rogerson and J.W. Leftley. 2000. The abundance of heterotrophic protists associated with intertidal seaweeds. Estuarine, Coastal and Shelf Science 50: 415–424.

Arroyo, N.L., M. Maldonado, R. Perez-Portela and J. Benito. 2004. Distribution patterns of meiofauna associated with a sublittoral *Laminaria* bed in the Cantabrian Sea (north-eastern Atlantic). Marine Biology 144: 231–242.

Asmus, H. 1987. Secondary production of an intertidal mussel bed community related to its storage and turnover compartment. Marine Ecology Progress Series 39: 251–266.

Baden, S., A. Emanuelsson, L. Pihl, C.J. Svensson and P. Aberg. 2012. Shift in seagrass food web structure over decades is linked to overfishing. Mar. Ecol. Prog. Ser 451: 61–73.

Barron, C., N. Marba, C.M. Duarte, M.F. Pedersen, C. Lindblad, K. Kersting, F. Moy and T. Bokn. 2003. High organic carbon export precludes eutrophication responces in experimental rocky shore communities. Ecosystems 6: 144–153.

Bartsch, I., C. Wiencke, K. Bischof, C.M. Buchholz, B.H. Buck, A. Eggert, P. Feuerpfeil, D. Hanelt, S. Jacobsen, R. Karez, U. Karsten, M. Molis, M.Y. Roleda, H. Schubert, R. Schumann, K. Valentin, F. Weinberger and J. Wiese. 2008. The genus *Laminaria sensu lato*: recent insights and developments. Eur. J. Phyc. 43: 1–86.

Bengtsson, M.M., K. Sjøtun and L. Øvreås. 2010. Seasonal dynamics of bacterial biofilms on the kelp *Laminaria* hyperborean. Aquatic Microbial Ecology 60: 71–83.

Bengtsson, M.M., K. Sjøtun, A. Lanzén and L. Øvreås. 2012. Bacterial diversity in relation to secondary production and succession on surfaces of the kelp *Laminaria hyperborea*. The ISME J. 6: 2188–2198.

Benke, A. 1984. Secondary production of aquatic insects. pp. 289–322. *In*: Resh, V.H. and D.M. Rosenberg (eds.). The Ecology of Aquatic Insects. Praeger, New York, USA.

Benke, A.C., J.B. Wallace, J.W. Harrison and J.W. Koebel. 2001. Food web quantification using secondary production analysis: predaceous invertebrates of the snag habitat in a subtropical river. Freshwater Biology 46: 329–346.

Brey, T. 1990. Estimating productivity of macrobenthic invertebrates from biomass and mean individual weight. Meeresforschung 32: 329–343.

Brey, T., A. Jarre-Teichmann and O. Borlich. 1996. Artificial neural network versus multiple linear regression: predicting P/B ratios from empirical data. MEPS 140: 251–256.

Bustamente, R.H. and G.M. Branch. 1996. The dependence of intertidal consumers on kelp-derived organic matter on the west coast of South Africa. Journal of Experimental Marine Biology and Ecology 196: 1–28.

Carroll, R.L. 1988. Vertebrate Paleontology and Evolution. Freeman, New York, USA.

Cartes, J.E., T. Brey, J.C. Sorbe and F. Maynou. 2001. Comparing production-biomass ratios of benthos and suprabenthos in macrofaunal marine crustaceans. Canadian Journal of Fishery and Aquatic Science 59: 1616–1625.

Christianen, M.J.A., L.L. Govers, T.J. Bouma, W. Kiswara, J.G.M. Roelofs, L.P.M. Lamers and M.M. van Katwijk. 2012. Marine megaherbivore grazing may increase seagrass tolerance to high nutrient loads. Journal of Ecology 100: 546–560.

Christie, H. and P. Kraufvelin. 2004. Mechanisms regulating amphipod population density within macroalgal communities with restricted predator impact. Scientia Marina 68: 189–198.

Christie, H., N.M. Jørgensen, K.M. Norderhaug and E. Waage-Nielsen. 2003. Species distribution and habitat exploitation of fauna associated with kelp (*Laminaria hyperborea*) along the Norwegian coast. J. Mar. Biol. Ass. U.K. 83: 687–699.

Christie, H., N.M. Jørgensen and K.M. Norderhaug. 2007. Bushy or smooth, high or low; Importance of habitat architecture and vertical level for distribution of fauna on kelp. Journal of Sea Research 58: 198–208.

Christie, H., K.M. Norderhaug and S. Fredriksen. 2009. Macrophytes as habitat for fauna. Mar. Ecol. Prog. Ser. 396: 221–233.

Cowles, A., J.E. Hewitt and R.B. Taylor. 2009. Density, biomass and productivity of small mobile invertebrates in a wide range of coastal habitats. Marine Ecology Progress Series 384: 175–185.

Cruz-Rivera, E. and M.E. Hay. 2000a. The effect of diet mixing on consumer fitness: macroalgae, epiphytes, and animal matter as food for marine amphipods. Oecologia 123: 252–264.

Cruz-Rivera, E. and M.E. Hay. 2000b. Can quantity replace quality? Food choice, compensatory feeding, and fitness of marine mesograzers. Ecology 81: 201–219.

Cusson, M. and E. Bourget. 2005. Global patterns of macroinvertebrate production in marine benthic habitats. Marine Ecology Progress Series 297: 1–14.

Dayton, P.K., M.J. Tegner, P.E. Parnell and P.B. Edwards. 1992. Temporal and spatial patterns of disturbance and recovery in a kelp forest community. Ecological Monographs 62: 421–445.

De Troch, M., F. Fiers and M. Vincx. 2003. Niche segregation and habitat specialisation of harpacticoid copepods in a tropical seagrass bed. Marine Biology 142: 345–355.

Duarte, C.M. 1995. Submerged aquatic vegetation in relation to different nutrient regimes. Ophelia 41: 87–112.

Duarte, C.M. and J. Cebrián. 1996. The fate of marine autotrophic production. Limnology and Oceanography 41: 1758–1766.

Duffy, J.E. and M.E. Hay. 2000. Strong impacts of grazing amphipods on the organization of a benthic community. Ecological Monographs 70: 237–263.

Duffy, J.E., K.S. Macdonald, J.M. Rhode and J.D. Parker. 2001. Grazer diversity, functional redundancy, and productivity in seagrass beds: an experimental test. Ecology 82: 2417–2434.

Duggins, D.O. and J.E. Eckman. 1997. Is kelp detritus a good food for suspension feeders? Effects of kelp species, age and secondary metabolites. Marine Biology 128: 489–495.

Duggins, D.O., C.A. Simenstad and J.A. Estes. 1989. Magnification of secondary production by kelp detritus in coastal marine ecosystems. Science 245: 170–173.

Dunton, K. 2001. δ15N and δ13C measurements of Antarctic peninsula fauna: trophic relationships and assimilation of benthic seaweeds. American Zoologist 41: 99–112.

Edgar, G.J. 1990a. The use of the size structure of benthic macrofaunal communities to estimate faunal biomass and secondary production. Journal of Experimental Marine Biology and Ecology 137: 195–214.

Edgar, G.J. 1990b. The influence of plant structure on the species richness, biomass and secondary production of macrofaunal assemblages associated with Western Australian seagrass beds. Journal of Experimental Marine Biology and Ecology 137: 215–240.

Edgar, G.J. 1993. Measurements of the carrying capacity of benthic habitats using a metabolic-rate based index. Oecologia 95: 115–121.

Edgar, G.J., C. Shaw, G.F. Watson and L.S. Hammond. 1994. Comparison of species richness, size-structure and production of benthos in vegetated and unvegetated habitats in Western Port, Victoria. Journal of Experimental Marine Biology and Ecology 176: 201–226.

Eriksson, B.K., L. Ljunggren, A. Sandström, G. Johansson, J. Mattila, A. Rubach, S. Råberg and M. Snickars. 2009. Declines in predatory fish promote bloom-forming macroalgae. Ecological Applications 19: 1975–1988.

Eriksson, B.K., K. Sieben, J. Eklof, L. Ljunggren, J. Olsson, M. Casini and U. Bergstrom. 2011. Effects of altered offshore food webs on coastal ecosystems emphasize the need for cross-ecosystem management. AMBIO 40: 786–797.

Fredette, T.J., R.J. Diaz, J. van Montfrans and R.J. Orth. 1990. Secondary production within a seagrass bed (*Zostera marina* and *Ruppia maritima*) in Lower Cheasapeake Bay. Estuaries 13: 431–440.

Fredriksen, S. 2003. Food web studies in a Norwegian kelp forest based on stable isotope (d^{13}C and d^{15}N) analysis. Marine Ecology Progress Series 260: 71–81.

Fredriksen, S., H. Christie and C. Bostrom. 2004. Deterioration of eelgrass (*Zostera marina* L.) through destructive grazing by the gastropod Rissoa membranacea (J. Adams). Sarsia 89: 218–222.

Fredriksen, S., H. Christie and B.A. Sætre. 2005. Species richness in macroalgae and macrofauna assemblages on *Fucus serratus* L. (Phaeophyceae) and *Zostera marina* L. (Angiospermae) in Skagerrak, Norway. Marine Biology Research 1: 2–19.

Gismervik, I. 2004. Podite carrying ciliates dominate the benthic ciliate community in the kelp forest. Aquatic Microbial Ecology 36: 305–310.

Hacker, S.D. and R.S. Steneck. 1990. Habitat architecture and the abundance and body-size-dependent habitat selection of a phytal amphipod. Ecology 71: 2269–2285.

Hamilton, A.L. 1969. On estimating annual production. Limnology and Oceanography 14: 771–782.

Hagerman, L. 1966. The macro- and microfauna associated with *Fucus serratus* L., with some ecological remarks. Ophelia 3: 1–43.

Hay, M.E., J.E. Duffy, W. Fenical and K. Gustafson. 1988. Chemical defense in the seaweed Dictyopteris delicatula: differential effects against reef fishes and amphipods. Marine Ecology Progress Series 48: 185–192.

Hynes-Hamilton, H.B.N. and M. Coleman. 1968. A simple method for assessing the annual production of stream benthos. Limnology and Oceanography 13: 569–573.

Jepson, T., P. Nystrom, P.O. Moksnes and S.P. Baden. 2008. Trophic Interactions in *Zostera marina* Beds along the Swedish Coast. Marine Ecology Progress Series 369: 63–76.

Jeong, S.J., O.H. Yu and H.L. Suh. 2006. Secondary production of *Jassa slatteryi* (Amphipoda, Ischyroceridae) on a *Zostera marina* seagrass bed in southern Korea. Marine Ecology Progress Series 309: 205–211.

Jørgensen, N.M. and H. Christie. 2003. Diurnal, horizontal and vertical dispersal of kelp-associated fauna. Hydrobiologia 503: 69–76.

Kaiser, M.J., M.J. Attrill, S. Jennings, D.N. Thomas, D.K.A. Davis Barnes, A.S. Brierly, J.G. Hiddink, H. Kaartokallio, N.V.C. Polunin and D.G. Raffaelli. 2011. Marine Ecology: Processes, Systems, and Impacts. Oxford University Press, New York, 501 pp.

Krumhansl, K.A. and R.E. Scheibling. 2012. Production and fate of kelp detritus. Marine Ecology Progress Series 467: 281–302.

Lawrence, J.M. 1975. On the relationship between marine plants and sea urchins. Oceanogr. Mar. Biol., Annu. Rev. 13: 213–286.

Ling, S.D. 2008. Range expansion of a habita-modifying species leads to loss of taxonomic diversity; a new and impoverished reef state. Oecologia 156: 883–894.

Mann, K.H. 1982. Ecology of coastal waters: a systems approach. University of California Press, Berkeley. USA.

Moksnes, P.O., M. Gullström, K. Tryman and S. Baden. 2008. Trophic cascades in a temperate seagrass community. Oikos 117: 763–777.

Nair, K.K.C. and K. Anger. 1979a. Life cycle of *Corophium insidiosum* (Crustacea, Amphipoda) in laboratory culture. Helgoländer wissenschaftliche Meeresuntersuchungen. 32: 279–294.

Nair, K.K.C. and K. Anger. 1979b. Experimental studies on the life cycle of *Jassa falcata* (Crustacea, Amphipoda). Helgoländer wissenschaftliche Meeresuntersuchungen. 32: 444–452.

Nellemann, C., E. Corcoran, C.M. Duarte, L. Valdes, C. De Young, L. Fonseca and G. Grimsditch (eds.). 2009. Blue Carbon. The Role of Healthy Oceans in Binding Carbon. A Rapid Response Assessment, United Nations Environmental Programme. GRID-Arendal, Arendal, Norway.

Newell, R.C. and J.G. Field. 1983. The contribution of bacteria and detritus to carbon and nitrogen flow in a benthic community. Mar. Biol. Lett. 4: 23–36.

Newell, R.C., J.G. Field and C.L. Griffiths. 1982. Energy balance and significance of micro-organisms in a kelp bed community. Marine Ecology Progress Series 8: 103–113.

Norderhaug, K.M. and H. Christie. 2009. Sea urchin grazing and kelp re-vegetation in the NE Atlantic. Marine Biology Research 5: 515–528.

Norderhaug, K.M. and H. Christie. 2011. Secondary production in a *Laminaria hyperborea* kelp forest and variation according to wave exposure. Estuarine Coast Shelf science 95: 135–144.

Norderhaug, K.M., S. Fredriksen and K. Nygaard. 2003. The trophic importance of *Laminaria hyperborea* to kelp forest consumers and the importance of bacterial degradation for food quality. Marine Ecology Progress Series 255: 135–144.

Norderhaug, K.M., K. Nygaard and S. Fredriksen. 2006. Importance of phlorotannin content and C:N ratio of *Laminaria hyperborea* in determining its palatability as food for consumers. Marine Biology Research 2: 367–371.

Norderhaug, K.M., H. Christie and S. Fredriksen. 2007. Is habitat size an important factor for faunal abundances on kelp (*Laminaria hyperborea*)? Journal of Sea Research 58: 120–124.

Norström, A.V., M. Nyström, J. Lokrantz and C. Folke. 2009. Alternative states on coral reefs: beyond coral-macroalgaal phace shifts. Marine Ecology Progress Series 376: 295–306.

Nyström, M., A.V. Norström, T. Blenckner, M. de la Torre-Castro, J.S. Eklöf, C. Folke, H. Österblom, R.S. Steneck, M. Thyresson and M. Troell. 2012. Confronting feedbacks of degraded marine ecosystems. Ecosystems 15: 695–710.

Padilla, D.K. 1985. Structural resistance of algae to herbivores. A biomechanical approach. Marine Biology 90: 103–109.

Paul, V.J., E. Cruz-Rivera and R.W. Thacker. 2001. Chemical mediation of macroalgal-herbivore interactions: ecological and evolutionary perspectives. pp. 227–265. *In*: McClintock J.B. and B.J. Baker (eds.). Marine Chemical Ecology. CRC Press, London, UK.

Pauly, D. and V. Christensen. 1995. Primary production required to sustain global fisheries. Nature 374: 255–257.

Peterson, C.H. and J.A. Estes. 2001. Conservation and management of marine communities. pp. 469–508. *In*: Bertness, M.D., S.D. Gaines and M.E. Hay (eds.). Marine Community Ecology. Sinauer, Sunderland, USA.

Ricciardi, A. and E. Bourget. 1999. Global patterns of macroinvertebrate biomass in marine intertidal communities. Marine Ecology Progress Series 185: 21–35.

Scheibling, R.E. and A.W. Hennigar. 1997. Recurrent outbreaks of disease in sea urchins Strongylocentrotus droebachiensis in Nova Scotia: evidence for a link with large-scale meteorological and oceanographic events. Marine Ecology Progress Series 152: 155–165.

Schramm, W. and P.H. Nienhuis. 1996. Marine Benthic Vegetation. Recent Changes and Effects of Eutrophication. Springer, Heidelberg, Germany.

Schultze, K., K. Janke, A. Krüß and W. Weidemann. 1990. The macrofauna and macroflora associated with *Laminaria digitata* and *L. hyperborea* at the island of Helgoland (German Bight, North Sea). Helgoländer Meeresuntersuchungen 44: 39–51.

Seiderer, L.J. and R.C. Newell. 1985. Relative significance of phytoplankton, bacteria and plant detritus as carbon and nitrogen resources for the kelp bed filter-feeder *Choromytilus meridionalis*. Marine Ecology Progress Series 22: 127–139.

Sejr, M.K., M.K. Sand, T. Jensen, J.K. Petersen, P.B. Christensen and S. Rysgaard. 2002. Growth and production of *Hiatella arctica* (Bivalvia) in a high-Arctic fjord (Young Sound, Northeast Greenland). Marine Ecology Progress Series 244: 163–169.

Shillaker, R.O. and P.G. Moore. 1987. The feeding habits of the amphipods Lembos websteri Bate and *Corophium bonellii* Milne Edwards. Journal of Experimental Marine Biology and Ecology 110: 93–112.

Skjoldal, H.R. 2004. The Norwegian Sea ecosystem. Tapir Academic Press, Trondheim. Norway.

Steinarsdottir, M.B., A. Ingolfsson and E. Olafsson. 2009. Trophic relationships on a fucoid shore in South-Western Iceland as revealed by stable isotope analyses, laboratory experiments, field observations and gut analyses. J. Sea Res. 61: 206–215.

Steinarsdottir, M.B., A. Ingolfsson and E. Olafsson. 2010. Field evidence of different food utilization of phytal harpacticoids collected from *Fucus serratus* indicated by δ13C and δ15N stable isotopes. Estuarine Coastal and Shelf Science 88: 160–164.

Steinberg, P.D. 1992. Geographical variations in the interaction between marine herbivores and brown algal secondary metabolites. pp. 51–92. *In*: Paul, V.J. (ed.). Ecological Roles of Marine Natural Products. Cornell University Press, London, UK.

Steneck, R.S., M.H. Graham, B.J. Bourque, D. Corbett, J.M. Erlandson, J.A. Estes and M.J. Tegner. 2002. Kelp forest ecosystem: biodiversity, stability, resilience and future. Environmental Conservation 29: 436–459.

Steneck, R.S., J. Vavrinec and A.V. Leland. 2004. Accelerating trophic-level dysfunction in kelp forest ecosystems of the Western North Atlantic. Ecosystems 7: 323–332.

Steneck, R.S., A. Leland, D.C. McNaught and J. Vavrinec. 2013. Ecosystem flips, locks, and feedbacks: the lasting effects of fisheries on Maine's kelp forest ecosystems. Bulletin of Marine Science 89: 31–55.

Thibaut, L.M., S.R. Conolly and P.A. Sweatman. 2012. Diversity and stability of herbivorous fishes on coral reefs. Ecology 93: 891–901.

Toth, G.B. and H. Pavia. 2002. Lack of phlorotannin induction in the kelp *Laminaria hyperborea* in response to grazing by two gastropod herbivores. Marine Biology 140: 403–409.

Tumbiolo, M.A. and J.A. Downing. 1994. An empirical model for the prediction of secondary production in marine benthic invertebrate populations. Marine Ecology Progress Series 114: 165–174.

Valentine, J.F. and K.L. Heck. 1993. Mussels in seagrass meadows: their influence on Macroinvertebrate abundance and secondary production in the northern Gulf of Mexico. Marine Ecology Progress Series 96: 63–74.

Van Alstyne, K.L., M.N. Dethier and D.O. Duggins. 2001. Spatial patterns in macroalgal chemical defences. pp. 301–324. *In*: McClintock, J.B. and B.J. Baker (eds.). Marine Chemical Ecology. CRC Press, London, UK.

Vetter, E.W. 1995. Detritus-based patches of high secondary production in the nearshore benthos. Marine Ecology Progress Series 120: 251–262.

Vetter, E.W. 1996. Secondary production of a Southern California *Nebalia* (Crustacea: Leptrostraca). Marine Ecology Progress Series 137: 95–101.

Waters, T.F. 1977. Secondary production in inland waters. Advances in Ecological Research 10: 91–164.

Wong, M.C., C.H. Peterson and M.F. Piehler. 2011. Evaluating estuarine habitats using secondary production as a proxy for food web support. Marine Ecology Progress Series 440: 11–25.

Worm, B. and H.K. Lotze. 2006. Effects of eutrophication, grazing, and algal blooms on rocky shores. Limnology and Oceanography 51: 569–579.

III. Human Threats to Macrophytic Foundation Species

8

Eutrophication and the Challenge of Changing Biotic Interactions

Eva Rothäusler[a] and *Veijo Jormalainen*[b]

Introduction to eutrophication

Since industrialization, gradually increasing manmade activities have caused significant alterations in the structure and function of coastal ecosystems. One prominent mode of human disturbance is the surplus of nutrients (i.e., cultural eutrophication) into waters across the globe (Vitousek et al. 1997, Boesch 2002). Scientists have recognized cultural eutrophication since the early 1970s (Cloern 2001, Boesch 2002), where it has been invoked as one of the major contributors to macrophyte decline (Duarte 1995, Benedetti-Cecchi et al. 2001, Lotze et al. 2006, Burkholder et al. 2007, Waycott et al. 2009, Schmidt et al. 2012).

Nitrogen flux from land to the coastal zone has increased to many times its natural level (e.g., Bennett et al. 2001, Tilman et al. 2001, Galloway and Cowling 2002, Foley et al. 2005) with stronger effects in areas characterized by restricted water exchange (e.g., bays, estuaries, lagoons and fjords). As a result, algal biomass has increased much faster as would occur under natural circumstances, followed by changes in the abundance and community structure of long-living macrophytes. Eutrophication-induced shifts of macrophytes are prominent worldwide but only a relatively limited number of studies (on average 4% of all marine environmental research based on ISI Web of Science) in the last two decades have focused on the

Section of Ecology, Department of Biology, 20014 University of Turku, Finland.
[a] Email: eva.rothausler@utu.fi
[b] Email: veijo.jormalainen@utu.fi

vulnerability of macrophytes in the light of cultural eutrophication (Fig. 1A). The proportion of studies addressing this research increased during the twentieth century but ceased when coming to the twenty-first century (Fig. 1A). In absolute numbers, eutrophication studies have increased 5-fold during the last two decades together with the overall increase in the scientific publication activity.

Indirect effects of nutrient pollution, such as increased turbidity, sedimentation of organic matter, increased competition and epibiotism, and alterations in grazer densities and susceptibility to grazing (Korpinen et al. 2007a and references therein), have been inferred to be the most important factors affecting macrophyte communities. Slow-

A

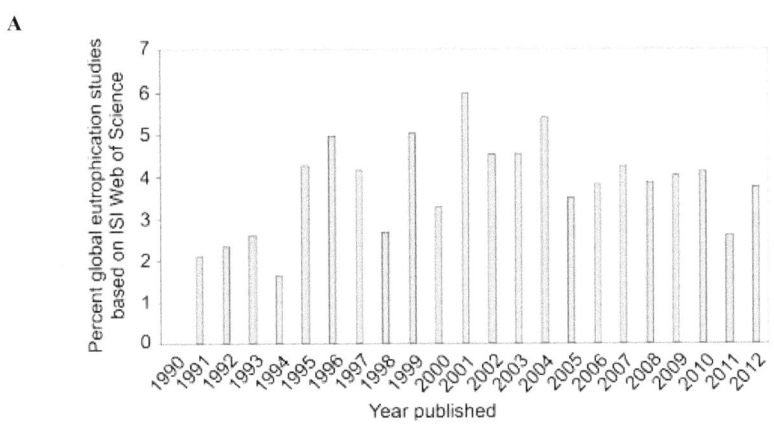

B

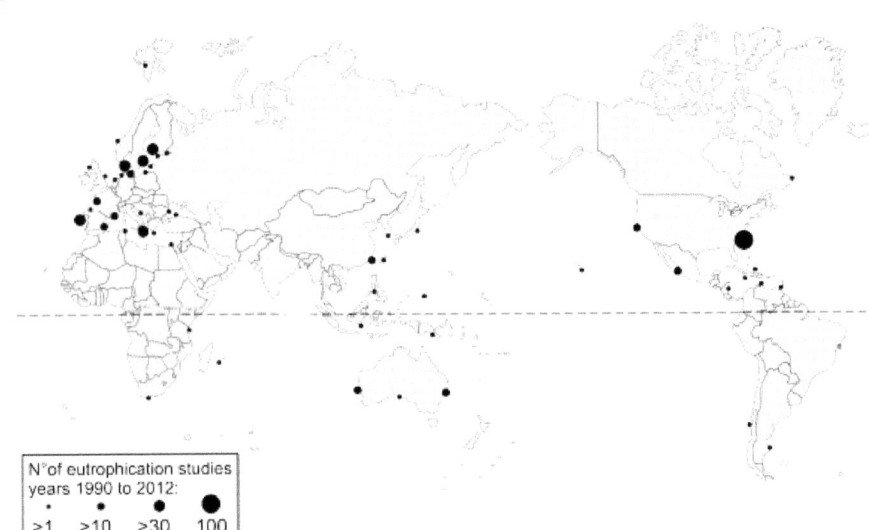

Figure 1. Percent global marine eutrophication studies out of total marine studies published between years 1990 and 2012, based on ISI Web of Science (A). Geographic scope of the number of marine eutrophication studies (black dots) conducted within the last two decades until now (B).

growing large macroalgae in the orders Laminariales (kelps) and Fucales (fucoids), as well as marine flowering seagrass meadows are declining worldwide due to reduced light penetration, siltation and overgrowth by epiphytic and filamentous or sheet-like macroalgae (Vogt and Schramm 1991, Schories et al. 1997, Hauxwell et al. 2001, Eriksson et al. 2002, Orth et al. 2006, Burkholder et al. 2007, Coleman et al. 2008, Connell et al. 2008, Phillips and Blackshaw 2011, Moy and Christie 2012).

The increase of these bloom forming opportunistic macro- and microalgae is influenced not only by nutrient supply but also by herbivory. Blooms can provide the nutritional base for large grazer populations and elevated nutrients improve macrophyte quality for herbivores as a food source, which can trigger an overall increased grazing (e.g., McGlathery 1995, Hemmi and Jormalainen 2002, Bokn et al. 2003) and consequently exert control over abundance of species. In order to understand the state and fate of macrophytes in nutrient over-enriched coastal habitats it is important to ask questions such as the following: Where does eutrophication take place? What biotic interactions are modified by eutrophication and how? Does eutrophication alter cascading regulatory effects on the producer community? How has eutrophication changed the trophic structure of the littoral community? How does the overall reshuffling of the biotic interactions due to eutrophication affect benthic macrophyte communities?

Eutrophication as a global threat to macrophyte assemblages

Loss and degradation of habitat forming macrophytes and their associated communities due to eutrophication has been reported for a number of bays and estuaries worldwide (Benedetti-Cecchi et al. 2001, Cloern 2001, Gorgula and Connell 2004). This leads to loss of ecosystem services provided by macrophytes, such as carbon or nutrient storage capacity or habitat for thousands of species (Waycott et al. 2009, Schmidt et al. 2012). Eutrophication can be found on both hemispheres but within the last two decades the northern hemisphere has been subject of particular attention (Fig. 1B). Along the European coasts in numerous places, including the Mondego Estuary (Portugal), Venice Lagoon (Italy), the Baltic Sea (Sweden, Finland, Germany, and Denmark, to a lesser extent Russia, Lithuania, Poland and Estonia), the North Sea (Germany, Denmark, Netherlands, and Great Britain), the Atlantic shores (France, Spain, Great Britain, and Norway), the Mediterranean and the Black Sea, changes within the macrophyto-benthos communities have been reported. Many small bays, estuaries, and lagoons along the northeastern US Atlantic coast, such as, e.g., Waquoit Bay, Chesapeake Bay, and Long Island Sound have been significantly modified by nutrient over-enrichment but also estuaries from the Pacific Ocean, Gulf of Mexico and the Caribbean Sea. In the southern hemisphere, most studies come from Australia (e.g., Great Barrier Reef lagoons, Moreton Bay, Port Phillip Bay), while few studies have been conducted so far in Central/South America, and Africa. Overall, most of the studies were conducted in temperate regions while to a lesser extent in tropical waters. Consequences of eutrophication are likely to be more persuasive in temperate shallow littoral ecosystems as they are naturally more productive than the tropical seas (Huston and Wolverton 2009) and the runoff from land-based activities and ample application of fertilizers has a longer history there (Galloway and Cowling 2002).

Nutrient pollution, negatively affects numerous kelp and fucoid species as well as seagrass meadows, and even some of these species have seen to become squeezed out of their habitats entirely (Orth and Moore 1983, Cardoso et al. 2004, Waycott et al. 2009, Kavanaugh et al. 2009, Phillips and Blackshaw 2011, Moy and Christie 2012). For instance, along the south and west coast of Norway there was a shift from the sugar kelp *Saccharina latissima* to filamentous ephemeral algae (Moy and Christie 2012). A similar trend was reported for the fucoid species *Phyllospora camosa*, which disappeared from the urbanized east coast of Australia (Coleman et al. 2008). Observations of reducing, disappearing, or retreats from deeper to shallow parts of *Fucus vesiculosus* and *F. serratus* belts have been made in the Baltic Sea in Finnish (Kangas et al. 1982, Ronnberg 1984), Swedish (Kautsky et al. 1986, Nilsson et al. 2004), Polish (Plinski and Florczyk 1984) and German (Vogt and Schramm 1991) coastal waters and estuaries. Large-scale losses of seagrasses due to nutrient over-enrichment associated with major overgrowth of algae and thus shading of seagrass beds have been reported for *Posidonia* species from Western Australia (Walker et al. 2006) and the Mediterranean Sea (Marba et al. 1996, Marba et al. 2005). For *Zostera* species, similar observations have been made in the Baltic Sea (Krause-Jensen et al. 2008, Baden et al. 2012) the Mondego Estuary (Cardoso et al. 2004), Chesapeake Bay (Orth and Moore 1983), the Seto Inland Sea (Tamaki et al. 2002), and the Western Wadden Sea (van Katwijk et al. 2010). Likewise, along tropical shorelines of Belize and Brazil human induced developmental impacts are the possible cause for seagrass disappearance (Short et al. 2006 and references therein). Overall, of the known global coverage of seagrass habitats, 29% has disappeared mainly due to degraded water quality (Waycott et al. 2009).

Multifaceted ramifications of eutrophication

Macrophytes typically are in a role of foundation species for the associated community, indicating that they affect the community structure and function and harbor a variety of associated species. Therefore macrophyte performance largely depends on biotic interactions even in pristine stands. Accordingly, eutrophication in these ecosystems takes place in a community context through altering biotic interactions, i.e., increasing competition for space, light, nutrients, and modifying grazing pressure (e.g., Valentine and Duffy 2006, Krause-Jensen et al. 2008, Moy and Christie 2012), emphasizing the importance of understanding ecology of macrophyte interactions with competitors and natural enemies. The indirect eutrophication effects are not limited to direct competition or herbivory but extend to development of phytoplankton blooms and to consequent light attenuation and increased organic sedimentation (Fig. 2). There are only two potentially direct effects of increased nutrient availability on macrophytes through physiological changes, either in processes affecting growth or in those affecting resistance to natural enemies. Both of them are affected by indirect effects through attenuation of light as well (Fig. 2). We will first discuss the responses of growth and resistance to nutrient enrichment and light attenuation, and then highlight the community context and the relevant interactions of the indirect causation.

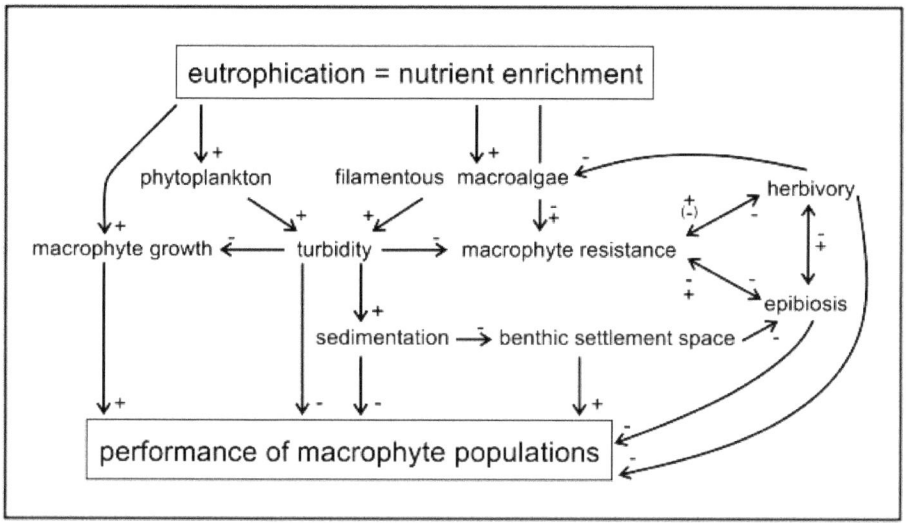

Figure 2. Conceptual illustration of the main abiotic and biotic factors causing macrophyte decline in eutrophied coastal estuaries and bays. Plus and minus signs indicate positive and negative effects, respectively, which one factor is having on another.

Responses of growth and defense to shifting resource levels

The resource that first starts limiting photosynthesis will be the main determinant of the physiological performance of macrophytes. Traditionally, light is considered to be a plentiful resource, whereas inorganic nutrients and carbon are expected to be the limiting factors. However, this understanding is probably biased because photosynthesis has typically been measured in ecologically non-realistic conditions, where the community context was excluded (Sand-Jensen et al. 2007). By definition the foundational macrophyte species form dense stands with lots of thallus/leaf biomass and strong three-dimensional structure. Within such stands self-shading is common and light attenuates efficiently, thus growth becomes seldom saturated by light. Even the maximum irradiances at noon may not saturate photosynthesis within the communities (Sand-Jensen et al. 2007). This implies that light limitation caused by eutrophication is the single most important mechanism hampering macrophyte performance.

Dissolved nitrogen is an essential element for much of the macrophyte physiology and represents the most limiting macronutrient for macrophyte growth (Lobban and Harrison 1994). Nitrogen assimilation is a metabolically expensive process that requires photosynthesis to form the carbon skeletons to build up amino acids (Turpin et al. 1991). On the other hand, as far as macrophytes can fix photosynthetic carbon, elevated nitrogen availability may boost photosynthesis. Initially, a common response of macrophytes in moderate eutrophied areas is a relative increase in photosynthesis, inherent growth, and tissue nitrogen content (Kenworthy and Fonseca 1992, Agawin et al. 1996, Udy and Dennison 1997, Valiela et al. 1997, Jaschinski and Sommer 2008a). For instance, *Fucus vesiculosus* and *Zostera marina* from Danish waters showed high tissue nitrogen content under enriched conditions relative to control plants, followed by a slight increase in their photosynthetic capacity (Pedersen 1995). The positive effect of nutrient availability on the photosynthetic capacity of *Posidonia oceanica* from the Spanish Mediterranean Sea was shown by Alcoverro et al. (2001a) under field conditions (Fig. 3).

In the course of increasing nutrient loading, the photosynthetic performance of seagrasses and perennial macroalgae is curtailed (Diaz-Pulido and McCook 2005, Lee et al. 2007, Rohde et al. 2008), which is mainly linked to a decrease in light penetration caused by overgrowth of periphyton and opportunistic algae (Tomasko et al. 1996, Brush and Nixon 2002). To sustain growth during low light conditions macrophytes are able to respond with photoacclimation, which may be achieved through increased pigment content and/or changes in electron transfer chain components, thereby enhancing their light harvesting efficiencies (Ralph et al. 2002, Lamote and Dunton 2006). These processes are relevant because they permit macrophytes to persist under light deprivation. However, with increased shading, photoacclimation processes become too costly, which is reflected in low growth rates and eventually leads to

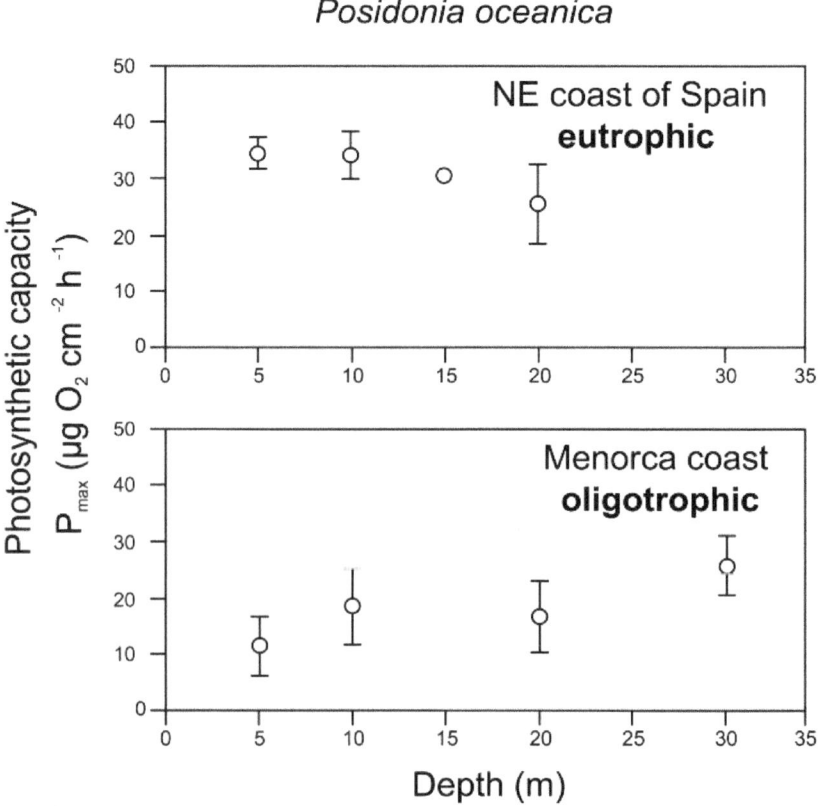

Figure 3. Photosynthetic capacity of leaves from *Posidonia oceanica* along a depth gradient within a eutrophic and oligotrophic coastal area (N = 3 sites for each area, dots represent site specific means ± STD). Modified after Alcoverro et al. (2001a).

macrophyte deterioration (Lee and Dunton 1997, Ruiz and Romero 2001). For instance, in the eelgrass *Zostera marina* increased nitrogen loads lead to an increase of the standing stock of opportunistic algae causing severe light limitation (Hauxwell et al. 2003). Light availability largely determines shoot density as well as the lower depth distributional limit, thus light attenuation effectively leads to diminished seagrass coverage (Krause-Jensen et al. 2000). Similarly, shading by phytoplankton blooms along the open US west coast are capable to cause light attenuation of a magnitude that is detrimental to growth of macrophytes such as kelp and surf grasses and may lead to their large-scale reduction (Kavanaugh et al. 2009). As a consequence, a movement of the lower depth-distributional limit of macrophytes towards shallow depths has commonly been found in highly eutrophied areas (Krause-Jensen et al. 2008).

Macroalgae and seagrasses not just grow and reproduce but they also need to survive under the grazing pressure of often highly abundant herbivores. Evolutionary history with strong grazing pressure has selected for defense adaptations, evolution of traits that decrease the probability of becoming eaten or diminish losses to herbivory. Defenses are particularly important for perennial, slow growing species as they need to survive for long periods and they cannot tolerate losses by growing fast (Coley et al. 1985, Endara and Coley 2011). Aquatic macroalgae, including brown algae contain a wide range of constitutive and/or inducible chemical defenses (reviewed by Cronin 2001, Paul et al. 2006, Toth and Pavia 2007, Jormalainen and Honkanen 2008). Less is known about chemical defenses of seagrasses but they commonly contain lots of phenolics (McMillan 1984), and chemical deterrence of phenolic seagrass extracts to a range of herbivores has been recently documented (Vergés et al. 2007).

Because of a fundamental physiological trade-off between the needs for growth and defenses, macrophytes respond plastically to variations in resource availability. Shortage of light generates carbon limitation, and when somatic needs and growth are prioritized, allocation to defenses is expected to decrease (Tuomi et al. 1988, Herms and Mattson 1992). Similarly, when mineral nutrient availability is high enough, photosynthesized carbon is primarily allocated to growth. Consequently, eutrophication (resulting in an increase in nutrient availability and a decrease in light) is predicted to cause a shift in resource allocation from defenses to growth, which may lead to higher palatability and grazing pressure on macrophytes. For example, Yates and Peckol (1993) showed that under high nutrient conditions, defensive responses of *F. vesiculosus* were significantly lower than under low nutrient conditions, and similarly, seagrass leafs of *Thalassia testudium* with high nitrogen content also were significantly less defended (Goecker et al. 2005). The prediction has gained support particularly from the responses of brown-algal phlorotannins to nutrient and light availability, although interpretations are often complicated by the indirect effects of nutrient manipulation and among-species differences in the responses (reviewed by Pavia and Toth 2008).

Opportunistic winners and perennial losers

Eutrophication favors fast-growing ephemeral macroalgae (Rönnberg et al. 1992, Hein et al. 1995, Pedersen and Borum 1996, Lotze and Schramm 2000, Raven and

Taylor 2003) that can speed up their growth, leading to a decline and competitive exclusion of perennial macrophytes (Borum and Sand-Jensen 1996, Valiela et al. 1997, Hauxwell et al. 2001, Raven and Taylor 2003, Steen 2004, Rohde et al. 2008, Moy and Christie 2012). Ephemeral opportunistic macroalgae include members of red, brown but mainly green algal species, and all of them are thin, delicately branched or sheet-like. That is why they feature a relatively high surface area to volume ratio for the uptake of nutrients as compared to the voluminous, thick, and leathery thalli of, e.g., fucoids or kelps (Fig. 4a). Physiological traits such as nutrient requirements, nutrient uptake and growth are scaled to the surface area to volume ratio of the algae such that thin algae require more nutrients and grow faster than thicker ones (Karez et al. 2004). Thus frequent pulses of nutrients or high concentrations of nutrients can be quickly exploited by these macroalgae (Plus et al. 2005, Kennison et al. 2011).

On the other hand, perennial forms have a low surface area to volume ratio and tend to be slow growers (Fig. 4b). However, fucoids, kelps, and seagrasses can take up and store nitrogen assimilating them into organic forms (Gerard 1982 for *Macrocystis pyrifera*, Invers et al. 2004 for *Posidonia oceanica*). Also, brown algae can store large amounts of carbon as mannitol and/or laminaran, using them as an energy source to support growth and metabolic activity over periods of low light availability (Sjotun and Gunnarsson 1995, Lehvo et al. 2001). Similarly, seagrasses store carbohydrates using

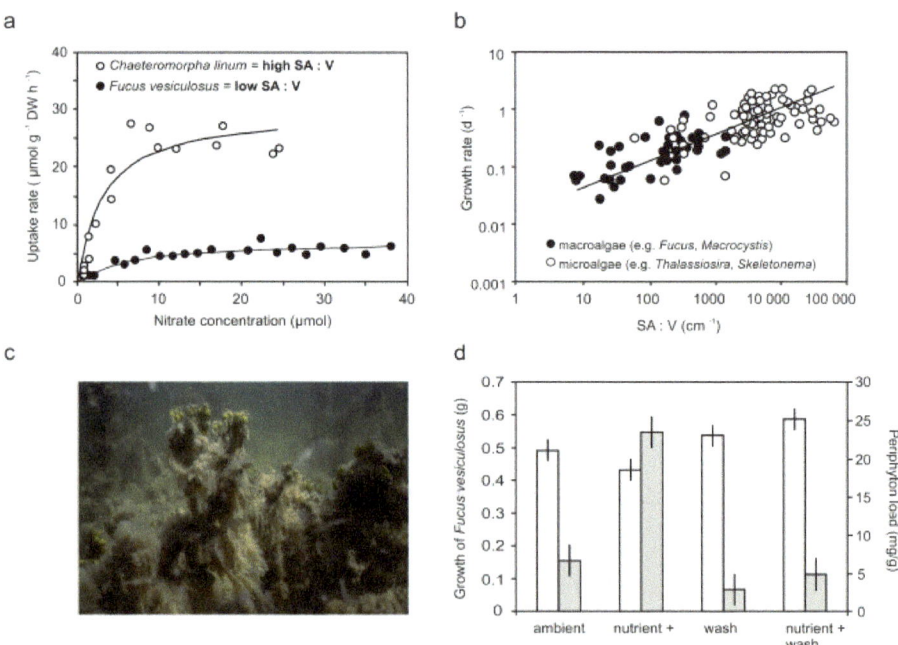

Figure 4. Nitrate uptake rate for filamentous (*Chaeteromorpha linum*) versus leathery macroalgae (*Fucus vesiculosus*) (a). Growth rate of macro- and micro-algae in relation to their surface area to volume ratio (SA : V) (b). Epiphytic overgrowth of *F. vesiculosus* in the Baltic Sea (c). Effect of different nutrient and cleaning treatments on the periphyton load (grey bars) and thus growth of adult *F. vesiculosus* (white bars) (d). Figures modified after Pedersen and Borum (1997), Rees et al. (2007) and Jormalainen et al. (2003).

them for overwintering and re-growth under conditions of negative carbon balance (Alcoverro et al. 2001b). Overall, remobilization of storage compounds from, e.g., old tissues and reallocation to new tissues (morphologically complex macroalgae), or translocation from, e.g., rhizomes and roots to shoots and leafs (seagrass), reduces the continuous demand of nutrient uptake from the environment. Consequently, slow-growing macrophytes are well adapted to low nutrient requirements but that asset is useless when nutrient supply that opportunistic macroalgae can exploit is constantly available. Moreover the broad tolerant ranges of opportunistic macroalgae to changes in abiotic factors such as radiation, temperature, desiccation, salinity and dissolved oxygen levels (Raffaelli et al. 1998, Taylor et al. 2001, Choi et al. 2010) makes them the winners over longer-living species.

Eutrophication promotes competition for settlement space owing to high biomass accumulations of filamentous algae and periphyton, sometimes exceeding 0.5 m in thickness (Peckol and Rivers 1996, Burkholder et al. 2007, Korpinen and Jormalainen 2008). Mats of decaying algae, originating from extensive growth of ephemeral algae may further limit colonization space particularly in the highly eutrophied Baltic Sea (Vahteri et al. 2000), but also elsewhere (e.g., Holmquist 1997, Merceron and Morand 2004), and can cause anoxia and have toxic effects on perennial macrophytes and animals (Eklund et al. 2005). Such drifting mats of filamentous algae have been observed to cover eelgrass meadows with a consequent 40% decrease in seedling survival (Valdemarsen et al. 2010). Settlement space does not become lost by opportunistic macroalgae alone but also by increased inorganic and organic sedimentation and siltation. Eriksson and Johansson (2003) showed experimentally that the ambient level of organic sedimentation in the Baltic Sea largely prevents colonization and hampers growth of *F. vesiculosus* propagules. Propagules were able to grow where the sediment was manually removed, showing that the lower distributional depth limit is affected by organic sedimentation. Sedimentation further affected the community composition because ephemeral species with continuous reproduction were more tolerant to sedimentation than the belt-forming perennial brown algae (Eriksson and Johansson 2005).

From space competition to epibiotism

Epibiotism can be considered as a form of space competition and has evolved as a consequence of surface limitation (Wahl 1989, Harder 2008), which increases with eutrophication. Algae and periphyton spreading epiphytically over thalli of habitat-forming macrophytes (Fig. 4c) may not only reduce the amount of light reaching macrophyte thalli (Sand-Jensen 1977, Brush and Nixon 2002, Mvungi et al. 2012), but also decrease the uptake rate of nutrients and hinder the exchange of gases at the leaf surface (Sand-Jensen et al. 1985, Hurd et al. 2000, Irlandi et al. 2004). The overall decrease of photosynthesis in optimal light conditions can be considerable; in eelgrass that was covered by a thick layer of epiphytic diatoms, Sand-Jensen (1977) measured a decrease of 45% in photosynthesis. In addition to reduced photosynthesis and consequent decrease in growth and reproduction, epibiotism may increase the risk of mechanical breakage and attract herbivores thereby increasing tissue loss

(Sand-Jensen 1977, Oswald et al. 1984, Wahl et al. 1997, Hurd et al. 2000, Brush and Nixon 2002, Jormalainen et al. 2003, Rohde et al. 2008, Krumhansl et al. 2011). These negative effects of epibiosis over their hosts can be even amplified in eutrophied regions (e.g., Worm and Sommer 2000, Hemmi et al. 2005, Russell et al. 2005). For example, high epibiotic load of calcified sessile animals due to increased nutrient availability has been suggested to lead to a higher frequency in ripping off algal parts and thus to a smaller size of the basibiont *F. vesiculosus* (Hemmi et al. 2005). Also, the amount of periphyton on *F. vesiculosus* was tripled when nutrients were added, thereby additionally reducing algal growth (Jormalainen et al. 2003). In contrast, nutrient addition can have a positive effect on the host growth, whenever periphyton became artificially removed, and consequently increased light availability (Fig. 4d).

Macrophytes have developed mechanisms to remove and/or suppress epibionts, via, e.g., surface-active chemical metabolites and/or the periodical shedding of their epidermis (Steinberg and de Nys 2002, Wahl 2008). These traits provide resistance to epibionts being thus ways to deal with their harmful consequences. Although epibiotism has increased with eutrophication and can negatively affect macrophyte performance (Sand-Jensen 1977, Oswald et al. 1984, Silberstein et al. 1986, Duarte 1995, Jaschinski and Sommer 2008a) it is not known whether this is solely due to epibionts capitalizing the chances of the increased nutrient availability or whether eutrophication hamper the resistance of macrophytes to epibiotism. For example, it has been suggested that the antifouling resistance of *F. vesiculosus* is compromised by light attenuation (Wahl et al. 2010), which may imply occurrence of negative feedbacks under eutrophied conditions (Wahl 2008).

Herbivory as the major regulating interaction

Herbivory is an important interaction in marine benthic habitats, where herbivores remove globally on average 68% of the abundance of benthic producers (Gruner et al. 2008, Poore et al. 2012). The strong plant-herbivore link in marine benthic habitats arises as a consequence of a good stoichiometric match between food and consumers, i.e., benthic producers in general provide a nutrient rich diet for grazers, and generalism being the dominant grazer strategy, i.e., grazing is spatially and temporally spread more evenly over the producer level (Shurin et al. 2006). The strong producer-herbivore link emphasizes the role of top-down regulation of producer abundance and it is also a precondition for the arousal of cascading effects from the higher trophic levels (see below). Although the grazing pressure on benthic producers is generally strong, a recent meta-analysis revealed lots of variation among different ecosystems and producer groups (Poore et al. 2012), with the strongest grazing effects found on intertidal and subtidal rocky reefs. Accordingly, grazing effects on the habitat-forming brown algae such as fucales, laminariales, and dictyotales were strong while those on seagrasses did not on average deviate from zero. Even though grazers had a minor direct effect on seagrasses, they played an important role in removing their epiphytic algae. The overwhelming role of herbivory in marine macrophytes infers that modulation of the producer-herbivore interaction may be an important indirect effect of eutrophication.

Grazers benefit from eutrophication

Eutrophication supports higher secondary production (Dolbeth et al. 2003, Worm and Lotze 2006, Korpinen et al. 2007c). Grazers may benefit from eutrophication through increasing food supply or increasing quality of the food. For example, Kangas et al. (1982) suggested that eutrophication by promoting growth of filamentous algae increases the amount of suitable juvenile habitat and food for the herbivorous isopod *Idotea balthica*, the populations of which have consequently increased to the level where their grazing pressure on *Fucus vesiculosus* decreased the abundance of the alga. The possible contribution of the quantitative increase of filamentous algae on population dynamics of *I. balthica* is still not known, but qualitative changes in the algae as food clearly play a role. Hemmi and Jormalainen (2004) showed that the quality of the filamentous green alga *Cladophora glomerata*, which is an important food source for juvenile *I. balthica*, increases when the alga grows in nutrient enriched conditions. The juveniles gained twice the weight of the control isopods in shorter time when fed with the nutrient enriched algae. Furthermore, adult *I. balthica* that was fed with *F. vesiculosus* reared in nutrient enriched conditions, gained more weight and laid more and heavier eggs than those fed with control algae (Hemmi and Jormalainen 2002). Similar results have been found for the amphipod *Gammarus locusta* (Kraufvelin et al. 2006) and the gastropod *Turbo castanea* (Baggett et al. 2013). Such eutrophication induced improvements in herbivore performance will increase the potential growth rate of herbivore populations, thus raising grazing pressure on producers.

Increasing herbivore abundance and grazing pressure on macrophytes and their recruits as a response to nutrient enrichment has been found in field experiments. In the heavily grazed seagrass habitats in the Caribbean Sea, nutrient enrichment leads to 30% increase in consumption of macroalgae (Boyer et al. 2004). In subtropical turtle grass meadows, grazing by herbivorous fish (Olsen and Valiela 2010) is stimulated by nutrient enrichment, and the epifaunal density of shrimp, isopods and amphipods can be twice the one in control plots (Gil et al. 2006). In the Baltic Sea, isopod herbivores have been observed to be capable to completely defoliate fucoid canopies thereby profoundly altering the littoral ecosystem (Engkvist et al. 2000, personal observation by V. Jormalainen). The effect of grazing can vary over small spatial scales: In the eutrophied Finnish Archipelago Sea, grazing pressure by the isopod *Idotea balthica* strongly decreased the number of apical tips of *F. vesiculosus* at 3 m depth but not at 1 m depth (Fig. 5a). Authors suggested that at 1 m depth wave disturbance prevents efficient grazing and algae can better compensate for grazing losses due to higher light availability (Jormalainen and Ramsay 2009). Grazing pressure also clearly reduced the colonization success of *F. vesiculosus*, as did the increased space competition with other algae due to nutrient enrichment (Fig. 5b, Korpinen et al. 2007c). Also, profound herbivory effects have been reported in kelp forests, where sea urchins (Estes et al. 2004), amphipods (Tegner and Dayton 1991, Graham 2002) or gastropods (Krumhansl and Scheibling 2011) may cause large biomass losses by direct consumption and indirectly by increasing blade tearing and loss. The role of eutrophication in boosting grazing pressure remains uncertain in some of these cases, but they clearly indicate the potential magnitude of the effects on macrophyte abundance arising through increasing herbivory.

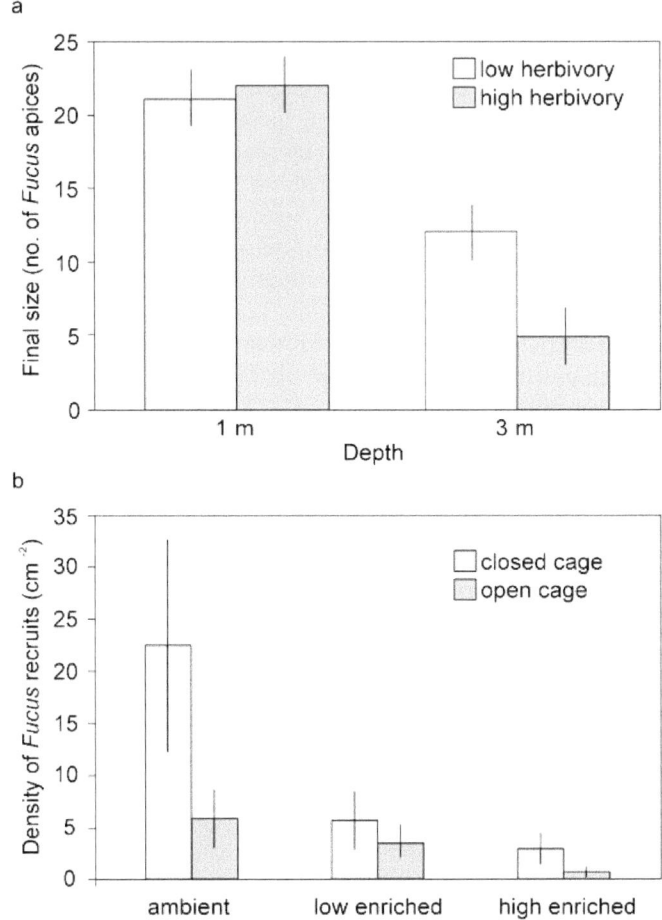

Figure 5. Effect of herbivore pressure on growth of adult *Fucus vesiculosus* along a depth gradient (a). Density of *F. vesiculosus* recruits under different nutrient and grazing treatments in shallow water areas (b). Herbivore treatments were obtained via cage enclosure for both experiments. Modified after Jormalainen and Ramsay (2009) and Korpinen et al. (2007c), respectively.

Herbivory counteracting eutrophication effects

The top-down regulation by herbivores and the bottom-up regulation by nutrient availability are both acting in concert and in an interactive way. Recent meta-analyses have shown that the magnitude of the herbivore effect may exceed that of the nutrient enrichment (although the comparison is difficult depending on the magnitude of manipulations), and most interestingly, the presence of herbivores typically dampens the nutrient enrichment effect (Burkepile and Hay 2006, Gruner et al. 2008). Herbivory, if strong enough can suppress blooms of opportunistic macroalgae. It may thereby counteract the increasing dominance of opportunistic competitors, thus benefitting perennial macrophytes by preventing competitive exclusion and by pre-emptying space for new settlers (Worm et al. 2000, Heck and Valentine 2007 and references therein). Grazers, thus, have an even more important role in eutrophied conditions. However, under high enough nutrient availability, grazer control has been found to collapse but the threshold when this will happen varies among systems differing in abundance and composition of the grazer community (Worm and Lotze 2006, Korpinen et al. 2007c).

Herbivores may have a particularly important role in removing epibiota from the surface of macrophytes thereby counteracting harmful effects of increased epibiotism. For example, in *Fucus vesiculosus* removal of periphyton from the thallus by small-sized gastropod grazers doubled the growth of the alga (Honkanen and Jormalainen 2005). Epiphyte removal has been found to have a particularly important role in seagrass habitats. A recent study by Verhoeven et al. (2012) showed for the seagrass *Posidonia australis* in plots with elevated nutrient levels and high manipulated grazer densities that the trochid gastropod *Calthalotia fragum* significantly reduced epiphytic growth, while the shoot growth of the seagrass and its photosynthesis increased. The opposite was true when grazer densities were low and the seagrass became shaded by epiphytes. A similar pattern has been reported for several other seagrass species (Neckles et al. 1993, Fong et al. 2000, reviewed by Hughes et al. 2004, Heck and Valentine 2006, Jaschinski and Sommer 2008b, Baden et al. 2010), implying that the magnitude of grazing in suppressing epiphytic biomass is an important factor for the productivity and persistence of habitat-forming seagrasses in eutrophied areas.

Changing trophic structures: Cascading top-down regulation versus eutrophication?

Eutrophication acting as a key factor from the bottom can closely interact with another global threat, the decline of apex predators (Cloern 2001, Myers and Worm 2003, Eriksson et al. 2009, Sieben et al. 2011a, Sieben et al. 2011b), thereby provoking a change of the relative role of different control mechanisms on macrophyte communities (Fig. 6a). The strength of top-down control is decreasing in many regions of the world and by removing apex predators intermediate predators are released, which can reduce the abundance and/or composition of mesograzers and subsequently increase the bloom of opportunistic macroalgae (Cloern 2001, Heck and Valentine 2007, Fox et al. 2012). Thus, human induced changes in cascading top-down trophic regulation may lead to a release of primary producer control by grazers. It has even been suggested that many negative effects on benthic communities that have conventionally been interpreted as consequences of eutrophication may, in fact, result from indirect effects of changes in upper trophic levels in the food webs (Heck and Valentine 2007). Rather than being a question of whether it is the top-down or bottom-up regulation that dominates, both are acting in concert. Thus eutrophication can promote the propagation of trophic cascades from apex predators to algal mass occurrence and outcompeting macrophytes (Fig. 6b). Along the coastline of the western Baltic Sea, mesograzer assemblages were strongly depressed by the release of three-spined sticklebacks, and this depression cascaded down to enhance the recruitment and consequently the biomass of opportunistic macroalgae (Fig. 6c). The mass development of these ephemeral macroalgae has been attributed to the additive effects of high nutrient levels and a removing of

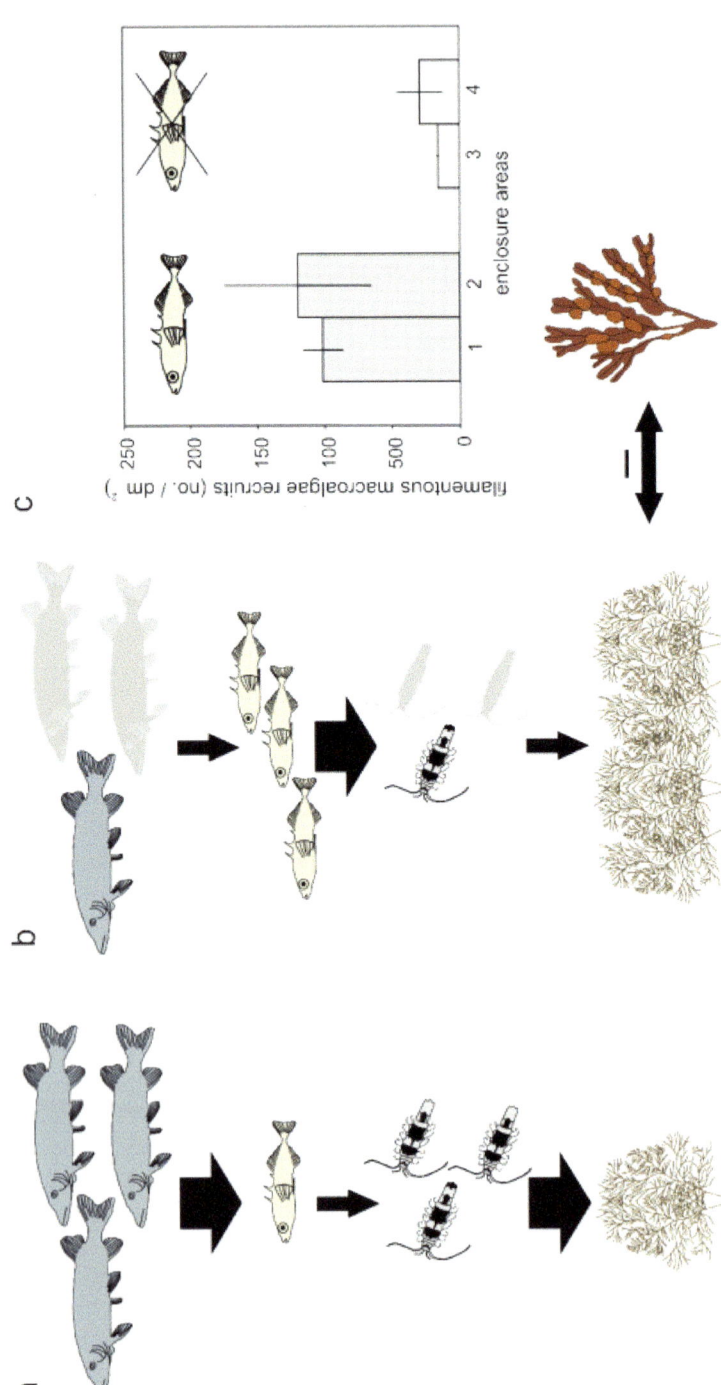

Figure 6. Hypothetical cascading top-down effects of apex predators on filamentous macroalgae and thus persistence of *Fucus vesiculosus* under (a) historical and (b) current apex fish densities in eutrophied areas of the Baltic Sea. Grey faded icons represent decreasing species and thick/thin arrows represent strong/low consumption pressure on species. Minus sign indicates competition effects between filamentous macroalgae and *F. vesiculosus*. Recruit densities of the filamentous *Cladophora* and *Ulva* for intermediate predator (three-spined stickleback) addition (grey bars) and removal (white bars) enclosure areas in a field experiment by Sieben et al. (2011a) (c). Figure (c) modified after Sieben et al. (2011a).

apex predatory fish (Eriksson et al. 2009, Sieben et al. 2011a, Sieben et al. 2011b). Subsequently, in regions where almost no invertebrate grazer control is present due to an intermediate predator release, the risk of a decline in the foundation species *Fucus vesiculosus* is intensified. Similarly, a shift in food web structure has been suggested to cause the decline of seagrass habitats in the west coast of Sweden (Baden et al. 2012). Overfishing of cod has led to high abundance of intermediate predatory fish (gobids, sticklebacks) and subsequently to the disappearance of idoteid and gammarid herbivores. In turn, the lack of mesoherbivores together with eutrophication has resulted in an overgrowth of seagrasses by filamentous algae and ultimately to a disappearance of seagrass meadows.

Nowadays fishing pressure on higher trophic levels and consequent high abundances of intermediate predators can play a decisive role in reducing invertebrate mesograzer densities, which in turn can release opportunistic algae from grazer control and let them bloom if nutrients are excessive and light is available (Eriksson et al. 2009, Sieben et al. 2011a, Sieben et al. 2011b, Baden et al. 2012) (Fig. 5c). Therefore, the degree of top-down control in structuring macrophyte assemblages is at the same time mediated by bottom-up mechanisms (Fox et al. 2012), i.e., the interactive effect of removal of apex predators and nutrient enrichment on standing biomass of the opportunistic species is synergistic.

Synthesis and outreach

Nutrient pollution can be found on both hemispheres from the tropics to temperate seas, particularly in coastal estuaries and bays, which are characterized by long water residence times. There it drives the growth of opportunistic macroalgal species and replaces important perennial foundation species such as kelps, rockweeds and seagrasses. The worldwide removal of apex predators (Dayton et al. 1995, Myers and Worm 2003) and their trophic regulatory function can act in concert with coastal eutrophication, thereby altering the overall regulation over macrophyte assemblages. Even though the present day effects of apex predators are greatly reduced from what they once were, they continue to shape and determine species abundance and composition (Heck and Valentine 2007). Therefore in order to preserve coastal macrophyte habitats and their ecosystem services, it is important to consider both the bottom-up and top-down regulatory factors.

To avoid local extinction and shrinking distributional ranges, macrophytes need to have adaptive potential, i.e., genetic variation in the traits relevant for eutrophication tolerance. Adaptive responses to selection arising from eutrophication will enable their persistence in eutrophied coastal estuaries and bays. In order to increase our predictive capability of eutrophication responses in macrophyte dominated areas, future research should examine the adaptive potential of macrophytes on population level, particularly their traits providing ecophysiological tolerance to multiple stressors. Of particular use will be studies about macrophyte resistance and tolerance to natural enemies and their interaction with nutrient and light availability.

Conservation efforts are needed not only in controlling anthropogenic nutrient inputs but also in preserving the still existent structure and functioning of the trophic network (Casini et al. 2008). The ecosystem responses to eutrophication will percolate through the network of biotic interactions and will therefore be functions of the organization of the community as a whole (Worm and Lotze 2006, Myers et al. 2007, Eriksson et al. 2009). For example, global variation in the effect of herbivory on macrophyte abundance may to some extent be related to seasonal variation in herbivore densities and their ability to temporarily escape from the predator control, i.e., cascading effects (Poore et al. 2012). Eutrophication and trophic release of mesopredators due to fisheries, especially when occurring jointly, can lead to ecosystem shifts from relatively stable perennial habitats supporting diverse communities to less diverse, and less stable communities of bloom forming macroalgae. While the pronounced role of herbivory in marine macrophyte assemblages is well documented, experiments about the importance of predation in structuring mesograzer assemblages in tropical and temperate systems, and their cascading impacts on primary producers remain an important challenge for future research. Based on such research, developing more accurate predictions for the joint and interactive effects of eutrophication and loss of apex predators will be highly valuable for an effective conservation and management of coastal macrophyte assemblages in this modern anomalous world.

References

Agawin, N.S.R., C.M. Duarte and M.D. Fortes. 1996. Nutrient limitation of Philippine seagrasses (Cape Bolinao, NW Philippines): *In situ* experimental evidence. Mar. Ecol. Prog. Ser. 138: 233–243.

Alcoverro, T., E. Cerbian and E. Ballesteros. 2001a. The photosynthetic capacity of the seagrass *Posidonia oceanica*: influence of nitrogen and light. J. Exp. Mar. Biol. Ecol. 261: 107–120.

Alcoverro, T., M. Manzanera and J. Romero. 2001b. Annual metabolic carbon balance of the seagrass *Posidonia oceanica*: the importance of carbohydrate reserves. Mar. Ecol. Prog. Ser. 211: 105–116.

Baden, S., C. Bostrom, S. Tobiasson, H. Arponen and P.O. Moksnes. 2010. Relative importance of trophic interactions and nutrient enrichment in seagrass ecosystems: A broad-scale field experiment in the Baltic-Skagerrak area. Limnol. Oceanogr. 55: 1435–1448.

Baden, S., A. Emanuelsson, L. Pihl, C.J. Svensson and P. Aberg. 2012. Shift in seagrass food web structure over decades is linked to overfishing. Mar. Ecol. Prog. Ser. 451: 61–73.

Benedetti-Cecchi, L., F. Pannacciulli, F. Bulleri, P.S. Moschella, L. Airoldi, G. Relini and F. Cinelli. 2001. Predicting the consequences of anthropogenic disturbance: large-scale effects of loss of canopy algae on rocky shores. Mar. Ecol. Prog. Ser. 214: 137–150.

Bennett, E.M., S.R. Carpenter and N.F. Caraco. 2001. Human impact on erodable phosphorus and eutrophication: A global perspective. Bioscience 51: 227–234.

Boesch, D.F. 2002. Challenges and opportunities for science in reducing nutrient over-enrichment of coastal ecosystems. Estuaries 25: 886–900.

Bokn, T.L., C.M. Duarte, M.F. Pedersen, N. Marbà, F.E. Moy, C. Barron, B. Bjerkeng, J. Borum, H. Christie, S. Engelbert, F.L. Fotel, E.E. Hoell, R. Karez, K. Kersting, P. Kraufvelin, C. Lindblad, M. Olsen, K.A. Sanderud, U. Sommer and K. Sorensen. 2003. The response of experimental rocky shore communities to nutrient additions. Ecosystems 6: 577–594.

Borum, J. and K. Sand-Jensen. 1996. Is total primary production in shallow coastal marine waters stimulated by nitrogen loading? Oikos 76: 406–410.

Boyer, K.E., P. Fong, A.R. Armitage and R.A. Cohen. 2004. Elevated nutrient content of tropical macroalgae increases rates of herbivory in coral, seagrass, and mangrove habitats. Coral Reefs 23: 530–538.

Brush, M.J. and S.W. Nixon. 2002. Direct measurements of light attenuation by epiphytes on eelgrass *Zostera marina*. Mar. Ecol. Prog. Ser. 238: 73–79.

Burkepile, D.E. and M.E. Hay. 2006. Herbivore vs. nutrient control of marine primary producers: Context-dependent effects. Ecology 87: 3128–3139.

Burkholder, J.M., D.A. Tomasko and B.W. Touchette. 2007. Seagrasses and eutrophication. J. Exp. Mar. Biol. Ecol. 350: 46–72.

Cardoso, P.G., M.A. Pardal, A.I. Lillebo, S.M. Ferreira, D. Raffaelli and J.C. Marques. 2004. Dynamic changes in seagrass assemblages under eutrophication and implications for recovery. J. Exp. Mar. Biol. Ecol. 302: 233–248.

Casini, M., J. Lovgren, J. Hjelm, M. Cardinale, J.C. Molinero and G. Kornilovs. 2008. Multi-level trophic cascades in a heavily exploited open marine ecosystem. Proc. R. Soc. Lond. B 275: 1793–1801.

Choi, T.S., E.J. Kang, J. Kim and K.Y. Kim. 2010. Effect of salinity on growth and nutrient uptake of *Ulva pertusa* (Chlorophyta) from an eelgrass bed. Algae 25: 17–26.

Cloern, J.E. 2001. Our evolving conceptual model of the coastal eutrophication problem. Mar. Ecol. Prog. Ser. 210: 223–253.

Coleman, M.A., B.P. Kelaher, P.D. Steinberg and A.J.K. Millar. 2008. Absence of a large brown macroalga on urbanized rocky reefs around Sydney, Australia, and evidence for historical decline. J. Phycol. 44: 897–901.

Coley, P.D., J.P. Bryant and I.I.I.F. Chapin. 1985. Resource Availability and Plant Antiherbivore Defence. Science 230: 895–899.

Connell, S.D., B.D. Russell, D.J. Turner, S.A. Shepherd, T. Kildea, D. Miller, L. Airoldi and A. Cheshire. 2008. Recovering a lost baseline: missing kelp forests from a metropolitan coast. Mar. Ecol. Prog. Ser. 360: 63–72.

Cronin, G. 2001. Resource allocation in seaweeds and marine invertebrates: chemical defense patterns in relation to defense theories. pp. 325–353. *In*: McClintock, J.B. and B.J. Baker (eds.). Marine Chemical Ecology. CRC Press.

Dayton, P.K., S.F. Thrush, M.T. Agardy and R.J. Hofman. 1995. Environmental effects of marine fishing. Aquat. Conserv. -Mar. Freshw. Ecosyst. 5: 205–232.

Diaz-Pulido, G. and L.J. McCook. 2005. Effects of nutrient enhancement on the fecundity of a coral reef macroalga. J. Exp. Mar. Biol. Ecol. 317: 13–24.

Dolbeth, M., M.A. Pardal, A.I. Lillebo, U. Azeiteiro and J.C. Marques. 2003. Short- and long-term effects of eutrophication on the secondary production of an intertidal macrobenthic community. Mar. Biol. 143: 1229–1238.

Duarte, C.M. 1995. Submerged aquatic vegetation in relation to different nutrient regimes. Ophelia 41: 87–112.

Eklund, B., A.P. Svensson, C. Jonsson and T. Malm. 2005. Toxic effects of decomposing red algae on littoral organisms. Estuar. Coast. Shelf Sci. 62: 621–626.

Endara, M.a.J. and P.D. Coley. 2011. The resource availability hypothesis revisited: a meta-analysis. Funct. Ecol. 25: 389–398.

Engkvist, R., T. Malm and S. Tobiasson. 2000. Density dependent grazing effects of the isopod *Idotea baltica* Pallas on *Fucus vesiculosus* L. in the Baltic Sea. Aquat. Ecol. 34: 253–260.

Eriksson, B.K. and G. Johansson. 2003. Sedimentation reduces recruitment success of *Fucus vesiculosus* (Phaeophyceae) in the Baltic Sea. Eur. J. Phycol. 38: 217–222.

Eriksson, B.K. and G. Johansson. 2005. Effects of sedimentation on macroalgae: species-specific responses are related to reproductive traits. Oecologia 143: 438–448.

Eriksson, B.K., G. Johansson and P. Snoeijs. 2002. Long-term changes in the macroalgal vegetation of the inner Gullmar Fjord, Swedish Skagerrak coast. J. Phycol. 38: 284–296.

Eriksson, B.K., L. Ljunggren, A. Sandstrom, G. Johansson, J. Mattila, A. Rubach, S. Raberg and M. Snickars. 2009. Declines in predatory fish promote bloom-forming macroalgae. Ecol. Appl. 19: 1975–1988.

Estes, J.A., E.M. Danner, D.F. Doak, B. Konar, A.M. Springer, P.D. Steinberg, M.T. Tinker and T.M. Williams. 2004. Complex trophic interactions in kelp forest ecosystems. Bull. Mar. Sci. 74: 621–638.

Foley, J.A., R. DeFries, G.P. Asner, C. Barford, G. Bonan, S.R. Carpenter, F.S. Chapin, M.T. Coe, G.C. Daily, H.K. Gibbs, J.H. Helkowski, T. Holloway, E.A. Howard, C.J. Kucharik, C. Monfreda, J.A. Patz, I.C. Prentice, N. Ramankutty and P.K. Snyder. 2005. Global consequences of land use. Science 309: 570–574.

Fong, C.W., S.Y. Lee and R.S.S. Wu. 2000. The effects of epiphytic algae and their grazers on the intertidal seagrass *Zostera japonica*. Aquat. Bot. 67: 251–261.

Fox, S.E., M. Teichberg, I. Valiela and L. Heffner. 2012. The relative role of nutrients, grazing, and predation as controls on macroalgal growth in the Waquoit Bay estuarine system. Estuar. Coast. 35: 1193–1204.

Galloway, J.N. and E.B. Cowling. 2002. Reactive nitrogen and the world: 200 years of change. Ambio 31: 64–71.

Gerard, V.A. 1982. Growth and utilization of internal nitrogen reserves by the giant kelp *Macrocystis pyrifera* in a low-nitrogen environment. Mar. Biol. 66: 27–35.

Gil, M., A.R. Armitage and J.W. Fourqurean. 2006. Nutrient impacts on epifaunal density and species composition in a subtropical seagrass bed. Hydrobiologia 569: 437–447.

Goecker, M.E., K.L. Heck and J.F. Valentine. 2005. Effects of nitrogen concentrations in turtlegrass *Thalassia testudinum* on consumption by the bucktooth parrotfish Sparisoma radians. Mar. Ecol. Prog. Ser. 286: 239–248.

Gorgula, S.K. and S.D. Connell. 2004. Expansive covers of turf-forming algae on human-dominated coast: the relative effects of increasing nutrient and sediment loads. Mar. Biol. 145: 613–619.

Graham, M.H. 2002. Prolonged reproductive consequences of short-term biomass loss in seaweeds. Mar. Biol. 140: 901–911.

Gruner, D.S., J.E. Smith, E.W. Seabloom, S.A. Sandin, J.T. Ngai, H. Hillebrand, W.S. Harpole, J.J. Elser, E.E. Cleland, M.E.S. Bracken, E.T. Borer and B.M. Bolker. 2008. A cross-system synthesis of consumer and nutrient resource control on producer biomass. Ecol. Lett. 11: 740–755.

Harder, T. 2008. Marine Epibiosis: Concepts, Ecological Consequences and Host Defence. Springer Berlin Heidelberg, pp. 1–13.

Hauxwell, J., J. Cebrian, C. Furlong and I. Valiela. 2001. Macroalgal canopies contribute to eelgrass (*Zostera marina*) decline in temperate estuarine ecosystems. Ecology 82: 1007–1022.

Hauxwell, J., J. Cebrian and I. Valiela. 2003. Eelgrass *Zostera marina* loss in temperate estuaries: relationship to land-derived nitrogen loads and effect of light limitation imposed by algae. Mar. Ecol. Prog. Ser. 247: 59–73.

Heck, K.L. and J.F. Valentine. 2006. Plant-herbivore interactions in seagrass meadows. J. Exp. Mar. Biol. Ecol. 330: 420–436.

Heck, K.L. and J.F. Valentine. 2007. The primacy of top-down effects in shallow benthic ecosystems. Estuar. Coast. 30: 371–381.

Hein, M., M.F. Pedersen and K. SandJensen. 1995. Size-dependent nitrogen uptake in micro- and macroalgae. Mar. Ecol. Prog. Ser. 118: 247–253.

Hemmi, A. and V. Jormalainen. 2002. Nutrient enhancement increases performance of a marine herbivore via quality of its food alga. Ecology 83: 1052–1064.

Hemmi, A. and V. Jormalainen. 2004. Genetic and environmental variation in performance of a marine isopod: Effects of eutrophication. Oecologia 140: 302–311.

Hemmi, A., A. Makinen, V. Jormalainen and T. Honkanen. 2005. Responses of growth and phlorotannins in *Fucus vesiculosus* to nutrient enrichment and herbivory. Aquat. Ecol. 39: 201–211.

Herms, D.A. and W.J. Mattson. 1992. The dilemma of plants: to grow or defend? Q. Rev. Biol. 67: 283–335.

Holmquist, J.G. 1997. Disturbance and gap formation in a marine benthic mosaic: influence of shifting macroalgal patches on seagrass structure and mobile invertebrates. Mar. Ecol. Prog. Ser. 158: 121–130.

Honkanen, T. and V. Jormalainen. 2005. Genotypic variation in tolerance and resistance to fouling in the brown alga *Fucus vesiculosus*. Oecologia 144: 196–205.

Hughes, A.R., K.J. Bando, L.F. Rodriguez and S.L. Williams. 2004. Relative effects of grazers and nutrients on seagrasses: a meta-analysis approach. Mar. Ecol. Prog. Ser. 282: 87–99.

Hurd, C.L., K.M. Durante and P.J. Harrison. 2000. Influence of bryozoan colonization on the physiology of the kelp *Macrocystis integrifolia* (Laminariales, Phaeophyta) from nitrogen-rich and -poor sites in Barkley Sound, British Columbia, Canada. Phycologia 39: 435–440.

Huston, M.A. and S. Wolverton. 2009. The global distribution of net primary production: resolving the paradox. Ecol. Monogr. 79: 343–377.

Invers, O., G.P. Kraemer, M. Perez and J. Romero. 2004. Effects of nitrogen addition on nitrogen metabolism and carbon reserves in the temperate seagrass *Posidonia oceanica*. J. Exp. Mar. Biol. Ecol. 303: 97–114.

Irlandi, E.A., B.A. Orlando and P.D. Biber. 2004. Drift algae-epiphyte-seagrass interactions in a subtropical *Thalassia testudinum* meadow. Mar. Ecol. Prog. Ser. 279: 81–91.

Jaschinski, S. and U. Sommer. 2008a. Top-down and bottom-up control in an eelgrass-epiphyte system. Oikos 117: 754–762.

Jaschinski, S. and U. Sommer. 2008b. Top-down and bottom-up control in an eelgrass-epiphyte system. Oikos 117: 754–762.

Jormalainen, V. and T. Honkanen. 2008. Macroalgal chemical defenses and their roles in structuring temperate marine communities. pp. 57–89. *In*: Amsler, C.D. (ed.). Algal Chemical Ecology. Springer.

Jormalainen, V. and T. Ramsay. 2009. Resistance of the brown alga *Fucus vesiculosus* to herbivory. Oikos 118: 713–722.

Jormalainen, V., T. Honkanen, R. Koivikko and J. Eränen. 2003. Induction of phlorotannin production in a brown alga: defense or resource dynamics? Oikos 103: 640–650.

Kangas, P., H. Autio, G. Hällfors, H. Luther, Å. Niemi and H. Salemaa. 1982. A general model of the decline of *Fucus vesiculosus* at Tvärminne, south coast of Finland in 1977–81. Acta Bot. Fennica 118: 1–27.

Karez, R., S. Engelbert, P. Kraufvelin, M.F. Pedersen and U. Sommer. 2004. Biomass response and changes in composition of ephemeral macroalgal assemblages along an experimental gradient of nutrient enrichment. Aquat. Bot. 78: 103–117.

Kautsky, N., H. Kautsky, U. Kautsky and M. Waern. 1986. Decreased depth penetration of *Fucus vesiculosus* (L.) since the 1940's indicates eutrophication of the Baltic Sea. Mar. Ecol. Prog. Ser. 28: 1–8.

Kavanaugh, M.T., K.J. Nielsen, F.T. Chan, B.A. Menge, R.M. Letelier and L.M. Goodrich. 2009. Experimental assessment of the effects of shade on an intertidal kelp: Do phytoplankton blooms inhibit growth of open-coast macroalgae? Limnol. Oceanogr. 54: 276–288.

Kennison, R.L., K. Kamer and P. Fong. 2011. Rapid nitrate uptake rates and large short-term storage capacities may explain why opportunistic green macroalgae dominate shallow eutrophic estuaries. J. Phycol. 47: 483–494.

Kenworthy, W.J. and M.S. Fonseca. 1992. The use of fertilizer to enhance growth of transplanted seagrasses *Zostera marina* l and *Halodule wrightii* Aschers. J. Exp. Mar. Biol. Ecol. 163: 141–161.

Korpinen, S. and V. Jormalainen. 2008. Grazing and nutrients reduce recruitment success of *Fucus vesiculosus* L. (Fucales: Phaeophyceae). Estuar. Coast. Shelf Sci. 78: 437–444.

Korpinen, S., T. Honkanen, O. Vesakoski, A. Hemmi, R. Koivikko, J. Loponen and V. Jormalainen. 2007a. Macroalgal communities face the challenge of changing biotic interactions: Review with focus on the Baltic Sea. Ambio 36: 203–211.

Korpinen, S., V. Jormalainen and T. Honkanen. 2007b. Bottom-up and cascading top-down control of macroalgae along a depth gradient. J. Exp. Mar. Biol. Ecol. 343: 52–63.

Korpinen, S., V. Jormalainen and T. Honkanen. 2007c. Effects of nutrients, herbivory, and depth on the macroalgal community in the rocky sublittoral. Ecology 88: 839–852.

Kraufvelin, P., S. Salovius, H. Christie, F.E. Moy, R. Karez and M.F. Pedersen. 2006. Eutrophication-induced changes in benthic algae affect the behaviour and fitness of the marine amphipod *Gammarus locusta*. Aquat. Bot. 84: 199–209.

Krause-Jensen, D., A.L. Middelboe, K. Sand-Jensen and P.B. Christensen. 2000. Eelgrass, *Zostera marina*, growth along depth gradients: upper boundaries of the variation as a powerful predictive tool. Oikos 91: 233–244.

Krause-Jensen, D., S. Sagert, H. Schubert and C. Bostrom. 2008. Empirical relationships linking distribution and abundance of marine vegetation to eutrophication. Ecological Indicators 8: 515–529.

Krumhansl, K.A. and R.E. Scheibling. 2011. Spatial and temporal variation in grazing damage by the gastropod *Lacuna vincta* in Nova Scotian kelp beds. Aquat. Biol. 13: 163–173.

Krumhansl, K.A., J.M. Lee and R.E. Scheibling. 2011. Grazing damage and encrustation by an invasive bryozoan reduce the ability of kelps to withstand breakage by waves. J. Exp. Mar. Biol. Ecol. 407: 12–18.

Lamote, M. and K.H. Dunton. 2006. Effects of drift macroalgae and light attenuation on chlorophyll fluorescence and sediment sulfides in the seagrass *Thalassia testudinum*. J. Exp. Mar. Biol. Ecol. 334: 174–186.

Lee, K.S. and K.H. Dunton. 1997. Effects of *in situ* light reduction on the maintenance, growth and partitioning of carbon resources in *Thalassia testudinum* Banks ex Konig. J. Exp. Mar. Biol. Ecol. 210: 53–73.

Lee, K.S., S.R. Park and Y.K. Kim. 2007. Effects of irradiance, temperature, and nutrients on growth dynamics of seagrasses: A review. J. Exp. Mar. Biol. Ecol. 350: 144–175.

Lehvo, A., S. Back and M. Kiirikki. 2001. Growth of *Fucus vesiculosus* L. (Phaeophyta) in the northern Baltic proper: Energy and nitrogen storage in seasonal environment. Bot. Mar. 44: 345–350.

Lobban, C.S. and P.J. Harrison. 1994. Seaweed Ecology and Physiology. Cambridge University Press, Cambridge.

Lotze, H.K. and W. Schramm. 2000. Can ecophysiological traits explain species dominance patterns in macroalgal blooms? J. Phycol. 36: 287–295.

Lotze, H.K., H.S. Lenihan, B.J. Bourque, R.H. Bradbury, R.G. Cooke, M.C. Kay, S.M. Kidwell, M.X. Kirby, C.H. Peterson and J.B.C. Jackson. 2006. Depletion, degradation, and recovery potential of estuaries and coastal seas. Science 312: 1806–1809.

Marba, N., C.M. Duarte, J. Cebrian, M.E. Gallegos, B. Olesen and K. SandJensen. 1996. Growth and population dynamics of *Posidonia oceanica* on the Spanish Mediterranean coast: Elucidating seagrass decline. Mar. Ecol. Prog. Ser. 137: 203–213.

Marba, N., C.M. Duarte, E. Diaz-Almela, J. Terrados, E. Alvarez, R. Martiinez, R. Santiago, E. Gacia and A.M. Grau. 2005. Direct evidence of imbalanced seagrass (*Posidonia oceanica*) shoot population dynamics in the Spanish Mediterranean. Estuaries 28: 53–62.

McGlathery, K.J. 1995. Nutrient and grazing influences on a subtropical seagrass community. Mar. Ecol. Prog. Ser. 122: 239–252.

McMillan, C. 1984. The condensed tannins (proanthocyanidins) in seagrasses. Aquat. Bot. 20: 351–357.

Merceron, M. and P. Morand. 2004. Existence of a deep subtidal stock of drifting *Ulva* in relation to intertidal algal mat developments. J. Sea Research 52: 269–280.

Moy, F.E. and H. Christie. 2012. Large-scale shift from sugar kelp (*Saccharina latissima*) to ephemeral algae along the south and west coast of Norway. Marine Biology Research 8: 309–321.

Mvungi, E.F., T.J. Lyimo and M. Björk. 2012. When *Zostera marina* is intermixed with *Ulva*, its photosynthesis is reduced by increased pH and lower light, but not by changes in light quality. Aquat. Bot. 102: 44–49.

Myers, R.A. and B. Worm. 2003. Rapid worldwide depletion of predatory fish communities. Nature 423: 280–283.

Myers, R.A., J.K. Baum, T.D. Shepherd, S.P. Powers and C.H. Peterson. 2007. Cascading effects of the loss of apex predatory sharks from a coastal ocean. Science 315: 1846–1850.

Neckles, H.A., R.L. Wetzel and R.J. Orth. 1993. Relative effects of nutrient enrichment and grazing on epiphyte-macrophyte (*Zostera marina* L.) dynamics. Oecologia 93: 285–295.

Nilsson, J., R. Engkvist and L.-E. Person. 2004. Long-term decline and recent recovery of *Fucus* populations along rocky shores of southeast Sweden, Baltic Sea. Aquat. Ecol. 38: 587–598.

Orth, R.J. and K.A. Moore. 1983. Chesapeake Bay—An unprecedented decline in submerged aquatic vegetation. Science 222: 51–53.

Orth, R.J., T.J.B. Carruthers, W.C. Dennison, C.M. Duarte, J.W. Fourqurean, K.L. Heck, A.R. Hughes, G.A. Kendrick, W.J. Kenworthy, S. Olyarnik, F.T. Short, M. Waycott and S.L. Williams. 2006. A global crisis for seagrass ecosystems. Bioscience 56: 987–996.

Oswald, R.C., N. Telford, R. Seed and C.M. Happeywood. 1984. The effect of encrusting bryozoans on the photosynthetic activity of *Fucus serratus* l. Estuar. Coast. Shelf Sci. 19: 697–702.

Paul, V.J., M.P. Puglisi and R. Ritson-Williams. 2006. Marine chemical ecology. Nat. Prod. Rep. 23: 153–180.

Pavia, H. and G. Toth. 2008. Macroalgal models in testing and extending defense theories. pp. 147–172. *In*: Amsler, C.D. (ed.). Algal Chemical Ecology. Springer.

Peckol, P. and J.S. Rivers. 1996. Contribution by macroalgal mats to primary production of a shallow embayment under high and low nitrogen-loading rates. Estuar. Coast. Shelf Sci. 43: 311–325.

Pedersen, M.F. 1995. Nitrogen limitation of photosynthesis and growth—Comparison across aquatic plant-communities in a Danish estuary (Roskilde Fjord). Ophelia 41: 261–272.

Pedersen, M.F. and J. Borum. 1996. Nutrient control of algal growth in estuarine waters. Nutrient limitation and the importance of nitrogen requirements and nitrogen storage among phytoplankton and species of macroalgae. Mar. Ecol. Prog. Ser. 142: 261–272.

Pedersen, M.F. and J. Borum. 1997. Nutrient control of estuarine macroalgae: growth strategy and the balance between nitrogen requirements and uptake. Mar. Ecol. Prog. Ser. 161: 155–163.

Phillips, J.A. and J.K. Blackshaw. 2011. Extirpation of macroalgae (*Sargassum* spp.) on the subtropical East Australian coast. Conserv. Biol. 25: 913–921.

Plinski, M. and I. Florczyk. 1984. Changes in the phytobenthos resulting from the eutrophication of the Puck Bay. Limnologica 15: 325–327.

Plus, M., I. Auby, M. Verlaque and G. Levavasseur. 2005. Seasonal variations in photosynthetic irradiance response curves of macrophytes from a Mediterranean coastal lagoon. Aquat. Bot. 81: 157–173.

Poore, A.G.B., A.H. Campbell, R.A. Coleman, J.E. Duffy, G.J. Edgar, V. Jormalainen, P.L. Reynolds, E.E. Sotka, J.J. Stachowicz, R.B. Taylor and M.A. Vanderklift. 2012. Global patterns in the impact of marine herbivores on benthic primary producers. Ecol. Lett. 15: 912–922.

Raffaelli, D.G., J.A. Raven and L.J. Poole. 1998. Ecological impact of green macroalgal blooms. Oceanogr. Mar. Biol. Ann. Rev. 36: 97–125.

Ralph, P.J., S.M. Polk, K.A. Moore, R.J. Orth and W.O. Smith. 2002. Operation of the xanthophyll cycle in the seagrass *Zostera marina* in response to variable irradiance. J. Exp. Mar. Biol. Ecol. 271: 189–207.

Raven, J. and R. Taylor. 2003. Macroalgal growth in nutrient-enriched estuaries: A biogeochemical and evolutionary perspective. Water, Air, & Soil Pollution: Focus 3: 7–26.

Rees, T.A.V. 2007. Metabolic and ecological constraints imposed by similar rates of ammonium and nitrate uptake per unit surface area at low substrate concentrations in marine phytoplankton and macroalgae. J. Phycol. 43: 197–207.

Rohde, S., C. Hiebenthal, M. Wahl, R. Karez and K. Bischof. 2008. Decreased depth distribution of *Fucus vesiculosus* (Phaeophyceae) in the Western Baltic: effects of light deficiency and epibionts on growth and photosynthesis. Eur. J. Phycol. 43: 143–150.

Ronnberg, O. 1984. Recent changes in the distribution of *Fucus vesiculosus* around the Aland Islands (N-Baltic). Ophelia 189–193.

Rönnberg, O., K. Ådjers, C. Ruokolahti and M. Bondestam. 1992. Effects of fish farming on growth, epiphytes and nutrient content of *Fucus vesiculosus* L. in the Åland archipelago, northern Baltic Sea. Aquat. Bot. 42: 109–120.

Ruiz, J.M. and J. Romero. 2001. Effects of in situ experimental shading on the Mediterranean seagrass *Posidonia oceanica*. Mar. Ecol. Prog. Ser. 215: 107–120.

Russell, B.D., T.S. Elsdon, B.M. Gillanders and S.D. Connell. 2005. Nutrients increase epiphyte loads: broad-scale observations and an experimental assessment. Mar. Biol. 147: 551–558.

Sand-Jensen, K. 1977. Effect of epiphytes on eelgrass photosynthesis. Aquat. Bot. 3: 55–63.

Sand-Jensen, K., N.P. Revsbech and B.B. Jorgensen. 1985. Microprofiles of oxygen in epiphyte communities on submerged macrophytes. Mar. Biol. 89: 55–62.

Sand-Jensen, K., T. Binzer and A.L. Middelboe. 2007. Scaling of photosynthetic production of aquatic macrophytes—a review. Oikos 116: 280–294.

Schmidt, A.L., J.K.C. Wysmyk, S.E. Craig and H.K. Lotze. 2012. Regional-scale effects of eutrophication on ecosystem structure and services of seagrass beds. Limnol. Oceanogr. 57: 1389–1402.

Schories, D., A. Albrecht and H. Lotze. 1997. Historical changes and inventory of macroalgae from Koenigshafen Bay in the northern Wadden Sea. Helgolander Meeresuntersuchungen 51: 341.

Short, F.T., E.W. Koch, J.C. Creed, K.M. Magalhaes, E. Fernandez and J.L. Gaeckle. 2006. Seagrass Net monitoring across the Americas: case studies of seagrass decline. Marine Ecology-An Evolutionary Perspective 27: 277–289.

Shurin, J.B., D.S. Gruner and H. Hillebrand. 2006. All wet or dried up? Real differences between aquatic and terrestrial food webs. Proc. R. Soc. Lond. B 273: 1–9.

Sieben, K., L. Ljunggren, U. Bergstrom and B.K. Eriksson. 2011a. A meso-predator release of stickleback promotes recruitment of macroalgae in the Baltic Sea. J. Exp. Mar. Biol. Ecol. 397: 79–84.

Sieben, K., A.D. Rippen and B.K. Eriksson. 2011b. Cascading effects from predator removal depend on resource availability in a benthic food web. Mar. Biol. 158: 391–400.

Silberstein, K., A.W. Chiffings and A.J. Mccomb. 1986. The loss of Seagrass in Cockburn sound, Western Australia. 3. The effect of epiphytes on productivity of *Posidonia australis* Hook F. Aquat. Bot. 24: 355–371.

Sjotun, K. and K. Gunnarsson. 1995. Seasonal growth pattern of an Icelandic *Laminaria* population (section Simplices, Laminariaceae, Phaeophyta) containing solid and hollow-stiped plants. Eur. J. Phycol. 30: 281–287.

Steen, H. 2004. Interspecific competition between *Enteromorpha* (Ulvales : Chlorophyceae) and *Fucus* (Fucales : Phaeophyceae) germlings: effects of nutrient concentration, temperature, and settlement density. Mar. Ecol. Prog. Ser. 278: 89–101.

Steinberg, P.D. and R. de Nys. 2002. Chemical mediation of colonization of seaweed surfaces. J. Phycol. 38: 621–629.

Tamaki, H., M. Tokuoka, W. Nishijima, T. Terawaki and M. Okada. 2002. Deterioration of eelgrass, *Zostera marina* L., meadows by water pollution in Seto Inland Sea, Japan. Mar. Pollut. Bul. 44: 1253–1258.

Taylor, R., R.L. Fletcher and J.A. Raven. 2001. Preliminary studies on the growth of selected 'Green tide' algae in laboratory culture: Effects of irradiance, temperature, salinity and nutrients on growth rate. Bot. Mar. 44: 327–336.

Tegner, M.J. and P.K. Dayton. 1991. Sea urchins, El Ninos, and the long-term stability of southern California kelp forest communities. Mar. Ecol. Prog. Ser. 77: 49–63.

Tilman, D., J. Fargione, B. Wolff, C. D'Antonio, A. Dobson, R. Howarth, D. Schindler, W.H. Schlesinger, D. Simberloff and D. Swackhamer. 2001. Forecasting agriculturally driven global environmental change. Science 292: 281–284.

Tomasko, D.A., C.J. Dawes and M.O. Hall. 1996. The effects of anthropogenic nutrient enrichment on turtle grass (*Thalassia testudinum*) in Sarasota Bay, Florida. Estuaries 19: 448–456.

Toth, G.B. and H. Pavia. 2007. Induced herbivore resistance in seaweeds: a meta-analysis. J. Ecol. 95: 425–434.

Tuomi, J., P. Niemelä, F.S. Chapin III, J.P. Bryant and S. Sirén. 1988. Defensive responses of trees in relation to their carbon/nutrient balance. pp. 57–72. *In*: Mattson, W.J., J. Levieux and C. Bernard-Dagan (eds.). Mechanisms of Woody Plant Defenses against Insects: Search for Pattern. Springer.

Turpin, D.H., G.C. Vanlerberghe, A.M. Amory and R.D. Guy. 1991. The inorganic carbon requirements for nitrogen assimilation. Canadian Journal of Botany-Revue Canadienne de Botanique 69: 1139–1145.

Udy, J.W. and W.C. Dennison. 1997. Growth and physiological responses of three seagrass species to elevated sediment nutrients in Moreton Bay, Australia. J. Exp. Mar. Biol. Ecol. 217: 253–277.

Vahteri, P., A. Makinen, S. Salovius and I. Vuorinen. 2000. Are drifting algal mats conquering the bottom of the Archipelago Sea, SW Finland? Ambio 29: 338–343.

Valdemarsen, T., P. Canal-Verges, E. Kristensen, M. Holmer, M.D. Kristiansen and M.R. Flindt. 2010. Vulnerability of *Zostera marina* seedlings to physical stress. Mar. Ecol. Prog. Ser. 418: 119–130.

Valentine, J.F. and J.E. Duffy. 2006. The central role of grazing in seagrass ecology. pp. 463–501. *In*: Larkum, A.W.D., R.J. Orth and C.M. Duarte (eds.). Seagrasses: Biology, Ecology and Conservation. Springer, Netherlands.

Valiela, I., J. McClelland, J. Hauxwell, P.J. Behr, D. Hersh and K. Foreman. 1997. Macroalgal blooms in shallow estuaries: Controls and ecophysiological and ecosystem consequences. Limnol. Oceanogr. 42: 1105–1118.

van Katwijk, M.M., A.R. Bos, P. Kennis and R. de Vries. 2010. Vulnerability to eutrophication of a semi-annual life history: A lesson learnt from an extinct eelgrass (*Zostera marina*) population. Biological Conservation 143: 248–254.

Vergés, A., M.A. Becerro, T. Alcoverro and J. Romero. 2007. Experimental evidence of chemical deterrence against multiple herbivores in the seagrass *Posidonia oceanica*. Mar. Ecol. Prog. Ser. 343: 107–114.

Verhoeven, M.P.C., B.P. Kelaher, M.J. Bishop and P.J. Ralph. 2012. Epiphyte grazing enhances productivity of remnant seagrass patches. Austral. Ecology 37: 885–892.

Vitousek, P.M., J.D. Aber, R.W. Howarth, G.E. Likens, P.A. Matson, D.W. Schindler, W.H. Schlesinger and D. Tilman. 1997. Human alteration of the global nitrogen cycle: Sources and consequences. Ecol. Appl. 7: 737–750.

Vogt, H. and W. Schramm. 1991. Conspicuous decline of *Fucus* in Kiel Bay (Western Baltic)—What are the causes. Mar. Ecol. Prog. Ser. 69: 189–194.

Wahl, M. 1989. Marine epibiosis. 1. fouling and antifouling—Some basic aspects. Mar. Ecol. Prog. Ser. 58: 175–189.

Wahl, M. 2008. Ecological lever and interface ecology: epibiosis modulates the interactions between host and environment. Biofouling 24: 427–438.

Wahl, M., M.E. Hay and P. Enderlein. 1997. Effects of epibiosis on consumer-prey interactions. Hydrobiologia 355: 49–59.

Wahl, M., L. Shahnaz, S. Dobretsov, M. Saha, F. Symanowski, K. David, T. Lachnit, M. Vasel and F. Weinberger. 2010. Ecology of antifouling resistance in the bladder wrack *Fucus vesiculosus*: patterns of microfouling and antimicrobial protection. Mar. Ecol. Prog. Ser. 411: 33–U61.

Walker, D.I., G.A. Kendrick and A. McComb. 2006. Decline and recovery of seagrass ecosystems—the dynamics of change. pp. 551–565. *In*: Larkum, A.W.D., R.J. Orth and C.M. Duarte (eds.). Seagrasses: Biology, Ecology and Conservation. Springer.

Waycott, M., C.M. Duarte, T.J.B. Carruthers, R.J. Orth, W.C. Dennison, S. Olyarnik, A. Calladine, J.W. Fourqurean, K.L. Heck, A.R. Hughes, G.A. Kendrick, W.J. Kenworthy, F.T. Short and S.L. Williams. 2009. Accelerating loss of seagrasses across the globe threatens coastal ecosystems. Proc. Nat. Acad. Sci. USA 106: 12377–12381.

Worm, B. and U. Sommer. 2000. Rapid direct and indirect effects of a single nutrient pulse in a seaweed-epiphyte-grazer system. Mar. Ecol. Prog. Ser. 202: 283–288.

Worm, B. and H.K. Lotze. 2006. Effects of eutrophication, grazing, and algal blooms on rocky shores. Limnol. Oceanogr. 51: 569–579.

Worm, B., H.K. Lotze and U. Sommer. 2000. Coastal food web structure, carbon storage, and nitrogen retention regulated by consumer pressure and nutrient loading. Limnol. Oceanogr. 45: 339–349.

Yates, J.C. and P. Peckol. 1993. Effects of nutrient availability and herbivory on polyphenolics in the seaweed *Fucus vesiculosus*. Ecology 74: 1757–1766.

9

Threats to Ecosystem Engineering Macrophytes: Climate Change

Thomas Wernberg,[1,]* *Francisco Arenas,*[2] *Celia Olabarria,*[3]
Mads S. Thomsen[4] and *Margaret B. Mohring*[1]

Warming and other climate changes: an overview

Anthropogenic activities are impacting Earth's climate dynamics, causing rapid global climate change and increasing climate variability (IPCC 2007, 2012). Coastal marine ecosystems, which are among the most ecologically and socio-economically important on the planet (Costanza et al. 1997, Barbier et al. 2011), are already experiencing the effects of this physical forcing, with impacts on biodiversity and ecosystem functioning at multiple spatial and temporal scales (Harley et al. 2006, Hoegh-Guldberg and Bruno 2010, Wernberg et al. 2011b, Doney et al. 2012).

Although geological records show climatic changes throughout the history of the Earth, anthropogenic activities have been causing a rapid increase in atmospheric carbon dioxide (Hansen et al. 2006, IPCC 2007). This increase in CO_2, in combination with other greenhouse gases, has intensified the infrared opacity of the atmosphere and

---

[1] UWA Oceans Institute & School of Plant Biology, The University of Western Australia 35 Stirling Highway, Nedlands 6009 WA, Australia.
[2] Laboratory of Coastal Biodiversity - CIIMAR, University of Porto, Rua dos Bragas 289, 4050-123 Porto, Portugal.
[3] Departamento de Ecoloxía e Bioloxía Animal, Facultade de Ciencias Experimentais, Universidade de Vigo, Campus Lagoas-Marcosende, 26310, Vigo, Spain.
[4] Marine Ecology Research Group, School of Biological Sciences, University of Canterbury, Private Bag 4800, Christchurch, New Zealand.
* Coresponding author: thomas.wernberg@uwa.edu.au

is generally believed to be the major cause of warming on the Earth's surface (Sabine and Tanhua 2010). In May 2013, CO_2 levels reached 400 ppm for the first time in human history, in fact, the last time in Earth's history that CO_2 concentrations were this high was 4.5 million years ago (National Oceanic and Atmospheric Administration 2013). Since 1960, mean land temperatures have increased by as much as 0.24°C decade^{-1} (Hansen et al. 2006, Burrows et al. 2011). Oceans are currently taking up > 80% of the heat added to the atmosphere, and sea temperature consequently increases as the atmosphere warms. Over the past century, the mean global rate of ocean temperature increase has been 0.06–0.07°C decade^{-1} although there has been pronounced heterogeneity in time and space (Burrows et al. 2011, Lima and Wethey 2012, Wu et al. 2012). In addition, the changing atmospheric conditions are also forcing changes in global wind patterns that are driving ocean circulation. This has resulted in a general strengthening of western boundary currents (Gulf Stream, Kuroshio Current, East Australia Current, Brazil Current and Agulhas Current) which carry warm tropical water towards temperate latitudes, further contributing to the net warming. The fastest warming ocean regions (0.09–0.23°C decade^{-1}) have therefore been found within these poleward-flowing currents (Wu et al. 2012). The increase in global sea temperature is accompanied by changes to patterns and strength of winds and ocean currents, upwelling and the associated nutrient supply, ocean stratification, a rise in sea levels, and changes in rainfall dynamics.

Oceans play a critical role in buffering the effects of increasing atmospheric CO_2 concentrations. In the last two centuries, around 50% of anthropogenic CO_2 emissions (118 ± 19 Pg C, each Pg C is 10^{15} g of Carbon) have been absorbed by the oceans. The increase in dissolved CO_2 in the oceans has increased bicarbonate (HCO_3^-) concentrations, decreased the concentration of carbonate ions (CO_3^{2-}) and caused a decline in pH of 0.02 units per decade over the past three decades (Hepburn et al. 2011; Doney et al. 2012).

On top of the gradual shift in mean environmental conditions, there have been changes to the occurrence of extreme events (IPCC 2012, Trenberth 2012). In their recent analysis of satellite derived global sea surface temperatures for the last 30 years, Lima and Wethey (2012) found that 46% and 38% of the worlds' coastlines have experienced a significant decrease in the frequency of extremely cold days and an increase in extremely hot days, respectively. Also, in 36% of temperate coastal regions the onset of the warm season was found to be advancing earlier in the year. Increases in extreme rainfall occurrences have resulted from the rising amount of water vapour in a warmer atmosphere (Trenberth 2012), modifying coastal salinity and stratification as well as increasing terragenous sediment and nutrient runoff. Climate change models predict a further rise in the frequency and intensity of extreme events such as storms, including cyclones and hurricanes, and increasing wave heights, flooding and heat waves (IPCC 2012).

All of these physical and chemical changes interplay to impact marine macrophytes both directly and indirectly (Fig. 1) through changes in physiology, abundance, distribution, phenology and species interactions (Poloczanska et al. 2007, Kordas et al. 2011, Wernberg et al. 2011a, Gao et al. 2012, Harley et al. 2012, Richardson et al. 2012).

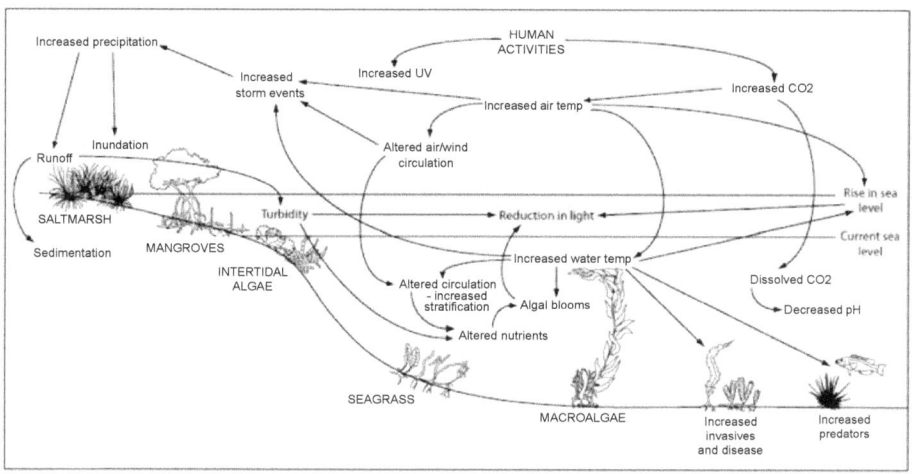

Figure 1. A conceptual model of the flow of impacts on ecosystem engineering macrophytes resulting from human activities. Anthropogenic activities have led to carbon dioxide accumulation in the atmosphere which has driven atmospheric and ocean warming, changes in circulation patterns, a rise in sea level, changes in ocean chemistry, and increased storm events. This diagram represents all major direct, and resulting indirect stressors, impacting marine macrophytes.

Climate change is not the only peril for marine macrophytes. The worlds' oceans are simultaneously affected by multiple human activities, such as eutrophication, fishing, habitat destruction, hypoxia, pollution, and species introductions (Halpern et al. 2008, Wernberg et al. 2011b). Climate change, therefore, cannot be understood in isolation from these multiple stressors because complex interactions often determine the responses of organisms and ecosystems (Crain et al. 2008a, Wernberg et al. 2011b). Effects of climate change stressors can be subtle, but synergistic effects often causes loss of resilience making ecosystems more vulnerable to changes that previously could be absorbed (Folke et al. 2004, Ling et al. 2009a, Connell and Russell 2010, Wernberg et al. 2010).

Ecosystem engineering macrophytes

Marine macrophytes, loosely defined as large primary producers in habitats substantially affected by salt water from the sea, play a critical role in most coastal ecosystems. Although marine macrophytes are very diverse taxonomically and functionally, they can be divided into four major groups: seaweeds, seagrasses, mangroves and salt marshes (Fig. 2).

Seaweeds (macroalgae) are found in most coastal habitats from tropical coral reefs to temperate and polar kelp forests (Mann 1973, Steneck et al. 2002). Usually seaweed habitats are associated with hard bottoms, although some soft bottom habitats can be dominated by unattached (e.g., *Ulva*, *Gracilaria*) or attached (e.g., *Caulerpa*) seaweeds (Thomsen et al. 2012). Even though seaweeds are generally restricted to coastal areas, they contribute about 50 Pg C year^{-1}, or approximately 5%, of the total global marine productivity (Duarte and Cebrian 1996, Israel et al. 2010). Seaweed

Figure 2. Examples of marine ecosystem engineers macrophytes. (A) Mangrove (M. Rule), (B) Salt marsh (C. Olabarria), (C) Intertidal seaweeds (M. Rule), (D) Seagrass bed (T. Wernberg), (E) Tropical Sargassum bed (T. Wernberg), (F) Kelp forest (T. Wernberg).

habitats are particularly sensitive to direct disturbances from high temperatures (Tanaka et al. 2012, Smale and Wernberg 2013) or storms (Filbee-Dexter and Scheibling 2012) as well as altered species interactions such as increased rates of herbivory (Steneck et al. 2002, Ling et al. 2009a, Franco et al. 2015, Verges et al. 2014). Currently, seaweeds are generally not carbon limited, and increasing CO_2 levels are likely to have little direct effect on many seaweeds (Beardall et al. 1998, but see Connell and Russell 2010). However, decreasing pH associated with increased CO_2 levels in seawater may have longer-term physiological, mechanical and structural effects on calcifying seaweeds (Kuffner et al. 2008).

Seagrasses are the dominant ecosystem engineers in many intertidal and shallow subtidal soft-bottom ecosystems from tropical to sub-polar environments (Larkum et al. 1989). Seagrass ecosystems are among the most ecologically and economically valuable in the world (Short and Neckles 1999, Connolly 2012). Seagrass are also highly productive and are recognized as significant ecosystems in most coastal zones. Since seagrasses inhabit shallow coastal waters they are highly sensitive to fluctuations in light, temperature, CO_2, nutrients and sediment variability (Short and Neckles 1999, Connolly 2012). Physical damage and mass declines in seagrass populations are evident worldwide as a result of anthropogenic climate change (Short and Neckles 1999, Duarte 2002, Orth et al. 2006, Waycott et al. 2009).

Mangroves are salt-tolerant terrestrial forests located at the interface between land and sea, predominantly along tropical and subtropical coastlines (Nagelkerken et al. 2008). These habitats occupy a total area of 181000 km^2 worldwide and are ecologically highly significant to marine systems (Alongi 2002). Their distribution is controlled by thermal regimes, tidal patterns and major ocean currents. Nevertheless, sea-level rise is probably the greatest climate change challenge that mangrove ecosystems face. Geological records indicate that previous sea-level fluctuations have created both crises and opportunities for mangrove communities (Field 1998). Mangroves can adapt to sea-level rise if it is slow enough, if adequate expansion space exists, and if other environmental conditions are met. Unfortunately, in many areas retreat is not possible due to urbanisation, increasing farmland and human utilisation of the coastal zone behind mangrove and saltmarsh communities. Consequently, mangroves experience a 'coastal squeeze' as sea levels rise. Loss of mangrove forest area will increase the threat to shoreline development, as well as reduce coastal water quality, biodiversity, nursery habitat, and result in major loss of coastal ecosystem function and damage to complex food-webs (Gilman et al. 2008).

Salt marshes are complex coastal systems made up of a diverse range of salt-tolerant plants which colonize wave-protected shores in temperate regions. The halophytic vegetation within a salt marsh bind sediments, limit water motion and provide vital nursery habitat and protection for many commercially and ecologically important species (Simas et al. 2001, Bertness et al. 2004). Salt marsh formation is dependent on a complex set of conditions, where the relationship between run-off, inundation, sediment type and protection from exposure must be well balanced. Salt marshes are particularly vulnerable to sea-level rise, but increased storms, precipitation and changed sediment and nutrient regimes will also limit salt marsh performance. Loss of salt marshes will have negative implications for inhabiting communities, as well as recruitment from nursery habitats to other marine areas. A reduction in salt marshes will also lead to a loss of protection of coastlines and human development from erosion and storm damage (Simas et al. 2001).

Marine macrophyte habitats have some of the highest rates of primary production on Earth (Fig. 3). Marine macrophytes are essential ecological engineers, with important structuring and functioning roles, serving as food resources and providing

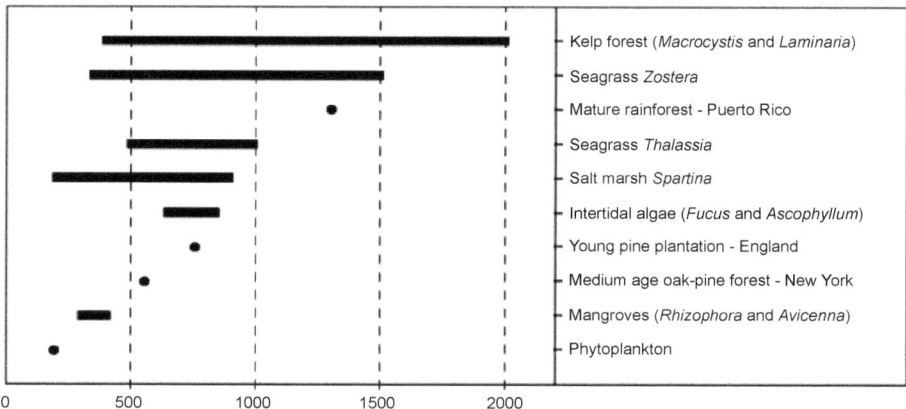

Figure 3. Net annual primary production (g C m⁻² year⁻¹) of the major marine macrophyte systems, compared with some terrestrial communities. Redrawn from Mann (1973).

physical habitat or nurseries for aquatic organisms, such as invertebrates, fish and birds (Beck et al. 2001, Hyndes et al. 2003, Bertness et al. 2004, Graham 2004, Wernberg et al. 2004, Nagelkerken et al. 2008, Thomsen et al. 2010). Analyses of the carbon budgets of marine vegetated systems indicate that they export significant amounts of organic carbon to adjacent ecosystems (Cebrian 2002, Krumhansl and Scheibling 2012) and also store great amounts of organic carbon in associated sediments (Duarte et al. 2005, Fourqurean et al. 2012). Alterations to these systems will have important consequences for the marine carbon budget, food-webs, coastal geomorphology and biodiversity.

Because of their ability to moderate environmental conditions (i.e., engineer the ecosystem) and reduce stress on associated organisms, ecosystem engineering macrophytes will become increasingly important to the maintenance of ecosystem function in a warmer, more stressful, future (Halpern et al. 2007).

Responses of marine macrophytes to climate change

Knowledge of how marine macrophytes have responded, and will continue to respond, to the changing climate is important in order to understand ecosystem level impacts and to provide guidelines to manage and conserve marine communities. Yet, despite their substantial ecological and socio-economic importance, only ~ 20% of the combined experimental effort in climate change experiments has been directed towards primary producers (Wernberg et al. 2012a). Consequently, there is a relatively poor mechanistic understanding behind observed climate impacts on marine macrophytes.

Temperature

Changes in temperature directly influence the physiology of marine macrophytes. However, all physiological processes are not equally responsive to temperature change. For example, respiration is typically more temperature sensitive than photosynthesis (e.g., Stæhr and Wernberg 2009), and so the metabolic 'costs of survival' increase faster than primary production. Where marine organisms live close to their thermal tolerance limits even relatively small increases in temperature can have dramatic effects on physiological processes, such as increased metabolic rates, cell protein damage and reduction in membrane fluidity (Eggert 2012), disruption to enzymatic processes, reduced nutrient uptake, photosynthetic inhibition, and respiratory stress (Davison and Pearson 1996). Prolonged exposure to increased temperatures can result in cellular and subcellular damage and reallocation of resources for protection and repair, leading to slowed growth, reduced productivity, delayed development, and eventually mortality (Harley et al. 2012). Extreme temperatures, if severe or prolonged, will lead to irreparable damage and death of macrophytes (i.e., disruptive stress *sensu* Eggert 2012). For instance, high temperatures associated with desiccation stress reduce recruitment of intertidal fucoids by directly influencing survival, but also indirectly by suppressing growth and reducing the likelihood of escape from grazing (Hawkins 1981). Similarly, physiological stress over several weeks during an extended marine heat wave was likely the cause of the local extirpation and 100 km range contraction of a subtidal habitat-forming fucoid seaweed in Australia (Smale and Wernberg 2013). While the physiological effects of thermal stress in macrophytes are relatively well established in the literature, the ecological relevance is mostly unknown. Few studies have focused on the relationship between the effects of sublethal thermal stress on the species fitness or their competitive abilities (but see Wernberg et al. 2010, Martínez et al. 2012a).

Beyond direct metabolic effects, water temperature is one of the main triggers for the timing of ontogenetic transitions such as spore production, flowering and seed germination in marine macrophytes (e.g., Diaz-Almela et al. 2007, Mohring et al. 2013). For example, temperature plays an important role in flowering induction and development in seagrass (*Zostera* sp.) on the Great Barrier Reef. In these populations the initiation of flowering occurs after a rapid rise in sea temperature, from the annual low, in conjunction with an increase in day length. Similarly, other species in this area flower in late spring before temperatures reach their summer peak. Therefore, changes to temperature regimes may disrupt these cues, and inhibit flowering and seed germination (Waycott et al. 2007). It is well known that different ontogenetic stages of both plants and animals have different temperature sensitivities. For example, kelp gametophytes are often described as more thermo-tolerant than sporophytes (Dieck 1993, Müller et al. 2012). A complete understanding of how temperature change might affect populations must therefore assess the thermal limits of all life stages (Russell et al. 2012).

Perhaps the most pervasive change due to rising ocean temperatures is the shift in distribution of many macrophytes (Lima et al. 2007, Fernández 2011, Wernberg et al. 2011a, Martínez et al. 2012b, Tanaka et al. 2012, Smale and Wernberg 2013).

Figure 4. Macrophytes impacted by climate change. (A) Direct physiological stress associated with record high temperatures during a marine heat wave caused a 100 km range contraction of the prominent fucoid seaweed *Scytothalia dorycarpa* in Western Australia (Smale and Wernberg 2013). (B) High summer temperatures have caused a change in flowering and increased mortality of the seagrass *Posidonia oceanica* in the Mediterranean Sea (Diaz-Almela et al. 2007, Jorda et al. 2012). (C) Altered species interactions following increased dispersal and survival of the herbivorous sea urchin *Centrostephanus rodgersii* as a consequence of a strengthening of the flow of the warm East Australia Current, has caused overgrazing of seaweeds (e.g., the kelp *Ecklonia radiata*) leading to the formation of extensive urchin barrens in Tasmania (Ling et al. 2009b, Johnson et al. 2011, photo: S.D. Ling).

However, some species occur along coastlines with physical barriers, including geological (e.g., substratum), geomorphological (e.g., depth, topography), oceanic (e.g., upwelling, off/alongshore currents), as well as human associated activities (e.g., around metropolitan areas). In these cases range shifts following the new temperature clines may not be possible and continued warming can therefore result in extinctions as species-ranges keep contracting until they eventually 'fall off the map' (Wernberg et al. 2011a). A study from Portugal showed differences between cool- and warm-water seaweeds in response to ocean warming. While warm-water species exhibited considerable range expansions cold-water species showed no clear patterns (Lima et al. 2007) perhaps illustrating the different population processes associated with range expansions and contractions. Vertical distribution shifts have also been observed in intertidal and shallow subtidal communities, with species shifting into deeper water where temperature fluctuations are less dramatic (Harley et al. 2012). Nevertheless, other factors not related (or indirectly related) to changes in temperature may also explain, in part, some of the distributional shifts (Lima et al. 2007). These include biotic interactions (e.g., grazing and competition, Ling et al. 2009b, Franco et al. 2015), and anthropogenic effects over the coast (e.g., pollution, coastal erosion, Coleman et al. 2008, Connell et al. 2008).

CO_2 and pH

Like all photosynthetic organisms, marine macrophytes use inorganic carbon during photosynthesis to create organic carbon-rich compounds. This does not, however, imply that the increasing CO_2 levels will benefit all marine macrophytes; large differences in the carbon uptake mechanisms between species prevent such clear-cut predictions.

Carbon dioxide diffusion rates in seawater are four orders of magnitude lower than in air. Thus, for macrophytes that rely solely on CO_2 diffusion, increases in CO_2 could enhance photosynthetic performance and growth. This is the case for seagrasses, which evolved from terrestrial plants during the Cretaceous (~ 90 mya) when CO_2 levels were considerably higher than at present, and so are often carbon limited (Beardall et al. 1998). Carbon dioxide addition experiments have found that seagrasses exposed to high CO_2 levels, combined with increased nutrient concentrations, can access the greater concentrations of CO_2 and boost their photosynthesis rates (Beardall et al. 1998). It has been demonstrated that mangroves will also perform better when exposed to elevated CO_2. In a long term study from the Caribbean, *Rhizophora mangle* showed enhanced total plant biomass, higher root:shoot ratios, as well as increases in seedling biomass, stem length, and total leaf area when subjected to CO_2 concentrations double the atmospheric average (Farnsworth et al. 1996).

For seaweeds, predictions are more complicated as different species have different adaptations and strategies to deal with increasing carbon availability. Macroalgae that rely exclusively on CO_2 uptake through diffusion are most likely to show enhanced photosynthesis under elevated CO_2 levels. Conversely, macroalgae that use carbon concentration mechanisms (CCM) are predicted to show little response to elevated CO_2, although down regulation of CCMs could be induced by elevated CO_2 and allow the energy to be used for other purposes (Hurd et al. 2009). There is increasing evidence for the existence of CCMs in many species of macroalgae and in many cases experimental studies have shown little growth response to CO_2 increases (Israel and Hophy 2002). However, studies have shown that a number of ephemeral, fast growing turf-forming algae respond positively to increased CO_2 when coupled with increases in other test-factors, such as, temperature and elevated nutrients (Russell et al. 2009a, Connell and Russell 2010). Moreover, evidence suggests that elevated CO_2 has variable, but often positive, effects on kelps like *Nereocystis luetkeana* and *Macrocystis pyrifera* (Thom 1996, Roleda et al. 2012).

Carbon availability for photosynthesis is, however, not the only issue. Higher CO_2-levels lead to lower seawater pH ('acidification'), and this can result in reduced photosynthesis, poor growth and even death of some algae (Israel and Hophy 2002, Martin and Gattuso 2009). Calcifying algae, such as encrusting or articulated coralline algae, are among the most ubiquitous macrophytes in marine environments. Calcareous algae play a critical role in the settlement of many ecologically important invertebrates (Pearce and Scheibling 1990) and act as the 'glue' that consolidates reef structures, protecting them from storm damage (Madin et al. 2012). Calcifying algae are particularly sensitive to reduced pH in the ocean because it changes the solubility of different CO_3-species, increases the metabolic costs of calcification and, eventually the likelihood of dissolution. As such, reduced growth rates in calcified algae have been linked to increasing ocean acidification (Harley et al. 2012, Wernberg et al. 2012b). These effects are likely to be exacerbated by increasing temperatures. For example in the Mediterranean Sea, communities on rocky shores close to volcanic CO_2 vents showed a significant reduction in abundance of coralline algae with decreasing pH (Hall-Spencer et al. 2008). On temperate reefs, calcifying algae can occupy up to 80% of hard substrate, dominating space beneath canopies (Steneck et al. 2002, Wernberg et al. 2012b). Increasing oceanic CO_2 could therefore reduce the competitiveness of calcifying algae over non-calcifying species, which may result in a reduction of reef diversity, as well as habitat and protection for a range of dependent communities (Beardall et al. 1998). Although ocean acidification has been considered a major threat to marine organisms, effects are highly variable across and within taxonomic groups. Furthermore, a recent meta-analysis suggests that marine biota may be more resistant to ocean acidification than previously suggested (Dupont et al. 2010, Hendriks et al. 2010).

Changes in circulation patterns

Warming of the upper layers of the ocean combined with alterations to wind patterns, drive changes to major ocean currents (Wu et al. 2012), greater stratification of the water column and a deepening of the thermocline (Harley et al. 2006). The results include greater intrusion of warm, nutrient-poor waters at higher latitudes and reduced upwelling of nutrient-rich waters (Johnson et al. 2011). Upwelled nutrients fuel growth and reproduction in benthic and planktonic algae, and future changes in upwelling regimes/dynamics could have important consequences for ecosystem productivity (Harley et al. 2006, Hoegh-Guldberg and Bruno 2010). For example, the decline of giant kelp (*Macrocystis pyrifera*) populations in Tasmania (Australia) has been attributed to an increase in the strength of the East Australia Current, now bringing warm and nutrient poor water onto the coast (Johnson et al. 2011). Altered circulation patterns also affect species distributions. Shifts from kelp (*Ecklonia*) to *Sargassum* dominated communities have been reported in Japan as a result of a strengthening of the Kuroshio current, leading to regional warming (Tanaka et al. 2012). The strengthening of warm poleward currents can also drive tropical herbivores into temperate macrophyte-dominated habitats where they can have strong negative effects on macrophytes (Verges et al. 2014, Bennett et al. 2015). In Portugal, Lima et al. (2007) reported poleward expansions of warm-water species of seaweeds presumably as a result of a weakening of coastal upwelling. There are also cases of intensified upwelling allowing cool-water species to expand into previously warmer areas. This was the case in South Africa, where a dominant canopy-forming kelp (*Ecklonia maxima*) expanded its distribution limit by ~ 70 km (Bolton et al. 2012).

Many marine macrophytes rely on water movement for dispersal, and changes to ocean currents and upwelling are, therefore, also likely to have implications for population connectivity and recruitment. For example, Coleman and colleagues (2011) found strong positive correlations between the strength of coastal currents and genetic connectivity among kelp (*Ecklonia*) populations around Australia. Dispersal and recruitment of seagrasses and mangroves are also highly dependent on currents (Kendrick et al. 2012). For example the viability of mangrove (*Avicennia marina*) propagules diminishes over time, and seeds can only successfully establish within 4–5 days of release. In New Zealand, southern populations of this species are controlled by limited transport of propagules from northern populations; thus, any climate induced changes to circulation patterns may lead to poor population replenishment, and ultimately loss of mangrove stands (Osunkoya and Creese 1997).

Sea level rise

The average global sea level is rising at a rate of 3.3 mm year^{-1} due to thermal expansion of the oceans and freshwater input from melting glaciers, and most models estimate that by 2100 sea levels will be between 0.5 to 1.2 m higher than today (Harley et al. 2006, Hoegh-Guldberg and Bruno 2010). The most obvious result of increased sea level is that current macrophyte habitats above or near the surface will be permanently submerged, and current subtidal habitats will deepen, causing changes in physical conditions (e.g., reduced light penetration and wave exposure). This is particularly important in regions where the rocky shore is mostly horizontal (e.g., eastern Mediterranean) and in lagoons and estuaries with extremely low slopes. Since more than 14% of worlds coastline can be categorised as low sloping rapid sea-level rise would cause an inundation of the intertidal zone by seawater, turning the platforms into subtidal reefs (Cazenave and Lovel 2010). Vaselli et al. (2008) showed that on NW Mediterranean rocky shores a rise in sea level in the range of 5 to 50 cm would increase the availability of steep substrata (> 40°) by almost 60%, drastically modifying the dominant assemblages. Because they are highly specialized, and often live close to their tolerance limits, mangroves are particularly sensitive to minor variation in hydrological or tidal regimes. In the laboratory, *Rhizophora mangle* seedlings subjected to simulated increased tidal inundation showed reduced growth and photosynthetic rates, which would eventually lead to reduced recruitment and increased mortality (Poloczanska et al. 2007). Marshes will be also affected by rising sea level. Numerical models suggest that an increase in the rate of sea level rise together with a reduction in sediment supply will lead to greater inundation, expansion of channel networks, and conversion of marshland into subtidal mudflats (Kirwan and Temmerman 2009). Many species are, however, expected to shift their distributions landward, and should be able to keep pace with predicted rates of sea level rise. However, a major hurdle for landward retreat is lack of habitat availability due to human development.

Ultra-violet radiation

Ultraviolet radiation is, strictly speaking, not a symptom of climate change. However, increased penetration of UV radiation from ozone depletion, as a result of human activities, also has a detrimental effect of marine systems on a global scale (Poloczanska et al. 2007, Harley et al. 2012). Ultraviolet (UV) radiation is known to harm physiological processes in marine macrophytes through damage to DNA and inhibition of photosynthesis and nutrient uptake. This can lead to a reduction in productivity and biomass (Beardall et al. 1998). For example, net photosynthesis of the mangrove seedlings (*Rhizophora apiculata*) decreased by 59% with a 40% increase in UV (Connolly 2012). Low levels of UV-A (long wave lengths, 315–400 nanometers) have been shown to enhance photosynthesis and repair photo-damaged molecules, whereas UV-B (short wave lengths, 280–315 nanometers) mostly has harmful effects. Importantly, responses to UV appear to be species specific, and are often determined by the environment that an organism inhabits. For example, the seaweed *Gracilaria lemaneiformis* showed reduced photosynthetic efficiency (on sunny days) with increased levels of UV-B, whereas photosynthetic efficiency (during the sunrise period) increased with elevated UV-A (Xu and Gao 2010). More generally, intertidal algae are better adapted, than subtidal algae to cope with the increased UV stress, and many species have evolved strategies such as (photo) repair mechanisms. Similarly, some seagrasses have adaptations to cope with UV-stress; for example, when exposed to increased UV shallow water *Halophila ovalis* shows less stress than the deeper water *Posidonia australis* (Beardall et al. 1998). As discussed for heat stress, increased UV typically has stronger negative effect on the juvenile phase of the life cycle, and recruits tend to be more sensitive to light fluctuations. For example, while mature *Rhizophora mangle* trees perform best in full sunlight, their seedlings do better in the understory environment because the ontogenetic development in mangroves produces changes in light adaptation (Farnsworth et al. 1996). Similar ontogenetic changes have been observed in some kelp species. Fejtek et al. (2011) found that adult *Macrocystis pyrifera* can survive in much shallower waters than populations occur, but the susceptibility of the juveniles to high light and UV levels limit its vertical range. This is also the case for many other species of kelp around the world (Wiencke et al. 2006). A reduction in recruitment has implications for the replenishment of damaged populations; thus, the effects of UV are often interactive, and magnify other climate change conditions.

Extreme events—storms and heat waves

Biological responses to environmental conditions are often determined by extremes in conditions rather than average levels (Gaines and Denny 1993). A recent report from the IPCC (IPCC 2012) concluded that the magnitude and frequency of extreme weather events is increasing as a consequence of climate change (see also Poloczanska et al. 2012, Trenberth 2012).

Storms and extreme heat wave events can cause mass mortality of marine macrophytes over short time periods. For instance, large losses of seagrass meadows have been observed in Australia following major storms and cyclones, as a result of physical disturbance to the substrate and undermining of roots and rhizomes (Connolly 2012), and in Canada, a hurricane severely decimated subtidal kelp beds (Filbee-Dexter and Scheibling 2012).

Sometimes the ultimate cause of impact associated with storms, are not the storms themselves. Storms are often also associated with disturbance to sediments and peak rainfall, which can result in large-scale flooding, the re-suspension of particulate matter, increased turbidity, sedimentation and nutrients from run-off, and fluctuations in salinity. All of these secondary effects can have detrimental effects on macrophytes through sedimentation and nutrient-fuelled blooms of algae causing reduced light penetration, smothering and hypoxia (Airoldi 2003, Thomsen et al. 2012). For example, in 1992 an intense tropical cyclone caused a pulse in re-suspension of particulate matter and turbidity which killed more than 1,000 km^2 of seagrass in Hervey Bay (Australia) in 1992 (Orth et al. 2006). Flooding and freshwater runoff also decrease salinity, and this can have dramatic effects on intertidal mangroves and seagrasses (UNESCO 1993, Short and Neckles 1999, Vaselli et al. 2008). Mangroves are particularly susceptible to fluctuations in salinity, and rely on both freshwater run-off and tidal inundation for survival. When salinity is too high or low, mangrove growth and survival can be compromised. Most marine seagrasses grow best within a narrow salinity range, and many species suffer reduced growth, productivity, and die-back during salinity fluctuations associated flooding and increased rainfall (Waycott et al. 2009, Riddin and Adams 2010).

Heat waves are becoming more frequent and more severe, and their local and regional scale impacts on marine macrophytes are dramatic (Poloczanska et al. 2012). For example, in the summer of 2003 a heat wave in the Mediterranean Sea began in early June 2003 and lasted until mid-August, with air temperature records about 3–6°C above the seasonal average in many parts of Europe and seawater temperature up to 28.8°C (Sparnocchia et al. 2006). As a consequence, large-scale seagrass mortality occurred (Garrabou et al. 2009), and shoot mortality rates were unusually high, exceeding recruitment rates in most areas (Marba and Duarte 2010). Similarly, large-scale seagrass dieback occurred in Spencer Gulf, South Australia in February 2003, where air temperatures exceeded 40°C, coupled with exceptionally low tides in the middle of the day, over several consecutive days. During this heat wave seagrass shoots suffered high damage to chloroplasts, causing high mortality rates, resulting in the loss of nearly 13,000 ha of seagrass (Connolly 2012). Another more recent heat wave occurred off the west coast of Australia in 2011, where temperatures exceeded anything recorded for > 140 years, remaining significantly elevated for ~ 10 weeks across ~ 2,000 km coastline causing widespread ecological impacts (Pearce et al. 2011, Wernberg et al. 2013). The soaring temperatures led to detrimental physiological stress to habitat-forming seaweeds, causing extensive loss of seaweed canopies and a 100 km range contraction of a prominent endemic species with likely ecosystem-wide flow-on effects (Smale and Wernberg 2012, Smale and Wernberg 2013, Wernberg et al. 2013). Seagrasses in the World Heritage lise Shark Bay were also severely impacted by the 2011 marine heat wave (Fraser et al. 2014).

Biological stressors—diseases, epiphytes and invasions

In addition to the physical drivers associated with climate change itself, biologically stressors such as diseases, epiphytes and invasive species will also impact marine macrophytes.

Micro-organisms are ubiquitous in seawater, and the way they interact with macrophytes is likely to change under future climate scenarios (Campbell et al. 2011). Climate change can influence the prevalence and severity of disease outbreaks in marine ecosystems. Increased disease can contribute to declines of populations or species, especially for generalist pathogens. Increasing environmental stress associated with climate changes will likely decreases host susceptibility to diseases (Campbell et al. 2011), increase disease virulence and abundance, and increase disease vectors and the expansion of pathogen ranges (Hoegh-Guldberg and Bruno 2010, Wernberg et al. 2012b). For example, fungi (*Fusarium* sp.) can cause dieback of salt marsh (*Spartina alterniflora*) when the plants are subjected to high temperatures and drought stress (Alber et al. 2008). In the past, seagrass communities have been severely affected by diseases. For example, the Atlantic-wide loss of eelgrass (*Zostera marina* and *Thalassia testudinum*) meadows resulted from a combination of environmental stress and the wasting disease associated with a fungus (*Labyrinthula zosterae*, Duarte 2002). In kelps, bacteria are the causative agents of red spot disease, whereas the known pathogens of the dark spot disease are endophytic brown filamentous algae. Field observations from different parts of the world document massive rates of infection prevalence by endophytic brown algae with mostly unknown pathogenic effects (Eggert et al. 2010). Heat stress in seaweeds also cause lowered chemical defenses (Campbell et al. 2011), which can lead to heavy epiphytism with detrimental consequences for kelp forests (Andersen et al. 2011, Smale and Wernberg 2012).

Global climate change can effect species distributions and resources availability thus, inducing biological invasions (Occhipinti-Ambrogi 2007). Climatic change affects non-indigenous species by altering environmental conditions, resource levels and native communities of the invaded habitat (Thomsen et al. 2011). Global warming coincides with an increase in human transport activities, which increases vectors for introductions (Hoegh-Guldberg and Bruno 2010). Changes in temperature may favour the settlement of invasive species where it was previously too cold. For example, warming conditions in the Mediterranean favoured the establishment of the green seaweed *Caulerpa taxifolia*, which produces toxic secondary metabolites potentially leading to *Posidonia oceanica* die-back (de Villèle and Verlaque 1995, but see Glasby 2012, Thomsen et al. 2012 for alternative interpretations). Increasing sea temperature is also putting pressure on endemic species, reducing potential competitive advantage over invaders. In Australia, summer heat waves events associated with global climate change have seen large scale die-backs in kelp beds. This mass mortality facilitated the establishment of the invasive macroalga, *Undaria pinnatifida* (Valentine and Johnson 2003, Valentine and Johnson 2004).

Increasing temperature is not the only climate change-related factor affecting the outcome of invasions. Changes in atmospheric circulation, storm frequency, and precipitation alter the dispersion and invasiveness of non-indigenous species. In a recent laboratory experiment, increasing temperature and pCO_2 altered the success and performance of the brown seaweed *Sargassum muticum* at multiple points in the life cycle (Vaz-Pinto et al. 2012). Finally, non-indigenous species that have become well-established in recipient communities are expected to alter community response to climate. For example, in rock pools of the western Atlantic coast of the Iberian Peninsula, macroalgal assemblages invaded by *S. muticum* might be more resistant increasing temperature and CO_2 in future climate scenarios (Olabarria et al. 2012).

Cumulative effects of multiple stressors

Many climate and non-climate physical, chemical and biological changes in the environment are occurring simultaneously (Halpern et al. 2008, Wernberg et al. 2011b). Consequently, marine macrophytes are subjected to multiple concurrent stressors at any one time, and these are likely to act in concert rather than in isolation (Crain et al. 2008a). To this end, almost all observed impacts on marine macrophytes possibly related to climate change have involved multiple stressors (Wernberg et al. 2011b, Harley et al. 2012). For example, losses or declines of habitat-forming macroalgae in Australia over the past decades were attributed to historical overfishing of herbivore predators (Ling et al. 2009a), pollution and reduced water quality associated with urbanization (Coleman et al. 2008, Connell et al. 2008). In the Gulf of Maine, kelp beds have suffered dramatic changes in the last twenty-five years mainly due to an increase of seawater temperatures in summer, overfishing and the introduction of non-indigenous species (Harris and Tyrrell 2001), where temperature mediated outbreaks of invasive bryozoans cause extensive defoliation of kelps and increases in the invasive *Codium* seaweeds (Scheibling and Gagnon 2009).

Higher CO_2 and nutrient levels boost the growth and productivity of turf-forming seaweeds, which can block the recruitment of kelps (Gorman et al. 2009), while acidification reduces the calcification and recruitment of coralline crusts (Russell et al. 2009b), and these effects are further exacerbated by increasing temperatures (Martin and Gattuso 2009, Diaz-Pulido et al. 2012). In addition, ocean acidification may disrupt the feeding biology of some herbivores, limiting their ability to control the climate-enhanced growth of opportunistic turfs (Russell et al. 2009b, Falkenberg et al. 2010). Temperature and CO_2 concentrations can also modulate the effects of UV radiation on the germination, growth and survivorship of kelps and fucoids (Hoffman et al. 2003, Swanson and Fox 2007).

For mangroves, increasing temperatures are causing stress, which is being exacerbated by increased inundation as a result of sea level rise and rainfall. Storms can cause damage to trees and defoliation, and the runoff can compound the pressure by introducing higher sediment loads which smother pneumatophores. While high rates of sedimentation are detrimental to mangroves, moderate sedimentation can enhance growth through increased availability of nutrients (Ellis et al. 2004, Lovelock et al. 2007). More toxic chemicals are also likely to wash from local watersheds to have negative effects of mangrove trees. Changes in salinity also co-occur with storm runoff, and this can have a negative or positive effect; populations existing in estuaries where salinity is low are likely to be put under further pressure as salinities decline. Populations occurring in oceanic waters, or arid areas where salinity is extreme may benefit from freshwater inflow and may exhibit increased growth rates, biodiversity and expansions in mangrove area (Gilman et al. 2008). Similarly, higher nutrient concentrations in runoff water may also have a positive effect on mangrove growth. Any one of these changing conditions has the potential to cause mortality, or at least reduced growth of mangroves, but in combination the effects are more severe (Gilman et al. 2008). Such effects may be exacerbated or ameliorated when acting with other anthropogenic stressors such as, nutrient loading or CO_2 (McKee and Rooth 2008, Cherry et al. 2009).

Seagrasses are often simultaneously influenced by multiple stressors at different temporal and spatial scales (Björk et al. 2008). For example, the negative effect associated with sea level rise and increased storm intensity/frequency is depth-dependant, and interacts with a reduction in the amount of available light reaching the seagrass canopy (Short and Neckles 1999). In combination with organic enrichment, extreme events such as floods, droughts and heat waves, may also impact seagrass meadows. In the Mondego estuary (Portugal), which has experienced significant eutrophication in the last 20 years, drought events caused salinity to increase above 35 ppt, increasing pressure on the seagrass, *Z. noltii*, and negatively affected meadow biomass (Cardoso et al. 2008).

In order to fully understand the effects of global climate change on marine macrophytes, it will be important to understand interactive effects and possible synergies (Crain et al. 2008b, Wernberg et al. 2012b). Until recently most marine climate change research involved single-stressor experiments focused on animals, highlighting an urgent need to conduct more studies with macrophytes, and in particular using multiple-stress approaches (Wernberg et al. 2012a).

Scaling up from physiological response to community— ecosystem level responses

Marine biota does not respond uniformly to climate change stressors. Experimental work has shown that even species with similar 'climate traits' (i.e., similar effect of climate change on one or several ecosystem functions) can respond very differently to the same environmental stressors. For example, responses of macroalgae to increases in temperature and CO_2 varies greatly, even among species belonging to the same functional group (Martin and Gattuso 2009). The lack of uniform easy-to-predict responses will have important implications for marine ecosystems where some species play disproportionately strong roles in community structure and function (Kroeker et al. 2010).

Most marine research on climate change published to date have used highly controlled single species experiments to look for specific responses and improve our understanding of physiological mechanisms (Wernberg et al. 2012a). Whilst these experiments are extremely important, they represent highly artificial experimental conditions, being isolated from common biological interactions (e.g., competition, predation, facilitation). Biological interactions may buffer or amplify individual responses thereby altering predicted assemblage or ecosystem-level responses (Kordas et al. 2011). For example, increasing CO_2 can cause seagrass photosynthetic rates to increase by 50% which may deplete the surrounding CO_2 pool, maintain an elevated pH and thus, protect associated calcifying organisms (Kroeker et al. 2010). Similarly, kelp loss might be exacerbated by the positive effects of increasing CO_2 and temperature on non-calcifying algal turfs, i.e., kelp inhibitors (Connell and Russell 2010).

Therefore, ecological change in coastal systems will reflect the combined influence of direct environmental impacts on individual species and indirect effects resulting from changes in interspecific interactions. Presently, there is little certainty of the actual effects of future climate change and how species will respond. When coupled with the intrinsic complexity of ecosystems, it is difficult to predict the consequences for biodiversity and ecosystem function. Ecosystem responses will depend on the pace and magnitude of climate changes and the longevity of species and their role in the ecosystem (Harley et al. 2012). Thus, it is important to conduct experiments with multiple species, simple assemblages or even complex subsets of whole ecosystems (Wernberg et al. 2012a). Such microcosm/mesocosm experiments have been used with great success to mimic pelagic ecosystems, but have yet to be fully exploited by researchers studying climate effects on benthic communities (but see Olabarria et al. 2012).

Climate related changes in species interactions are often a consequence of shifts in species ranges, resulting in new species interactions as well as disruptions of existing interactions. Higher trophic levels are often more sensitive to climate change than plants. Secondary consumers not only suffer their own specific climate related stress, but are generally dependent on these primary producers, which may be exhibiting reduced biomass, localised extinctions and rapidly altering their distributions. These bottom-up changes to food webs can lead to localised extinctions and important alterations to ecosystem function (Kordas et al. 2011, Wernberg et al. 2012b). Conversely, the addition of new consumers can have substantial impacts on macrophytes systems through changes top-down control, maintaining fundamentally changed systems (Verges et al. 2014, Bennett et al. 2015).

Research suggests that interspecific interactions shift from generally negative (e.g., competitive, predatory) when the environment is benign to generally positive (e.g., facilitative) when the environment is stressful (Bertness and Callaway 1994). For instance, in salt-marshes from New England (United States) the nature and intensity of interactions between neighbour plants depend on latitudinal and inter-annual variation in climate. In the south (where heat stress is common), species interactions tend to be facilitative, whereas in the north (where abiotic conditions are more benign) interactions are more competitive. Furthermore, interactions are more facilitative in warmer *versus* cooler years (Bertness and Ewanchuk 2002). Similar large-scale patterns have been found within a species; In Australia Wernberg et al. (2010) found that intra-specific facilitation dominated at warm northern latitudes (heat stressed) but intra-specific competition dominated at cool southern (more benign) latitudes. It is therefore expected that many interactions will shift from competitive to facilitative under increasing temperature stress associated with climate changes (Halpern et al. 2007). Additionally, important changes in interactions might result when the geographic range of species shifts, and when the seasonal timing of life history events of interacting species fall into, or out of, synchrony.

Integration adaption in the responses

Despite evidence illustrating the potential of microevolution to generate rapid phenotypic changes at short time scales (Pereyra et al. 2009), in species with lifespans like those of macrophytes, probably the most relevant component of adaptive potential to climate change are phenotypic plasticity and genetic variability. Both mechanisms are able to act within a generation, while evolutionary genetic changes involve multiple generations. Thus, understanding trait plasticity and limits to adaptation is essential to predict rates of extinction owing to climate change, as well as estimating the effect of environmental change on community ecology and ecosystem services (Bridle and Vines 2007). Together with phenotypic plasticity, genetic variability is the second immediate mechanism that allows a population to persist when its local environment changes (Chevin et al. 2010). Evidence of the importance of genetic diversity on the ability of macrophytes to cope climate change are already available in the literature. Ehlers et al. (2008) found a strong negative effect of warming and a positive effect of genotypic diversity on shoot densities of eelgrasses. Similarly, Pearson et al. (2009) provided evidences of mal-adaption to heat shocks and dehydration stress from low diverse edge population of the brown algae *Fucus serratus*.

Conclusions and future issues

Macrophytes are critical components of marine ecosystems, and while the evidence for substantial climate change impacts on marine macrophytes is increasing (Wernberg et al. 2011b, Harley et al. 2012), experimental work on macrophytes is substantially under-represented in the research effort (Wernberg et al. 2012a). A better understanding of interactions between macrophytes and climate change is critical because of the important contribution of this group to functioning of many ecosystems and its potential role on amelioration of environmental stress (Halpern et al. 2007). Despite the increase in the number of studies in the last decade, there are still significant gaps in our knowledge. Harley et al. (2012) summarizes these knowledge-gaps for macroalgae, but we contend that they apply generally to marine macrophytes and re-iterate them here: (1) the importance of rates, timing, magnitude, and duration of environmental change, (2) non-additive effects of multiple stressors, (3) population-level implications of variable environmental impacts among life-history stages, (4) the scope for population or species-level adaptation to environmental change, and

(5) ecological responses at the level of communities and ecosystems, including tipping points and sudden phase shifts.

Acknowledgements

TW was supported by a Future Fellows grant from the Australian Research Council. FA was supported by the European Regional Development Fund (ERDF) through the COMPETE—Operational Competitiveness Program and national funds through FCT—Foundation for Science and Technology, under the project "PEst-C/MAR/LA0015/2011" and the FCT funded projects PHYSIOGRAPHY (PTDC/MAR/105147/2008) and CLEF (PTDC-AAC-AMB-102866-2008). CO was supported by the Spanish Government through the Ministry of Science and Innovation-FEDER under the project CLINVA (CGL2009-07205).

References

Airoldi, L. 2003. The effects of sedimentation on rocky coast assemblages. Oceanography and Marine Biology 41: 161–236.

Alber, M., E.M. Swenson, S.C. Adamowicz and I.A. Mendelssohn. 2008. Salt Marsh Dieback: An overview of recent events in the US. Estuarine, Coastal and Shelf Science 80: 1–11.

Alongi, D.M. 2002. Present state and future of the world's mangrove forests. Environmental Conservation 29: 331–349.

Andersen, G.S., H. Steen, H. Christie, S. Fredriksen and F.E. Moy. 2011. Seasonal patterns of sporophyte growth, fertility, fouling, and mortality of *Saccharina latissima* in Skagerrak, Norway: implications for forest recovery. Journal of Marine Biology, ID 690375.

Barbier, E.B., S.D. Hacker, C.J. Kennedy, E.W. Koch, A.C. Stier and B.R. Silliman. 2011. The value of estuarine and coastal ecosystem services. Ecological Monographs 81: 169–193.

Beardall, J., S. Beer and J.A. Raven. 1998. Biodiversity of marine plants in an era of climate change: some predictions on the basis of physiological performance. Botanica Marina 41: 113–123.

Beck, M., K. Heck, K. Able, D. Childers, D. Eggleston, B. Gillanders, B. Halpern, C. Hays, K. Hoshino, T. Minello, R. Orth, P. Sheridan and M. Weinstein. 2001. The identification, conservation, and management of estuarine and marine nurseries for fish and invertebrates. BioScience 51: 633–641.

Bennett, S., T. Wernberg, E.S. Harvey, J. Santana-Garcon and B. Saunders. 2015. Tropical herbivores provide resilience to a climate mediated phase-shift on temperate reefs. Ecology Letters 18: 714–723.

Bertness, M.D. and R. Callaway. 1994. Positive interactions in communities. TRENDS in Ecology and Evolution 9: 191–193.

Bertness, M.D. and P.J. Ewanchuk. 2002. Latitudinal and climate-driven variation in the strength and nature of biological interactions in New England salt marshes. Oecologia 132: 392–401.

Bertness, M.D., B.R. Silliman and R. Jefferies. 2004. Salt marshes under siege: Agricultural practices, land development and overharvesting of the seas explain complex ecological cascades that threaten our shorelines. American Scientist 92: 54–61.

Björk, M., F. Short, E. Mcleod and S. Beer. 2008. Managing seagrasses for resilience to climate change. The International Union for the Conservation of Nature and Natural Resources.

Bolton, J.J., R.J. Anderson, A.J. Smit and M.D. Rothman. 2012. South African kelp moving eastwards: the discovery of *Ecklonia maxima* (Osbeck) Papenfuss at De Hoop Nature Reserve on the south coast of South Africa. African Journal of Marine Science 34: 147–151.

Bridle, J.R. and T.H. Vines. 2007. Limits to evolution at range margins: when and why does adaptation fail? Trends in Ecology and Evolution 22: 140–147.

Burrows, M.T., D.S. Schoeman, L.B. Buckley, P. Moore, E.S. Poloczanska, K.M.Brander, C. Brown, J.F. Bruno, C.M. Duarte, B.S. Halpern, J. Holding, C.V. Kappel, W. Kiessling, M.I. O'Connor, J.M. Pandolfi, C. Parmesan, F.B. Schwing, W.J. Sydeman and A.J. Richardson. 2011. The pace of shifting climate in marine and terrestrial ecosystems. Science 334: 652–655.

Campbell, A.H., T. Harder, S. Nielsen, S. Kjelleberg and P.D. Steinberg. 2011. Climate change and disease: bleaching of a chemically defended seaweed. Global Change Biology 17: 2958–2970.

Cardoso, P.G., D. Raffaelli and M.A. Pardal. 2008. The impact of extreme weather events on the seagrass *Zostera noltii* and related Hydrobia ulvae population. Marine Pollution Bulletin 56: 483–492.

Cazenave, A. and W. Lovel. 2010. Contemporary sea level rise. Annual Review of Marine Science 2: 145–173.

Cebrian, J. 2002. Variability and control of carbon consumption, export, and accumulation in marine communities. Limnol. Oceanogr. 47: 11–22.

Cherry, J.A., K.L. McKee and J.B. Grace. 2009. Elevated CO_2 enhances biological contributions to elevation change in coastal wetlands by offsetting stressors associated with sea-level rise. Journal of Ecology 97: 67–77.

Chevin, L.-M., R. Lande and G.M. Mace. 2010. Adaptation, plasticity, and extinction in a changing environment: towards a predictive theory. PLoS Biol. 8: e1000357.

Coleman, M.A., B.P. Kelaher, P.D. Steinberg and A.J.K. Millar. 2008. Absence of a large brown macroalga on urbanized rocky reefs around Sydney, Australia, and evidence for historical decline. Journal of Phycology 44: 897–901.

Coleman, M.A., J. Chambers, N.A. Knott, H.A. Malcolm, D. Harasti, A. Jordan and B.P. Kelaher. 2011. Connectivity within and among a network of temperate marine reserves. PLoS ONE 6: e20168.

Connell, S.D. and B.D. Russell. 2010. The direct effects of increasing CO_2 and temperature on non-calcifying organisms: increasing the potential for phase shifts in kelp forests. Proceedings of the Royal Society B: Biological Sciences.

Connell, S.D., B.D. Russell, D.J. Turner, S.A. Shepherd, T. Kildea, D. Miller, L. Airoldi and A. Cheshire. 2008. Recovering a lost baseline: missing kelp forests from a metropolitan coast. Marine Ecology Progress Series 360: 63–72.

Connolly, R. 2012. Seagrass. *In*: Poloczanska, E.S., A.J. Hobday and A.J. Richardson (eds.). A Marine Climate Change Impacts and Adaptation Report Card for Australia 2012. Retrieved from www. oceanclimatechange.org.au [25/11/12].

Costanza, R., R. d'Arge, R. de Groot, S. Farberk, M. Grasso, B. Hannon, K. Limburg, S. Naeem, R.V. O'Neill, J. Paruelo, R.G. Raskin, P. Sutton and M. van den Belt. 1997. The value of the worlds ecosystem services and natural capital. Nature 387: 253–260.

Crain, C.M., K. Kroeker and B.S. Halpern. 2008a. Interactive and cumulative effects of multiple human stressors in marine systems. Ecology Letters 11: 1304–1315.

Crain, C.M., K.J. Kroeker and B.S. Halpern. 2008b. Interactive and cumulative effects of multiple human stressors in marine systems. Ecology Letters 11: 1304–1315.

Davison, I.R. and G.A. Pearson. 1996. Stress tolerance in intertidal seaweeds. Journal of Phycology 32: 197–211.

de Villèle, X. and M. Verlaque. 1995. Changes and degradation in a *Posidonia oceanica* bed invaded by the introduced tropical alga *Caulerpa taxifolia* in the north western Mediterranean. Botanica Marina 38: 79–87.

Diaz-Almela, E., M. Nuria and C.M. Duarte. 2007. Consequences of Mediterranean warming events in seagrass (*Posidonia oceanica*) flowering records. Global Change Biology 13: 224–235.

Diaz-Pulido, G., K.R.N. Anthony, D.I. Kline, S. Dove and O. Hoegh-Guldberg. 2012. Interactions between ocean acidification and warming on the mortality and dissolution of coralline algae. Journal of Phycology 48: 32–39.

Dieck, I.T. 1993. Temperature tolerance and survival in darkness of kelp gametophytes (Laminariales, Phaeophyta): ecological and biogeographical implications. Marine Ecology Progress Series 100: 253–264.

Doney, S.C., M. Ruckelshaus, J. Emmett Duffy, J.P. Barry, F. Chan, C.A. English, H.M. Galindo, J.M. Grebmeier, A.B. Hollowed, N. Knowlton, J. Polovina, N.N. Rabalais, W.J. Sydeman and L.D. Talley. 2012. Climate change impacts on marine ecosystems. Annual Review of Marine Science 4: 11–37.

Duarte, C.M. 2002. The future of seagrass meadows. Environmental Conservation 29: 192–206.

Duarte, C.M. and J. Cebrian. 1996. The fate of marine autotrophic production. Limnology and Oceanography 41: 1758–1766.

Duarte, C.M., J.J. Middelburg and N. Caraco. 2005. Major role of marine vegetation on the oceanic carbon cycle. Biogeosciences 2.

Dupont, S., N. Doreya and M. Thorndyke. 2010. What meta-analysis can tell us about vulnerability of marine biodiversity to ocean acidification? Estuarine, Coastal and Shelf Science 89: 182–185.

Eggert, A. 2012. Seaweed responses to temperature. pp. 47–66. *In*: Wiencke, C. and K. Bischof (eds.). Seaweed Biology. Springer-Verlag, Berlin Heidelberg.

Eggert, A., A. Peters and F. Kupper. 2010. The potential impact of climate change on endophyte infections in kelp sporophytes. pp. 139–154. *In*: Seckbach, J., R. Einav and A. Israel (eds.). Seaweeds and their Role in Globally Changing Environments, Cellular Origin, Life in Extreme Habitats and Astrobiology. Springer Netherlands doi: 10.1007/978-90-481-8569-6_9.

Ehlers, A., B. Worm and T.B.H. Reusch. 2008. Importance of genetic diversity in eelgrass *Zostera marina* for its resilience to global warming. Marine Ecology Progress Series 355: 1–7.

Ellis, J., P. Nicholls, R. Craggs, D. Hofstra and J. Hewitt. 2004. Effects of terrigenous sedimentation on mangrove physiology and associated macrobenthic communities. Marine Ecology Progress Series 270: 71–82.

Falkenberg, L.J., O.W. Burnell, S.D. Connell and B.D. Russell. 2010. Sustainability in near-shore marine systems: promoting natural resilience. Sustainability 2: 2593–2600.

Farnsworth, E.J., A.M. Ellison and W.K. Gong. 1996. Elevated CO_2 alters anatomy, physiology, growth, and reproduction of red mangrove (*Rhizophora mangle* L.). Oecologia 108: 599–609.

Fejtek, S.M., M.S. Edwards and K.Y. Kim. 2011. Elk Kelp, *Pelagophycus porra*, distribution limited due to susceptibility of microscopic stages to high light. Journal of Experimental Marine Biology and Ecology 396: 194–201.

Fernández, C. 2011. The retreat of large brown seaweeds on the north coast of Spain: the case of Saccorhiza polyschides. European Journal of Phycology 46: 352–360.

Field, C.D. 1998. Rehabilitation of mangrove ecosystems: an overview. Marine Pollution Bulletin 37: 383–392.

Filbee-Dexter, K. and R. Scheibling. 2012. Hurricane-mediated defoliation of kelp beds and pulsed delivery of kelp detritus to offshore sedimentary habitats. Marine Ecology Progress Series 455: 51–64.

Folke, C., S. Carpenter, B. Walker, M. Scheffer, T. Elmqvist, L. Gunderson and C.S. Holling. 2004. Regime shifts, resilience, and biodiversity in ecosystem management. Annual Review of Ecology Evolution and Systematics 35: 557–581.

Fourqurean, J.W., C.M. Duarte, H. Kennedy, N. Marbà, M. Holmer, M.A. Mateo, E.T. Apostolaki, G.A. Kendrick, D. Krause-Jensen, K.J. McGlathery and O. Serrano. 2012. Seagrass ecosystems as a globally significant carbon stock. Nature Geoscience 5: 505–509.

Franco, J.N., T. Wernberg, I. Bertocci, P. Duarte, D. Jacinto, N. Vasco-Rodrigues and F. Tuya. 2015. Herbivory drives kelp recruits into 'hiding' in a warm ocean climate. Marine Ecology Progress Series 536: 1–9.

Fraser, M.W., G.A. Kendrick, J. Statton, R.K. Hovey, A. Zavala-Perez and D.I. Walker. 2014 Extreme climate events lower resilience of foundation seagrass at edge of biogeographical range. Journal of Ecology 102: 1528–1536.

Gaines, S.D. and M.W. Denny. 1993. The largest, smallest, highest, lowest, longest, and shortest: extremes in ecology. Ecology 74: 1677–1692.

Gao, K., E. Helbling, D.-P. Häder and D. Hutchins. 2012. Responses of marine primary producers to interactions between ocean acidification, solar radiation, and warming. Marine Ecology Progress Series 470: 167–189.

Garrabou, J., R. Coma, N. Bensoussan, M. Bally, P. Chevaldonne, M. Cigliano, D. Diaz, J.G. Harmelin, M.C. Gambi, D.K. Kersting, J.B. Ledoux, C. Lejeusne, C. Linares, C. Marschal, T. Perez, M. Ribes, J.C. Romano, N. Teixido, O. Torrents, M. Zabala, F. Zuberer and C. Cerrano. 2009. Mass mortality in Northwestern Mediterranean rocky benthic communities: Effects of the 2003 heat wave. Global Change Biology 15: 1090–1103.

Gilman, E.L., J. Ellison, N.C. Duke and C. Field. 2008. Threats to mangroves from climate change and adaptation options: A review. Aquatic Botany 89: 237–250.

Glasby, T.M. 2012. *Caulerpa taxifolia* in seagrass meadows: killer or opportunistic weed? Biological Invasions.

Gorman, D., B.D. Russell and S.D. Connell. 2009. Land-to-sea connectivity: linking human-derived terrestrial subsidies to subtidal habitat change on open rocky coasts. Ecological Applications 19: 1114–1126.

Graham, M.H. 2004. Effects of local deforestation on the diversity and structure of southern California giant kelp forest food webs. Ecosystems 7: 341–357.

Hall-Spencer, J.M., R. Rodolfo-Metalpa, S. Martin, E. Ransome, M. Fine, S.M. Turner, S.J. Rowley, D. Tedesco and M.-C. Buia. 2008. Volcanic carbon dioxide vents show ecosystem effects of ocean acidification. Nature 454: 96–99.

Halpern, B.S., B.R. Silliman, J.D. Olden, J.P. Bruno and M.D. Bertness. 2007. Incorporating positive interactions in aquatic restoration and conservation. Frontiers in Ecology and the Environment 5: 153–160.

Halpern, B.S., S. Walbridge, K.A. Selkoe, C.V. Kappel, F. Micheli, C. D'Agrosa, J.F. Bruno, K.S. Casey, C. Ebert, H.E. Fox, R. Fujita, D. Heinemann, H.S. Lenihan, E.M.P. Madin, M.T. Perry, E.R. Selig, M. Spalding, R. Steneck and R. Watson. 2008. A global map of human impact on marine ecosystems. Science 319: 948–952.

Hansen, J., M. Sato, R. Ruedy, K. Lo, D.W. Lea and M. Medina-Elizade. 2006. Global temperature change. Proc. Natl. Acad. Sci. USA 103: 14288–14293.

Harley, C.D.G., A. Randall Hughes, K.M. Hultgren, B.G. Miner, C.J.B. Sorte, C.S. Thornber, L.F. Rodriguez, L. Tomanek and S.L. Williams. 2006. The impacts of climate change in coastal marine systems. Ecology Letters 9: 228–241.

Harley, C.D.G., K.M. Anderson, K.W. Demes, J.P. Jorve, R.L. Kordas, T.A. Coyle and M.H. Graham. 2012. Effects of climate change on global seaweed communities. Journal of Phycology 48: 1064–1078.

Harris, L.G. and M.C. Tyrrell. 2001. Changing community states in the Gulf of Maine: synergism between invaders, overfishing and climate change. Biological Invasions 3: 9–21.

Hawkins, S.J. 1981. The influence of *Patella* grazing on the fucoidan barnacle mosaic on moderately exposed rocky shores. Kieler Meeresforsch 5: 537–544.

Hendriks, I.E., C.M. Duarte and M. Álvarez. 2010. Vulnerability of marine biodiversity to ocean acidification: a meta-analysis. Estuarine, Coastal and Shelf Science 86: 157–164.

Hepburn, C.D., D.W. Pritchard, C.E. Cornwell, R.J. McLeod, J. Beardall, J.A. Raven and C.L. Hurd. 2011. Diversity of carbon use strategies in a kelp forest community: implications for a high CO_2 ocean. Global Change Biology 17: 2488–2497.

Hoegh-Guldberg, O. and J.F. Bruno. 2010. The impact of climate change on the world's marine ecosystems. Science 328: 1523–1528.

Hoffman, J.R., L.J. Hansen and T. Klinger. 2003. Interactions between UV radiation and temperature limit inferences from single-factor experiments. Journal of Phycology 39: 268–272.

Hurd, C.L., C.D. Hepburn, K.I. Currie, J.A. Raven and K.A. Hunter. 2009. Testing the effects of ocean acidification on algal metabolism: recommendations for experimental designs. Journal of Phycology 45: 1236–1251.

Hyndes, G.A., A.J. Kendrick, L.D. MacArthur and E. Stewart. 2003. Differences in the species- and size-composition of fish assemblages in three distinct seagrass habitats with differing plant and meadow structure. Mar. Biol. 142: 1195–1206.

IPCC. 2007. Climate Change 2007, Synthesis Report. A Contribution of Working Groups I, II, and III to the Third Assessment Report of the Intergovernmental Panel on Climate Change. Cambridge University Press, Cambridge, UK.

IPCC. 2012. Summary for Policymakers. pp. 1–19. *In*: Field, C.B., V. Barros, T.F. Stocker, D. Qin, D.J. Dokken, K.L. Ebi, M.D. Mastrandrea, K.J. Mach, G.-K. Plattner, S.K. Allen, M. Tignor and P.M. Midgley (eds.). Managing the Risks of Extreme Events and Disasters to Advance Climate Change Adaptation. A Special Report of Working Groups I and II of the Intergovernmental Panel on Climate Change. Cambridge University Press, Cambridge, UK, and New York, NY, USA.

Israel, A. and M. Hophy. 2002. Growth, photosynthetic properties and Rubisco activities and amounts of marine macroalgae grown under current and elevated seawater CO_2 concentrations. Global Change Biology 8: 831–840.

Israel, A., R. Einav and J. Seckbach. 2010. Seaweeds and their Role in Globally Changing Environments. Springer, London.

Johnson, C.R., S.C. Banks, N.S. Barrett, F. Cazassus, P.K. Dunstan, G.J. Edgar, S.D. Frusher, C. Gardner, M. Haddon, F. Helidoniotis, K.L. Hill, N.J. Holbrook, G.W. Hosie, P.R. Last, S.D. Ling, J. Melbourne-Thomas, K. Miller, G.T. Pecl, A.J. Richardson, K.R. Ridgway, S.R. Rintoul, D.A. Ritz, D.J. Ross, J.C. Sanderson, S.A. Shepherd, A. Slotwinski, K.M. Swadling and N. Taw. 2011. Climate change cascades: shifts in oceanography, species' ranges and subtidal marine community dynamics in eastern Tasmania. Journal of Experimental Marine Biology and Ecology 400: 17–32.

Jorda, G., N. Marba and C.M. Duarte. 2012. Mediterranean seagrass vulnerable to regional climate warming. Nature Clim. Change, advance online publication.

Kendrick, G.A., M. Waycott, T. Carruthers, M.L. Cambridge, R. Hovey, S. Krauss, P. Lavery, D. Les, R. Lowe, O. Mascaró, J. Ooi Lean Sim, R.J. Orth, D. Rivers, L. Ruiz-Montoya, E.A. Sinclair, J. Statton, J.K. van Dijk and J. Verduin. 2012. The central role of dispersal in the maintenance and persistence of seagrass populations. Bioscience 62: 56–65.

Kirwan, M. and S. Temmerman. 2009. Coastal marsh response to historical and future sea-level acceleration. Quaternary Science Reviews 28: 1801–1808.

Kordas, R.L., C.D.G. Harley and M.I. O'Connor. 2011. Community ecology in a warming world: The influence of temperature on interspecific interactions in marine systems. Journal of Experimental Marine Biology and Ecology 400: 218–226.

Kroeker, K.J., R.L. Kordas, R.N. Crim and G.G. Singh. 2010. Meta-analysis reveals negative yet variable effects of ocean acidification on marine organisms. Ecology Letters 13: 1419–1434.

Krumhansl, K. and R. Scheibling. 2012. Production and fate of kelp detritus. Marine Ecology Progress Series 467: 281–302.

Kuffner, I.B., A.J. Andersson, P.L. Jokiel, K.u.S. Rodgers and F.T. Mackenzie. 2008. Decreased abundance of crustose coralline algae due to ocean acidification. Nature Geoscience 1: 114–117.

Larkum, A.W.D., A.J. McComb and S.A. Shepard. 1989. Biology of Seagrasses—A Treatise on the Biology of Seagrasses with Special Reference to the Australian Region. Elsevier, Amsterdam.

Lima, F.P. and D.S. Wethey. 2012. Three decades of high-resolution coastal sea surface temperatures reveal more than warming. Nat. Commun. 3: 704.

Lima, F.P., P.A. Ribeiro, N. Queiroz, S.J. Hawkins and A.M. Santos. 2007. Do distributional shifts of northern and southern species of algae match the warming pattern? Global Change Biology 13: 2592–2604.

Ling, S.D., C.R. Johnson, S.D. Frusher and K.R. Ridgway. 2009a. Overfishing reduces resilience of kelp beds to climate-driven catastrophic phase shift. Proceedings of the National Academy of Sciences of the United States of America 106: 22341–22345.

Ling, S.D., C.R. Johnson, K. Ridgway, A.J. Hobday and M. Haddon. 2009b. Climate-driven range extension of a sea urchin: inferring future trends by analysis of recent population dynamics. Global Change Biology 15: 719–731.

Lovelock, C.E., I.C. Feller, J. Ellis, A.M. Schwarz, N. Hancock, P. Nichols and B. Sorrell. 2007. Mangrove growth in New Zealand estuaries: the role of nutrient enrichment at sites with contrasting rates of sedimentation Oecologia 153: 633–641.

Madin, J.S., T.P. Hughes and S.R. Connolly. 2012. Calcification, storm damage and population resilience of tabular corals under climate change. PLoS ONE 7: e46637.

Mann, K.H. 1973. Seaweeds: their productivity and strategy for growth. Science 182: 975–981.

Marba, N. and C.M. Duarte. 2010. Mediterranean warming triggers seagrass (*Posidonia oceanica*) shoot mortality. Global Change Biology 16: 2366–2375.

Martin, S. and J.-P. Gattuso. 2009. Response of Mediterranean coralline algae to ocean acidification and elevated temperature. Global Change Biology 15: 2089–2100.

Martínez, B., F. Arenas, M. Rubal, S. Burgués, R. Esteban, I. García-Plazaola, F. Figueroa, R. Pereira, L. Saldaña, I. Sousa-Pinto, A. Trilla and R. Viejo. 2012a. Physical factors driving intertidal macroalgae distribution: physiological stress of a dominant fucoid at its southern limit. Oecologia 170: 341–353.

Martínez, B., R.M. Viejo, F. Carreño and S.C. Aranda. 2012b. Habitat distribution models for intertidal seaweeds: responses to climatic and non-climatic drivers. Journal of Biogeography 39: 1877–1890.

McKee, K.L. and J.E. Rooth. 2008. Where temperate meets tropical: multifactorial effects of elevated CO_2, nitrogen enrichment, and competition on a mangrove-salt marsh community. Global Change Biology 14: 971–984.

Mohring, M.B., T. Wernberg, G.A. Kendrick and M.J. Rule. 2013. Reproductive synchrony in a habitat-forming kelp and its relationship with environmental conditions. Mar. Biol. 160: 119–126.

Müller, R., C. Desel, F.S. Steinhoff, C. Wiencke and K. Bischof. 2012. UV-radiation and elevated temperatures induce formation of reactive oxygen species in gametophytes of cold-temperate/Arctic kelps (Laminariales, Phaeophyceae). Phycological Research 60: 27–36.

Nagelkerken, I., S.J.M. Blaber, S. Bouillon, P. Green, M. Haywood, L.G. Kirton, J.O. Meynecke, J. Pawlik, H.M. Penrose, A. Sasekumar and P.J. Somerfield. 2008. The habitat function of mangroves for terrestrial and marine fauna: A review. Aquatic Botany 89: 155–185.

National Oceanic and Atmospheric Administration. 2013. Carbon Dioxide at NOAA's Mauna Loa Observatory reaches new milestone: Tops 400 ppm. *In*: Reseach Matters published online. <http://researchmatters.noaa.gov/news/Pages/CarbonDioxideatMaunaLoareaches400ppm.aspx>.

Occhipinti-Ambrogi, A. 2007. Global change and marine communities: Alien species and climate change. Marine Pollution Bulletin 55: 342–352.

Olabarria, C., F. Arenas, R.M. Viejo, I. Gestoso, F. Vaz-Pinto, M. Incera, M. Rubal, E. Cacabelos, P. Veiga and C. Sobrino. 2012. Response of macroalgal assemblages from rockpools to climate change: effects of persistent increase in temperature and CO_2. Oikos.

Orth, R.J., T.J.B. Carruthers, W.C. Dennison, C.M. Duarte, J.W. Fourqurean, K.L. Heck, Jr., A.R. Huges, G.A. Kendrick, W.J. Kenworthy, S. Olyarnik, F.T. Short, M. Waycott and S.L. Williams. 2006. A global crisis for seagrass ecosystems. Bioscience 56: 987–996.

Osunkoya, O. and R.G. Creese. 1997. Population structure, spatial pattern and seedling establishment of the Grey Mangrove, *Avicennia marina* var. *australasica*, in New Zealand. Australian Journal of Botany 45: 707–725.

Pearce, A., R. Lenanton, G. Jackson, J. Moore, M. Feng and D. Gaughan. 2011. The "marine heat wave" off Western Australia during the summer of 2010/11. Fisheries Research Report No. 222. Department of Fisheries, Western Australia, 40 pp.

Pearce, C.M. and R.E. Scheibling. 1990. Induction of metamorphosis of larvae of the green sea urchin, *Strongylocentrotus droebachiensis*, by coralline red algae. Biol. Bull. 179: 304–311.

Pearson, G.A., A. Lago-Leston and C. Mota. 2009. Frayed at the edges: selective pressure and adaptive response to abiotic stressors are mismatched in low diversity edge populations. Journal of Ecology 97: 450–462.

Pereyra, R.T., L. Bergstroem and L. Kautsky. 2009. Rapid speciation in a newly opened postglacial marine environment, the Baltic Sea. BMC Evolutionary Biology 9: 70.

Poloczanska, E.S., R.C. Babcock, A. Butler, A.J. Hobday, O. Hoegh-Guldberg, T.J. Kunz, R. Matear, D.A. Milton, T.A. Okey and A.J. Richardson. 2007. Climate change and Australian marine life. Oceanography and Marine Biology: An Annual Review 45: 407–478.

Poloczanska, E.S., A.J. Hobday and A.J. Richardson. 2012. Extreme events and climate change. pp. xx-xx. *In*: Poloczanska, E.S., A.J. Hobday and A.J. Richardson (eds.). A Marine Climate Change Impacts and Adaptation Report Card for Australia 2012. Retrieved from www.oceanclimatechange.org.au [25/11/12].

Richardson, A.J., C.J. Brown, K. Brander, J.F. Bruno, L. Buckley, M.T. Burrows, C.M. Duarte, B.S. Halpern, O. Hoegh-Guldberg, J. Holding, C.V. Kappel, W. Kiessling, P.J. Moore, M.I. O'Connor, J.M. Pandolfi, C. Parmesan, D.S. Schoeman, F. Schwing, W.J. Sydeman and E.S. Poloczanska. 2012. Climate change and marine life. Biology Letters 8: 907–909.

Riddin, T. and J.B. Adams. 2010. The effect of a storm surge event on the macrophytes of a temporarily open/closed estuary, South Africa. Estuarine, Coastal and Shelf Science 89: 119–123.

Roleda, M.Y., J.N. Morris, C.M. McGraw and C.L. Hurd. 2012. Ocean acidification and seaweed reproduction: increased CO_2 ameliorates the negative effect of lowered pH on meiospore germination in the giant kelp *Macrocystis pyrifera* (Laminariales, Phaeophyceae). Global Change Biology 18: 854–864.

Russell, B.D., J.-A.I. Thompson, L.J. Falkenberg and S.D. Connell. 2009a. Synergistic effects of climate change and local stressors: CO_2 and nutrient-driven change in subtidal rocky habitats. Global Change Biology 15: 2153–2162.

Russell, B.D., J.A.I. Thompson, L.J. Falkenberg and S.D. Connell. 2009b. Synergistic effects of climate change and local stressors: CO(2) and nutrient-driven change in subtidal rocky habitats. Global Change Biology 15: 2153–2162.

Russell, B.D., C.D.G. Harley, T. Wernberg, N. Mieszkowska, S. Widdicombe, J.M. Hall-Spencer and S.D. Connell. 2012. Predicting ecosystem shifts requires new approaches that integrate the effects of climate change across entire systems. Biology Letters 8: 164–166.

Sabine, C.L. and T. Tanhua. 2010. Estimation of anthropogenic CO_2 inventories in the ocean. Annual Review of Marine Science 2: 175–198.

Scheibling, R.E. and P. Gagnon. 2009. Temperature-mediated outbreak dynamics of the invasive bryozoan *Membranipora membranacea* in Nova Scotian kelp beds. Marine Ecology Progress Series 390: 1–13.

Short, F.T. and H.A. Neckles. 1999. The effects of global climate change on seagrasses. Aquatic Botany 63: 169–196.

Simas, T., J.P. Nunes and J.G. Ferreira. 2001. Effects of global climate change on coastal salt marshes. Ecological Modelling 139: 1–15.

Smale, D. and T. Wernberg. 2012. Ecological observations associated with an anomalous warming event at the Houtman Abrolhos Islands, Western Australia. Coral Reefs 31: 441–441.

Smale, D. and T. Wernberg. 2013. Extreme climatic event drives range contraction of a habitat-forming species. Proceedings of the Royal Society B 280: 2012–2829.

Sparnocchia, S., M.E. Schiano, P. Picco, R. Bozzano and A. Cappelletti. 2006. The anomalous warming of summer 2003 in the surface layer of the Central Ligurian Sea (Western Mediterranean). Annales Geophysicae 24: 443–452.

Stæhr, P.A. and T. Wernberg. 2009. Physiological responses of *Ecklonia radiata* (Laminariales) to a latitudinal gradient in ocean temperature. Journal of Phycology 45: 91–99.

Steneck, R.S., M.H. Graham, B.J. Bourque, D. Corbett, J.M. Erlandson, J.A. Estes and M.J. Tegner. 2002. Kelp forest ecosystems: biodiversity, stability, resilience and future. Environmental Conservation 29: 436–459.

Swanson, A.K. and C.H. Fox. 2007. Altered kelp (Laminariales) phlorotannins and growth under elevated carbon dioxide and ultraviolet-B treatments can influence associated intertidal food webs. Global Change Biology 13: 1696–1709.

Tanaka, K., S. Taino, H. Haraguchi, G. Prendergast and M. Hiraoka. 2012. Warming off southwestern Japan linked to distributional shifts of subtidal canopy-forming seaweeds. Ecology and Evolution 2: 2854–2865.

Thom, R.M. 1996. CO_2-Enrichment effects on eelgrass (*Zostera marina* L.) and bull kelp (*Nereocystis luetkeana* (mert.) P. & R.). Water, Air, and Soil Pollution 88: 383–391.

Thomsen, M.S., T. Wernberg, A. Altieri, F. Tuya, D. Gulbransen, K.J. McGlathery, M.Holmer and B.R. Silliman. 2010. Habitat cascades: the conceptual context and global relevance of facilitation cascades via habitat formation and modification. Integr. Comp. Biol. 50: 158–175.

Thomsen, M.S., J.D. Olden, T. Wernberg, J.N. Griffin and B.R. Silliman. 2011. A broad framework to organize and compare ecological invasion impacts. Environmental Research 111: 899–908.

Thomsen, M.S., T. Wernberg, A.H. Engelen, F. Tuya, M.A. Vanderklift, M. Holmer, K.J. McGlathery, F. Arenas, J. Kotta and B.R. Silliman. 2012. A meta-analysis of seaweed impacts on seagrasses: generalities and knowledge gaps. PLoS ONE 7: e28595.

Trenberth, K. 2012. Framing the way to relate climate extremes to climate change. Climatic Change 115: 283–290.

UNESCO. 1993. Impact of expected climate change on mangroves. *In*: UNEP-UNESCO Task Team, Report of the First Meeting, Rio de Janeiro, 1–3 June 1992, II. UNESCO Reports in Marine Science 61.

Valentine, J.F. and C.R. Johnson. 2004. Establishment of the introduced kelp *Undaria pinnatifida* following dieback of the native macroalga *Phyllospora comosa* in Tasmania, Australia. Marine and Freshwater Research 55: 223–235.

Valentine, J.P. and C.R. Johnson. 2003. Establishment of the introduced kelp *Undaria pinnatifida* in Tasmania depends on disturbance to native algal assemblages. Journal of Experimental Marine Biology and Ecology 295: 63–90.

Vaselli, S., I. Bertocci, E. Maggi and L. Benedetti-Cecchi. 2008. Assessing the consequences of sea level rise: effects of changes in the slope of the substratum on sessile assemblages of rocky sea-shores. Marine Ecology Progress Series 368: 9–22.

Vaz-Pinto, F., C. Olabarria and F. Arenas. 2012. Role of top-down and bottom-up forces on the invasibility of intertidal macroalgal assemblages. Journal of Sea Research, in press.

Vergés, A., P.D. Steinberg, M.E. Hay et al. 2014. The tropicalization of temperate marine ecosystems: climate-mediated changes in herbivory and community phase shifts. Proceedings of the Royal Society B: Biological Sciences, 281: 20140846.

Waycott, M., C. Collier, K. McMahon, P. Ralph, L. McKenzie, J. Udy and A. Grech. 2007. Vulnerability of seagrasses in the Great Barrier Reef to climate change. pp. 193–299. *In*: Johnson, J.E. and P.A. Marshall (eds.). Climate Change and the Great Barrier Reef. Great Barrier Reef Marine Park Authority and Australian Greenhouse Office Australia.

Waycott, M., C.M. Duarte, T.J.B. Carruthers, R.J. Orth, W.C. Dennison, S. Olyarnik, A. Calladine, J.W. Fourqurean, K.L. Heck, A.R. Hughes, G.A. Kendrick, W.J. Kenworthy, F.T. Short and S.L. Williams. 2009. Accelerating loss of seagrasses across the globe threatens coastal ecosystems. Proceedings of the National Academy of Sciences 106: 12377–12381.

Wernberg, T., M.S. Thomsen, P.A. Staehr and M.F. Pedersen. 2004. Epibiota communities of the introduced and indigenous macroalgal relatives *Sargassum muticum* and Halidrys siliquosa in Limfjorden (Denmark). Helgoland Marine Research 58: 154–161.

Wernberg, T., M.S. Thomsen, F. Tuya, G.A. Kendrick, P.A. Staehr and B.D. Toohey. 2010. Decreasing resilience of kelp beds along a latitudinal temperature gradient: potential implications for a warmer future. Ecology Letters 13: 685–694.

Wernberg, T., B. Russell, M. Thomsen, C. Gurgel, C. Bradshaw, E. Poloczanska and S. Connell. 2011a. Seaweed communities in retreat from ocean warming. Current Biology 21: 1828–1832.

Wernberg, T., B.D. Russell, P.J. Moore, S.D. Ling, D.A. Smale, A. Campbell, M.A. Coleman, P.D. Steinberg, G.A. Kendrick and S.D. Connell. 2011b. Impacts of climate change in a global hotspot for temperate marine biodiversity and ocean warming. Journal of Experimental Marine Biology and Ecology 400: 7–16.

Wernberg, T., D.A. Smale and M.S. Thomsen. 2012a. A decade of climate change experiments on marine organisms: procedures, patterns and problems. Global Change Biology 18: 1491–1498.

Wernberg, T., D.A. Smale, A. Verges, A. Campbell, B.D. Russell, M.A. Coleman, S.D. Ling, P.D. Steinberg, C.R. Johnson, G.A. Kendrick and S.D. Connell. 2012b. Macroalgae and temperate rocky reefs. pp. xx-xx. *In*: Poloczanska, E.S., A.J. Hobday and A.J. Richardson (eds.). Report Card of Marine Climate Change for Australia; detailed scientific assessment NCCARF Publication.

Wernberg, T., D.A. Smale, F. Tuya, M.S. Thomsen, T.J. Langlois, T. de Bettignies, S. Bennett and C.S. Rousseaux. 2013. An extreme climatic event alters marine ecosystem structure in a global biodiversity hotspot. Nature Climate Change 3: 78–82.

Wiencke, C., M.Y. Roleda, A. Gruber, M.N. Clayton and K. Bischof. 2006. Susceptibility of zoospores to UV radiation determines upper depth distribution limit of Arctic kelps: Evidence through field experiments. Journal of Ecology 94: 455–463.

Wu, L., W. Cai, L. Zhang, H. Nakamura, A. Timmermann, T. Joyce, M.J. McPhaden, M. Alexander, B. Qiu, M. Visbeck, P. Chang and B. Giese. 2012. Enhanced warming over the global subtropical western boundary currents. Nature Climate Change 2: 161–166.

Xu, J. and K. Gao. 2010. Use of UV-A energy for photosynthesis in the red macroalga *Gracilaria lemaneiformis*. Photochemistry and Photobiology 86: 580–585.

10

Ecological Interactions between Marine Plants and Alien Species

Mads Solgaard Thomsen,[1,2,]* *Thomas Wernberg,*[2,3]
Peter A. Staehr[4] and *David Schiel*[1]

Introduction

Marine 'plants' (here including macroalgal protists and seagrass angiosperms) comprise thousands of species distributed in the photic zone along the world's coastlines. These play key ecological and biogeochemical roles in marine ecosystems, often controlling biodiversity, ecosystem functioning and energy flow. Many marine plants are foundations species (Dayton 1972) and function as ecosystem engineers (Jones et al. 1994), affecting the availability of resources to other species by modifying, maintaining and creating habitats. In this chapter we discuss two related aspects of ecological interactions between alien species and marine plants: (a) *how alien plants and animals impact native plants* (Fig. 1) and (b) *how alien plants impact native plants or animals* (Fig. 2). In the first scenario the emphasis is on *the native plants that are affected* by alien plants or animals, whereas in the second part the focus shifts to how *invasive plants drive ecological change* both of native plants and animals. For

---
[1] Marine Ecology Research Group, School of Biological Sciences, University of Canterbury, Private Bag 4800, Christchurch 8140, New Zealand.
[2] UWA Oceans Institute & School of Plant Biology, The University of Western Australia, Crawley WA 6009, Australia.
[3] Australian Institute of Marine Science, 39 Fairway, Crawley 6009 WA, Australia.
Email: thomas.wernberg@uwa.edu.au
[4] Department of Bioscience, Aarhus University, Frederiksborgvej 399, 4000, Roskilde.
Email: pst@bios.au.dk
* Corresponding author: mads.solgaard.thomsen@gmail.com

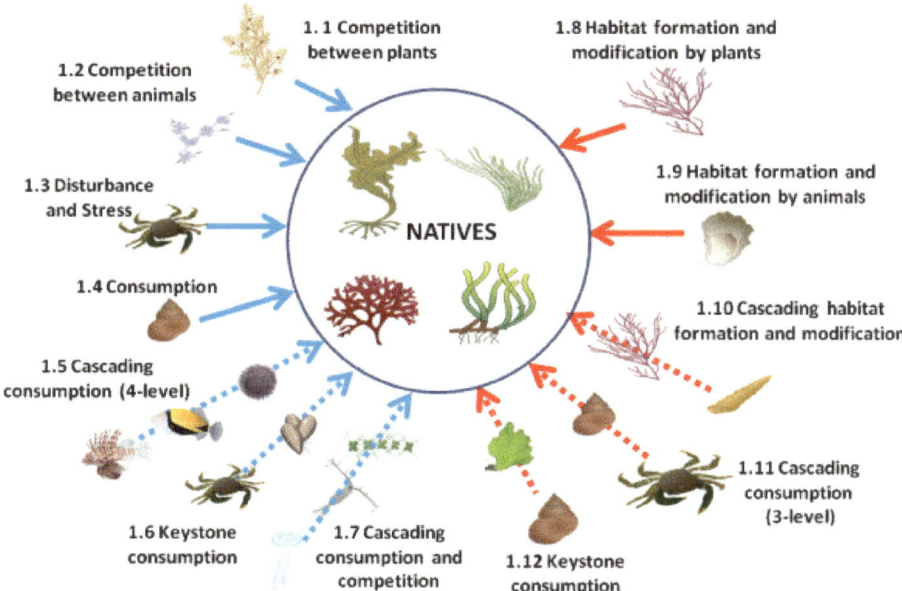

Figure 1. Common ecological interactions whereby alien marine plants and animals impact native marine 'plants' (seaweeds and seagrasses). Note minor overlap with Fig. 2; processes 1.1 and 1.8 correspond to processes 2.1 and 2.4 respectively. Direction of arrow = effect direction (impact is on the centered native plants); blue arrows = negative effects, red arrows = positive effects, solid arrows = direct effects; dashed arrows = indirect effects. See text for details on each interaction. Images and symbols used in the figure courtesy of the Integration and Application Network (http://ian.umces.edu/).

each scenario we provide a conceptual overview of typical ecological interactions (Figs. 1 and 2) with examples of how invaders affect the performance and abundance of local species and thereby influence energy flow, ecosystem functioning, biodiversity and structure of local marine communities. Our main focus is on the second scenario for which we compile a list of published experiments that have tested whether alien marine plants impact the diversity and structure of native communities. We focus solely on marine plants that perform best in saltwater; that is, we do not review studies that focus on terrestrial angiosperms (e.g., *Spartina alterniflora, Phragmites australis, Rhizophora mangle*) that can invade marine intertidal zones, or freshwater angiosperms (e.g., *Myriophyllum spicatum, Potamogeton crispus, Elodea canadensis*) that can invade brackish habitats.

Effects of alien plants and animals on native plants

Marine plants can control ecosystem function, biodiversity and community structure, but are often stressed by human activities, in particular climate change, eutrophication, pollution, habitat alteration, fisheries, and invasive alien species. Specifically, alien species can have strong impacts on native marine plants, and thereby impact ecosystem function and biodiversity (Thomsen et al. 2014b). Over 1700 marine alien or cryptogenic species, that is species that are not demonstrably native or introduced,

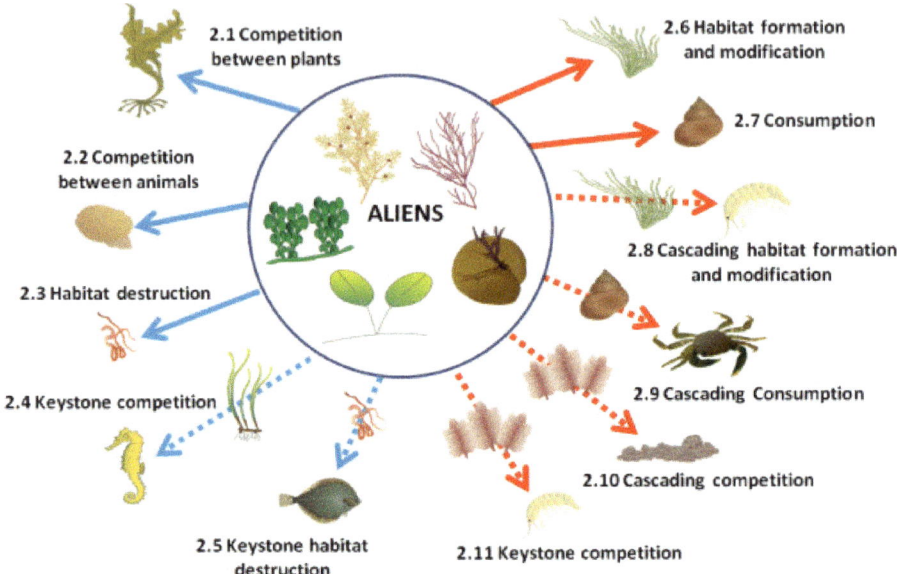

Figure 2. Common ecological interactions whereby alien marine plants impact native marine plants and animals. Note minor overlap with Fig. 1; processes 2.1 and 2.4 correspond to processes 1.1 and 1.8, respectively. Direction of arrow = effect direction (impact is caused by centered alien plants); blue arrows = negative effects, red arrows = positive effects, solid arrows = direct effects; dashed arrows = indirect effects. See text for details on each interaction. Images and symbols used in the figure courtesy of the Integration and Application Network (http://ian.umces.edu/).

have been identified around the world (Hewitt and Campbell 2010), and many of these have invaded shallow coastal habitats where they interact with and affect native seagrasses and seaweeds, often with cascading impacts on associated animal communities (Thomsen et al. 2010a). These effects are typically classified as positive or negative (*effect direction*; does the invader benefit or harm a native species?), small or large (*effect magnitude*; how much is the native species affected?), the type of data from which effects are evaluated (*effect data type*; are effects evaluated from speculation, anecdotes, or mensurative or manipulative experiments?), and if they are of a direct or indirect nature (*effect pathway*; do effects involve only interactions between the invader and the native species or do effects work through interactions with additional species). Below we provide examples of the different types of effects that invasive plants or animals can have on marine plants.

Negative effects on native marine plants

Invasion impact studies have traditionally focused on how terrestrial native plants have been negatively impacted by invaders (Rodriguez 2006) and this also appears to be valid for marine plants (Thomsen et al. 2011b). Below we describe direct (see Figs. 1.1 to 1.4) and indirect (see Figs. 1.5 to 1.7) species interactions whereby native plants can be negatively impacted by marine invaders.

Competition. Invasive plants can compete directly with native plants as they require the same basic resources for survival, growth and reproduction, in particular light, space, and nutrients (we consider both inferential and exploitive competition as direct effects because only the invader and native plant species are involved in this interaction (Wootton 1994)). Competition by plant invaders is one of the most frequent reported type of effects on native marine macrophytes (Schaffelke and Hewitt 2007, Williams and Smith 2007, Thomsen et al. 2009b). For example, native kelps and seagrasses can be negatively affected by alien *Sargassum muticum* (Fig. 1.1) (Staehr et al. 2000, Britton-Simmons 2004), *Codium fragile* (Levin et al. 2002) and several *Caulerpa* species (Ceccherelli and Cinelli 1997). However, in these cases (and many others), the underlying inhibitory mechanisms are unknown, but probably occur through competition for light and space. Native plants may also compete with alien animals because many of these are sessile biofoulers (Fig. 1.2). These aliens may compete for space, reduce light levels or, in extreme cases, excrete excessive amounts of organic material leading to toxic conditions for the native plants. For example, native kelps are negatively affected by the alien fouling bryozoan *Membranipora membranacea* that lives on the surface of kelp blades (Levin et al. 2002, Saunders and Metaxas 2008), thereby shading the kelp but also increasing the rigidity of the frond and possibly increasing the likelihood of frond breakage during storms.

Invaders as agents of disturbance and stress. Invasive animals can negatively impact native plants by creating biological disturbances. Native seagrasses (*Zostera marina* and *Halophila ovalis*) can be uprooted and their seeds buried by alien polychaetes (*Marenzelleria viridis*), snails (*Batillaria australis*) and crabs (*Carcinus maenas*) (Fig. 1.3). These invasive bioturbators can break seagrass rhizomes and leaves, or expose rhizomes and roots above the sediment surface and thereby increase susceptibility to storm-disturbances (Davis et al. 1998, Hoeffle et al. 2012). Invasive bioturbators can also bury seeds below a germination threshold, but burial could alternatively also increase protection from herbivores and storms (Delefosse and Kristensen 2012). These disturbance effects have been observed in laboratory and mesocosm experiments (Davis et al. 1998, Hoeffle et al. 2012) but effects by alien bioturbators may be less common in natural dense seagrass meadows where the rhizomes are interconnected, and high densities of common bioturbators (*C. maenas* and *B. australis*) are known to co-exist with dense seagrasses (Polte et al. 2005, Thomsen et al. 2010b). Disturbance effects from invasive bioturbaders are probably most important around seagrass edges and when new small patches establish from drifting clones or seeds. Information about invaders as 'disturbance agents' could therefore be important in conservation and for seagrass transplantation (Davis et al. 1998). In addition to uprooting (physical disturbances), invasive animals can also modify habitats to make environmental conditions hostile for marine plants (chemical disturbances), such as when *Marenzelleria viridis* decreases oxygen levels in sediments (Kristensen et al. 2011), potentially limiting seagrass expansion and recovery (Norkko et al. 2012).

Consumption. Alien animals can also have direct negative impacts on native plants through consumption (Fig. 1.4), although few field experiments have documented this (Thomsen et al. 2014b), perhaps because relatively few marine herbivores are invasive. The best known example is from the Northwest Atlantic where the invasive snail *Littorina littorea* can have large grazing effects on rocky intertidal shores and in tidepools. This snail can remove recruits of *Fucus* species on smooth substrates (Lubchenco 1983) and ephemeral algae from tidepools (Lubchenco and Menge 1978). However, experimentally determined impacts may overestimate effects if the invader displaces functionally similar native grazers (e.g., *L. littorea* displace *L. saxatilis*). In short, in the absence of the invasive grazer, the native grazers would instead have removed the same *Fucus* recruits and ephemeral alga (Eastwood et al. 2007). Eastwood's study highlights that short-term experiments conducted in already invaded locations can fail to identify long-term impacts, like niche shifts or local extinctions. Grazing effects from *L. littorea* have been documented in detail, probably because the snails are relatively large, conspicuous, slow-moving, abundant, and have a wide distribution in the easy-to-sample intertidal zone. In another invasion grazer study, alien fish (*Siganus luridus* and *S. rivulatus*) were shown to consume large amounts of erect algae, thereby creating extensive 'barren' areas on the Mediterranean coast of Turkey (Sala et al. 2011). However, grazing effects from alien species can also be more subtle. For example, the small mobile alien amphipod *Ampithoe valida* can consume large quantities of the reproductive structure of the native seagrass *Zostera marina*, thereby removing seeds and compromising the maintenance of genetic diversity and long-term persistence of meadows (Reynolds et al. 2012).

Indirect effects; cascading effects and consumption of keystone species. Invasion impact studies have generally focused on direct effects, although numerous forms of indirect effects are possible (White et al. 2006). We do not describe all possible indirect effects because most are poorly documented, but focus on a few simpler examples where 3–4 species interact in 'chains of events'. In these interaction chains, a 'basal' (or 'primary') invader affects a focal native species but only when 1–2 additional ('intermediate' or 'secondary') species are present (see Box 1 for more details and punctuated arrows on Figs. 1 and 2 for examples). It is well established that second-order consumers can have positive effects on plants (Shurin et al. 2005). Following analogous food chain theory, third-order consumers should therefore have negative impacts on native plants (Daskalov et al. 2007, Casini et al. 2008). We are not aware of studies that have quantified linkages between alien top-predators and native marine plants, but it is probably relatively common (for a freshwater example, see Tronstad et al. 2010). For example, tropical invasive lionfish consume trigger fish (Albins and Hixon 2008, Albins 2013), a group of fish that consumes urchins, which again consume seaweeds (McClanahan et al. 1996, O'Leary and McClanahan 2010) and seagrasses (Alcoverro and Mariani 2004). Lionfish could therefore have indirect negative effects on tropical seaweeds in a 4-level consumption cascade (Fig. 1.5), although this process would be less important if lionfish also consume large amounts of herbivorous reef fish. Keystone consumption (see Box 1) may also cause indirect negative effects on native plants. Invasive predators, like the crabs *Carcinus maenas* or *Hemigrapsus sanguineus*, consume large amounts of mussels and oysters (DeGraaf and Tyrrell 2004) that otherwise provide habitat for native seaweeds (Fig. 1.6) (Thomsen 2004, Lang and Buschbaum 2010). Invasive animals, therefore, by consuming habitat-formers, can have negative indirect effects on native plants. A final illustrative (hypothetical)

Box 1. Definitions

Alien, cryptogenic and invasive species. Aliens (= Non-indigenous, exotic, non-native, introduced species) are living outside their native distributional range, and have arrived at these locations aided by human activity, either deliberate or accidental. We do not include new species that arrived by their own dispersal mechanisms across natural barriers, but only survive due to climate changes. Aliens are typically called invasive if they are highly successful (e.g., with rapid spread and high abundance) or have a strong impact on resident ecosystems. In invasion biology, invaders are also aliens, but in biogeographical research invaders can be native species that arrive at new regions by natural dispersal mechanisms or by 'natural' breakdown of physical barriers. Cryptogenic species are of unknown origin; detailed taxonomic, biogeographical, and molecular analyses are required to determine if a cryptogenic species is a native or an alien.

Transport vectors. The pathway whereby aliens arrive to new systems through human activities. In marine systems the most important vectors are hull-fouling (found on the outside of ships), ballast water and sediments (found inside ships), aquaculture (intentional, escapees, and 'blind passengers'), and through aquarium trade.

Invasion impact studies. The invader is the causal agent of ecological change. Impact can be a synonym for 'effect', 'consequence', or 'cause of'. Impact studies focus on how invaders affect a property of a resident system. The invader is the independent variable, typically portrayed graphically on the x-axis. Impacts can be larger or smaller than a reference value, often defined as zero; the impact is then either positive (> 0) or negative (< 0). Positive or negative impacts are a statistical measure that differs from whatever impact is 'good' or 'bad' (the human interpretation of the statistical measure). Impacts can be reported on cultural (economics, health, societal) or natural (biotic, abiotic) properties. Biotic properties can be divided into impacts reported on or above the species level (the fundamental biological unit); negative impacts on species are simple to interpret whereas negative impacts reported above the species level can hide opposing effects (some species benefits, others are harmed).

Invasion impacts are evaluated from observations, hypotheses, models, and mensurative and manipulative experiments. No quantitative data are involved when an impact is evaluated from anecdotal observations (e.g., a researcher has noted, but not quantified, the occurrence of an abundant alien seaweed in the same location where native grazers are abundant) or hypothetical predictions/effects (e.g., it can predicted that any native generalist grazer can consume non-toxic invasive plants). Impacts evaluated from mensurative and manipulative experiments rely on quantitative data. In mensurative experiments, researchers have no control over the abundance of the invader, but compare responses between sample sites and periods that reflect different levels of invaders. Mensurative experiments can be conducted

Box 1. contd....

Box 1. contd.

on large and long spatio-temporal scales, with a wide range of invasive taxa and are without artifacts associated with cages, tethers or experimental disturbances. In manipulative experiments the researcher has (some) control over the abundance of the invader either by adding invaders to un-invaded plots or removing invaders from already invaded plots. In manipulative experiments causality can be inferred between the invader and the impacted resident species. Invader densities and the spatio-temporal context can here be controlled to form part of hypothesis testing. Impacts evaluated from models can be based on all kinds of combinations of the types of information above.

Cascade and keystone interaction chains. A 'cascade effect' is a 'disproportionally large ecological effect that occurs via a repeated ecological interaction', whereas a 'keystone effect' is a 'disproportionally large ecological effect'. 'Cascade' is a more precise term than 'keystone' because all processes in the interaction chain are the same. Net effects on focal organisms are positive in 3-level cascades, representing either 'a friend of my friend is my friend' (cascading habitat formation and mutualism) or 'an enemy of my enemy is my friend (cascading-consumption and competitions). The term 'keystone interactions' is less precise than cascade, and is defined by the first process in the interaction chain. In keystone consumption and keystone mutualism the first process is consumption and mutualism, respectively. The net effect on focal organisms from 3-level keystone interactions can be positive or negative depending on the direction of the processes involved (contrast, keystone consumption in Fig. 1.6 and Fig. 1.12).

example of a 'long consumption cascade' involves an invasive predator, the comb-jelly *Mnemiopsis leidii* (Fig. 1.7). *M. leidii* can consume large quantities of zooplankton and may therefore indirectly facilitate phytoplankton blooms (Riisgård et al. 2012) that reduce light and nutrient levels, resulting in negative net effects on benthic marine plants (Sand-Jensen and Borum 1991).

Positive effects on native marine plants

Alien plants and animals can have positive impacts on native plants through a variety of species interactions, such as habitat-formation, habitat-modification, and mutualism (direct effects, Figs. 1.8–1.9) as well as cascading habitat formation, cascading consumption, and keystone consumption (indirect effects, Figs. 1.10–1.12).

Habitat formation and modification. Habitat formation and habitat modification are common interactions whereby invasive plants and animals have positive direct effects on native plants. For example, the invasive red seaweed *Gracilaria vermiculophylla* creates habitat for native epiphytic plants, a particularly important process in soft-bottom estuaries that have little structural substrata for epibiotic organisms (Fig. 1.8) (Thomsen and McGlathery 2005, Thomsen et al. 2006a, Nyberg et al. 2009). Similar effects have been identified for many invasive animals that also create 3-dimensional structures for native plants to settle onto (Fig. 1.9), including invasive oysters (Lang and Buschbaum 2010), mussels (Sousa et al. 2009), snails (Thomsen et al. 2010b), tunicates (Castilla et al. 2004) and bryozoa (Sellheim et al. 2010). Of these species, oysters are particularly important because they have been transplanted to soft-bottom estuaries around the world (Padilla 2010). In addition to the physical creation of space for native species to live on (autogenic engineering), invasive habitat-formers also modify their environment, sometimes making abiotic conditions more benign for native species (allogenic engineering). For example, invasive oysters, mussels and tunicates can reduce desiccation stress and increase water clarity by filtering out plankton (Therriault and Herborg 2008, Rius and McQuaid 2009, Padilla 2010) and thereby increasing light levels for native benthic plants. In addition, invasive mobile bioturbators (e.g., *Batillaria australis*) can increase oxygen levels in sediments and thereby reduce stress on native seagrass rhizomes, although invasive snails can also uproot rhizomes and decrease oxygen levels if faecal products are deposited within the sediment (Hoeffle et al. 2012). These examples show that invaders often have direct positive effects on native plants by creating living space (habitat formation) and reducing environmental stress (habitat modification).

Indirect effects; cascading habitat formation, cascading consumption and keystone consumption. Invasive plants and animals can also have indirect positive effects on native plants. For example, invasive oysters and snails can provide habitat for native plants, that again provide habitat for native epiphytic plants, through cascading habitat formation (Fig. 1.10, Box 1) (Thomsen et al. 2010a). Indirect positive effects on native plants can also occur through consumption cascades (Fig. 1.11). Invasive crabs, such as *Carcinus maenas* or *Hemigrapsus sanguineus,* can reduce grazing on native plants by consuming (Eastwood et al. 2007) or 'scaring away' herbivores (Trussell et al. 2002, Trussell et al. 2004); that is, they reduce grazing through density- and trait-mediated consumption cascades, respectively. Finally, invasive grazers can also have positive effects on native plants, through keystone consumption, by preferentially consuming competitively dominant sessile species (Fig. 1.12). For example, invasive *L. littorea* preferentially consume fast-growing *Ulva* species and thereby reduce competition for slow-growing less palatable species like *Chondrus crispus* (Lubchenco 1978, 1983). The invasive grazer thereby provides the same function as a natural disturbance, opening up space (a limiting resource) that competitively inferior plants can colonize.

Effects of alien plants on native plants and animals

The previous section reviewed mechanisms whereby native marine plants can be impacted by invasive plant and animals. In this section we shift the focus to demonstrate that marine plants not only are affected by invaders, but also are invaders themselves causing major ecological changes to coastal ecosystems. This topic has been reviewed in detail (Schaffelke and Hewitt 2007, Williams 2007, Williams and Smith 2007, Thomsen et al. 2009b). In concert, these reviews conclude that: (a) impacts have been quantified from a small fraction (< 7%) of known introduced seaweeds; (b) many case studies reported strong impacts on native species; (c) few generalities of impacts have been shown; (d) underlying impact mechanisms are poorly known (but negative and positive impacts likely occur through competition and habitat formation and food provision, respectively); and (e) that major research gaps exist, particularly about how entire local communities are affected. Williams and Smith (2007) listed 277 marine alien seaweeds from around the world, represented by 45 Chlorophycea, 66 Phaeophyceae, 165 Rhodophytes (from a global pool of ca. 2400 Chlorophycea, 1800 Phaeophyceae and 6300 Rhodophytes). However, the status of several of the listed invaders is uncertain. For example, *Alaria esculenta*, and *Saccharina ochoensis* were listed as aliens in Denmark and *Gracilaria gracilis* as alien in Norway, but these species either do not have permanent populations there or are native (http://www.nobanis.org, Thomsen et al. 2007b). There are also at least three alien seagrasses (out of a pool of ca. 60 species (Short et al. 2007)): *Zostera japonica*, *Halophila stipulacea*, and *Zostera tasmanica* (Williams 2007, Castilla and Neill 2009). Williams (2007) suggests that *Halophila decipiens* may be an invader in Hawaii but other authors consider it native (McDermid et al. 2002, Russell et al. 2003). The most common transport vectors for these invaders are hull fouling and aquaculture (shellfish farming in particular), although a few species have also been transported in ballast water, through the Suez Canal, on fishing gear and with aquarium trade (Williams and Smith 2007). Marine plants have been introduced around the world, but some regions have more invaders than others; the Mediterranean Sea and the NE Atlantic alone account for around half of known introductions (Williams and Smith 2007). It should be noted there is a potential regional bias related to sampling effort, resource availability, and availability of historical data. Each of these 277 alien marine plants has had an impact on the invaded locality because each invader modifies both geochemical cycles (through metabolic activities) and community structure (through encounters with native species). However, many impacts are difficult to quantify, particularly if the impact magnitude is small or if the impact occurs in small areas, short time windows, and in inaccessible habitats. Below we review common mechanisms whereby marine alien plants affect native species.

Negative effects of alien marine plants

Typical forms of negative impacts by alien marine plants include competition, habitat-destruction/modification (direct effects, Figs. 2.1 to 2.3) and keystone competition and keystone habitat destruction (indirect effects, Figs. 2.4 to 2.5).

Direct negative effects; competition and habitat destruction. There are many examples of invasive plants competing with native plants for limited resources (light, space, nutrients). Invasive *Sargassum muticum* compete with native canopy-forming seaweeds (Fig. 2.1), including *Fucus vesiculosus* and *Halidrys siliquosa* (Staehr et al. 2000), *Macrocystis pyrifera* (Ambrose and Nelson 1982) and *Laminaria bongardiana* (Britton-Simmons 2004), although in some cases impacts appear to be minor, especially in the rocky intertidal zone (Sánchez and Fernández 2005, Olabarria et al. 2009). Another well-studied example is competition from the invasive kelp *Undaria pinnatifida* that may reduce native seaweed diversity (Casas et al. 2004) although other studies have found few impacts (Forrest and Taylor 2003, Valentine and Johnson 2005, South et al. in press). Importantly, both *S. muticum* and *U. pinnatifida* only have large canopies over a 4–5 month growth period where after the canopy is lost through erosion and fragmentation (Wernberg et al. 2001, Schiel and Thompson 2012, South et al. in press). This canopy loss may explain why these conspicuous invaders have relatively low impacts, particular in intertidal systems where light is abundant and desiccation stress is strong. Similar effects have been observed in soft-bottom systems where invasive *Caulerpa* species can compete with native seagrasses (de Villele and Verlaque 1995, Ceccherelli and Cinelli 1997), although other studies suggest that competition effects are limited (Jaubert et al. 1999, Ceccherelli and Sechi 2002, Jaubert et al. 2003, Thomsen et al. 2012). In short, competitive effects do occur but are context-dependent, such that the outcome depends on specific interactions between invasive and native species in conjunction with environmental conditions (Thomsen et al. 2011b). For example, *Gracilaria vermiculophylla* did not affect a native seagrass species under cold temperatures, disregarding its abundance, but had negative effects under high temperatures and when it was highly abundant (Hoeffle et al. 2011). This temperature-density interaction occurred because seagrasses are less stress-resistant near their upper temperature tolerance level and, at the same time, invader-effects are more severe at high abundance levels (in other words, elevated respiration and reduced oxygen conditions are more important at high densities and high temperatures) (Hoeffle et al. 2011).

Invasive marine plants may also compete with native animals, especially for limited space (Fig. 2.2). Invasive algal turf can compete with sessile encrusting animals (e.g., corals) on rocky substrata (Linares et al. 2012, Cebrian et al. 2013) and *Caulerpa racemosa* and *C. taxifolia* may compete for space with larger infaunal species that also depend on sandy substrata (but we are not aware of studies showing biotic resistance from soft-bottom animals to *Caulerpa* invasion). Thus, if abiotic conditions are benign, invasive plants typically invade soft-bottom systems, no matter what infaunal organisms live there (Berkenbusch et al. 2007, Tsai et al. 2010). Non-vegetated mud or sand habitats are thereby converted to vegetated meadows (= habitat destruction, cf. Fig. 2.3) and organisms that only can inhabit sand and mud are displaced. Only a few studies have documented negative effects from invasive marine plants on native infaunal species (Wright et al. 2007, Gribben et al. 2009b, Tsai et al. 2010), perhaps because many infaunal species are facultative mud-flat inhabitants and still survive (perhaps in lower densities) within the vegetated meadows (Gribben and Wright 2006, Wright and Gribben 2008, Gribben et al. 2009b, McKinnon et al. 2009, Byers et al. 2010, Klein and Verlaque 2011, Pacciardi et al. 2011). There are several examples of this type of habitat destruction, where invasive species convert mudflats to vegetated meadows, including the red alga *Gracilaria vermiculophylla* (Thomsen et al. 2007a, Thomsen et al. 2010a, Byers et al. 2012), green *Caulerpa* species (Gribben and Wright 2006, Wright and Gribben 2008, Gribben et al. 2009b, McKinnon et al. 2009, Byers et al. 2010, Klein and Verlaque 2011, Pacciardi et al. 2011), the brown alga *Sargassum muticum* (Strong et al. 2006) and the invasive seagrasses *Halophila stipulacea* and *Zostera japonica* (Posey 1988, Baldwin and Lovvorn 1994, Berkenbusch et al. 2007, Willette and Ambrose 2009, Ruesink et al. 2010, Willette and Ambrose 2012). Similar processes are also common for semi-marine invasive angiosperms, such as when invasive salt marshes and mangroves convert mudflats into vegetated meadows (Neira et al. 2007, Thomsen et al. 2009a, Wu et al. 2009, Demopoulos and Smith 2010).

Indirect negative effects; keystone competition and habitat-destruction. There has been little attention to these types of species interactions, but there are likely to be many indirect negative effects associated with marine plant invasions, particularly through keystone competition and keystone habitat destruction/modification. Invasive plants that negatively affect native plants can cause indirect negative effects on those animals that inhabit and forage among native plants (Fig. 2.4). For example, syngnathid and monacanthid fish are more abundant around native seagrass compared to invasive *Caulerpa* (York et al. 2006), juvenile fish are more abundant in native seagrass beds compared to invasive *Halophila stipulacea* (Willette and Ambrose 2012) and gastropods and seastars are more abundant on native kelp compared to invasive *Codium fragile* (Schmidt and Scheibling 2006). Still, many of these indirect effects associated with keystone competition and competitive displacement appear to be relatively minor, probably because marine animals often are generalists that can survive on different vegetation (Bell 1991). Similarly, many 'epibiota comparisons' show larger spatio-temporal differences in epibiota community structure, compared to community differences between the invasive and competitive native host plants (Viejo 1999, Wernberg et al. 2004, Gestoso et al. 2010, Soler-Hurtado and Guerra-García 2011, Gestoso et al. 2012, Guerra-García et al. 2012, Janiak and Whitlatch 2012). Indirect negative effects associated with keystone habitat destruction are conceptually similar to keystone competition; if invasive plants destroy mudflat habitats (converting them to meadows), this can lead to indirect negative impacts on the animals that inhabit/forage on the mudflat (Fig. 2.5). Native species likely to be indirectly and negatively affected by destruction of mudflats include burying fish like flounders, stargazers, and weeverfish, and wading birds that feed on mudflats, especially during migrations (although this has not yet been documented well for invasive marine plants) (Parks 2006).

Positive effects of marine alien plants

Invasions by marine plants provide some of the most compelling lines of evidence for positive impacts on native organisms (Wallentinus and Nyberg 2007, Thomsen et al. 2010a). Typical forms of positive impacts of alien marine plants include habitat formation and modification, mutualism, and consumption (direct effects, Figs. 2.6 and 2.7) as well as cascading habitat formation, cascading consumption, cascading competition and keystone competition (indirect effects) (Figs. 2.8 to 2.11).

Direct positive effects; habitat formation and modification, mutualism and consumption. Positive impacts of invasive plants through habitat formation has been documented many times, particularly for plants that invade mudflats (Fig. 2.6), such as *Caulerpa racemosa* (Klein and Verlaque 2011), *C. taxifolia* (McKinnon et al. 2009), *Gracilaria vermiculophylla* (Thomsen et al. 2006a, Thomsen et al. 2010a, Byers et al. 2012) (Hernández et al. 2012, Johnston and Lipcius 2012), *Halophila decepiens* (Willette and Ambrose 2012) and *Zostera japonica* (Posey 1988, Berkenbusch and Rowden 2007, Berkenbusch et al. 2007). Many positive effects are rather obvious as new physical structures provide living space and protection from predators for both sessile and interstitial mobile species (Huston 1994). In addition to creating habitat, habitat forming invaders also modify the environment (habitat modification), sometimes making it more benign for native species, for example by reducing intertidal desiccation stress (Bulleri et al. 2006). Positive effects can also be observed in invaded vegetated habitats when invaders 'add' structure and biomass, which support more interstitial organisms and epiphytes. For example, *S. muticum* invasion in Denmark has probably resulted in a system-wide increase in plant biomass and thereby more habitat for epiphytic plants (Thomsen et al. 2006b) and invertebrates (Wernberg et al. 2004). There has not been similar extensive research to evaluate how the epibiota (that benefit from added invasive structures) may impact the invader itself (except via consumption), but it is possible that mutualisms sometimes occur, for example if native mesograzers eat the epiphytes that shade the invader. Another common type of direct positive effect is through consumption (Fig. 2.7) whereby invasive plants provide food for native herbivores. Many studies have quantified grazing impact on invasive seaweeds, often trying to explain invasion success in the context of the enemy release theory (although this theory requires that the invader is top-down controlled in its native region, a prerequisite that is seldom investigated). It is possible that direct positive invasion effects are less important through consumption than habitat formation because many invasive plants are not particularly edible (e.g., *S. muticum*, Britton-Simmons 2004) or preferred food for dominant grazers like sea urchins (*C. fragile*, Scheibling and Anthony 2001). Other studies have shown that *S. muticum* is not preferred by mesograzers (Pedersen et al. 2005, Monteiro et al. 2009, Engelen et al. 2011), *Caulerpa* species have toxins making them relatively unpalatable (Boudouresque et al. 1996, Gollan and Wright 2006), *G. vermiculophylla* appear not to be impacted by grazers (Thomsen and McGlathery 2007, Nejrup and Pedersen 2010, Nejrup et al. 2012), and many mesograzers avoid invasive filamentous red algae (Cebrian et al. 2011, Tomas et al. 2011). Still, juvenile stages of invasive macroalgae and seagrasses may be edible and provide an important seasonal food supply for grazers (Thornber et al. 2004, Sjotun et al. 2007, Reynolds et al. 2012). Furthermore, siphonalian invaders, like *C. fragile* and various *Caulerpa* species, can have positive impacts on specialist saccoglossan grazers (Trowbridge and Todd 2001, Trowbridge 2002, Harris and Jones 2005) and these specialist grazers have therefore been suggested as potentially useful in control and eradication campaigns (Thibaut and Meinesz 2000). Similarly, grazing by *Littorina* species can limit *C. fragile* in tidepools (Scheibling et al. 2008), and waterfowl can consume invasive *Zostera japonica* (Baldwin and Lovvorn 1994). Finally, it is possible that the importance of consumption has been underestimated because edible opportunistic cosmopolitan seaweeds, like many ulvoid species, typically are considered (perhaps wrongly?) to be native. However, these species are notoriously difficult to identify and are well-known ship 'biofoulers' and are therefore likely to have been moved around in large quantities between biogeographical regions prior to scientific data collections (Carlton 1996).

Indirect effects; cascading habitat formation, cascading consumption, cascading competition, keystone consumption, and keystone competition. Invasive marine plants can have positive indirect effects through cascading habitat formation (Fig. 2.8). For example, invasive *S. muticum* and *G. vermiculophylla* provide primary habitat for structural sessile plants and animals, such as tunicates and filamentous seaweeds

(Wernberg et al. 2004, Thomsen et al. 2006b, Nyberg et al. 2009, Thomsen et al. 2010a, Gestoso et al. 2012, Engelen et al. 2013) that again provide a secondary habitat for many smaller hydrozoa, bryozoa, and small mobile invertebrates (Fig. 3). Invasive plants can also have indirect positive effects through cascading habitat modification, such as when *C. taxifolia* reduces sediment redox potential, thereby altering abiotic conditions. Native infaunal bivalves are thereby 'forced' to live at the sediment surface where they are exposed to sessile fouling organisms that settle on them (Gribben et al. 2009a). More common examples focus on positive aspects of plant invasions mediated through cascading consumption (Fig. 2.9) (Hairston et al. 1960, Estes and Palmisano 1974), where indirect facilitation follows from direct consumption by herbivores (1st order consumer), which again are consumed by predators (2nd or 3rd order consumers). Cascading consumption, where invasive plants provide the initial food source, could for example be common along the US east coast where *C. fragile* is consumed by (invasive) *Littorina* snails (Scheibling et al. 2008), which are consumed by (invasive) crabs (Trussell et al. 2002, Trussell et al. 2004, Eastwood et al. 2007), which finally are consumed by native crabs, seabirds and fish (deRivera et al. 2005). By contrast, we are not aware of studies on positive indirect effects associated with cascading competition (Fig. 2.10)—a process that typically occurs when competitors are limited by different resources (Levine 1999). It is possible that invasive *S. muticum,*

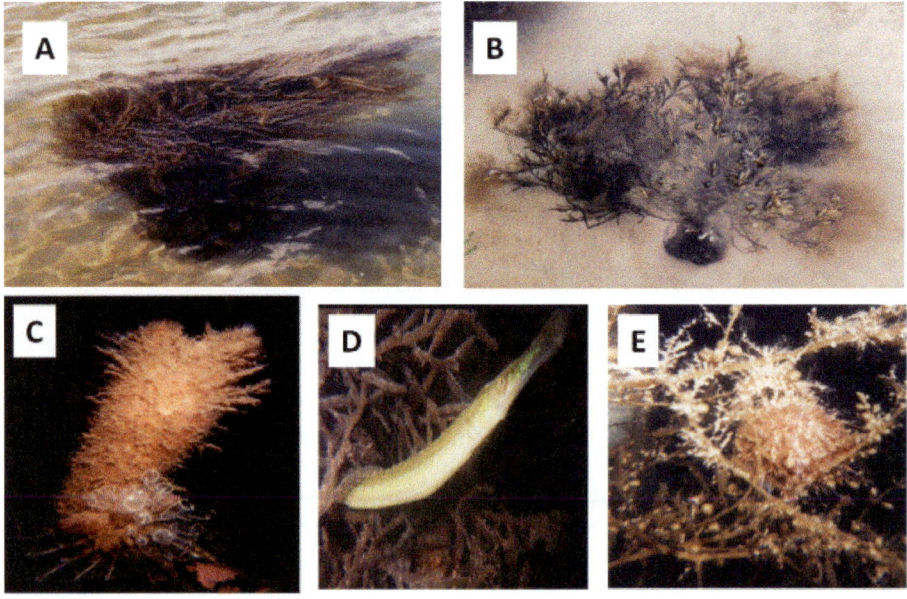

Figure 3. Impacts of the invasive brown alga *Sargassum muticum* on resident species. A–B. *Sargassum muticum* (A) attached to stone with a small holdfast but a large dense canopy, competes for light and substratum with the native morphologically similar species *Halidrys siliquosa* (B). (C) The invasive solitary tunicate *Styela clava* attaches to the invasive *S. muticum,* thereby providing more habitat for native hydrozoa and colonial tunicates through cascading habitat-formation. (D) The predatory pipefish (*Syngnathus*) feeds on mobile invertebrates inhabiting *S. muticum.* (E) Herbivorous sea urchins can have positive effects on *S. muticum* if they consume attached epiphytes, but negative effects if they consume *S. muticum* or if the fragile fronds are broken.

which reduces light levels but occupies little of the substratum, competes for light with small native turf algae (which have high light and substratum requirements), which in turn compete for substratum with encrusting plants and animals (which have high substratum but low light requirements). The invader may therefore have indirect positive impacts on encrusting species because it reduces the abundance of the strong substratum-competitors (and encrusting algae do appear to be common around *S. muticum*) (Staehr et al. 2000, Thomsen et al. 2006b). Finally, indirect positive effects can occur through keystone competition (Fig. 2.11) if invasive marine plants have negative impacts through competition on native plants that provide a poor habitat or food source for native species (e.g., for fish and mobile invertebrates).

Mini-review: community impact studies of alien plants

It has repeatedly been emphasized that studies documenting invasion impacts on entire communities are in short supply (Parker et al. 1999, Byers et al. 2002, Levine et al. 2003, Powell et al. 2011, Vilà et al. 2011, Pyšek et al. 2012, Thomsen et al. 2014b), as is also the case for studies that focus on impacts of marine plant invasions (Schaffelke and Hewitt 2007, Williams 2007, Williams and Smith 2007, Thomsen et al. 2009b, Thomsen et al. 2014a). We therefore conducted a brief review of marine invasion community impact studies up to September 2012 to investigate if this research gap is being addressed. Our objective is to identify published mensurative and manipulative experiments that report quantitative data on community impacts from marine plant invaders, and to highlight research gaps.

Methods. We searched for peer-reviewed manipulative and mensurative field experiments of marine plant invaders in Web of Science, Current Contents, and Google Scholar and by backtracking reference lists. Each study was classified according to study attributes (mensurative or manipulative data, reported community data), invader attributes (taxonomy) and attributes of the invaded system (latitude, ocean, country, habitat, tidal level, impacted focal organism). Several studies collected data from multiple locations, depths, and countries; for these studies we used the 'average value' for ordinal data (e.g., latitude) and 'multiple values' for categorical data (e.g., country).

Results. We identified 65 published papers addressing community impacts from 15 alien marine plants including a few studies that had results from pooled invaders (e.g., 'alien red turf' = *A. preissii* and *W. Setacea*; Table 1). These 15 species comprise 2 angiosperms, 4 green algae (in the order Caulerpales), 3 brown algae and 6 red algae, representing 66, 9, 5 and 4% of the registered aliens in each taxonomic group, respectively. The first study was published in 1988, but only 6 additional studies were published in the 1990s. However, community impact studies increased greatly in the 2000s (39 studies) and from 2010 to September 2012 we identified 19 new studies. This change in publication outputs corresponds to an exponential increase through time ($y = e^{0.1861x}$, $R^2 = 0.97$). Community impact was analyzed graphically in 39 studies, with multivariate inferential statistical tests in 41 studies, on taxonomic richness in 57 studies, on diversity in 19 studies and on evenness in 10 studies. We found a broad range of evidence from which community impacts were evaluated, including

Table 1. List of 65 published studies that document impacts of marine plant invaders on entire communities

Reference (by year)	Alien	Invaded habitat	Impact on (focal response)	Method
(Posey 1988)	*Zostera japonica*	Mudflat	Mobile animals	Man (Add)
	Zostera japonica	Mudflat	Mobile animals	Men (CI)
(Francour et al. 1995)	*Caulerpa taxifolia*	Seaweed-Mixed	Fish	Men (CI)
(Bellan-Santini et al. 1996)	*Caulerpa taxifolia*	Seaweed-Mixed	Epibiota	Men (CI)
(Viejo 1997)	*Sargassum muticum*	Seaweed-Mixed	Seaweeds	Man (Rem)
(Relini et al. 1998)	*Caulerpa taxifolia*	Seagrass-*Cymodocea nodosa*	Fish	Men (CI)
(Argyro et al. 1999)	*Caulerpa racemosa*	Seagrass-*Posidonia oceanica*	Epibiota/infauna	Men (BA)
(Viejo 1999)	*Sargassum muticum*	Seaweed-Seaweed *Cystoseira nodicaulis*	Epibiota	Men (TC)
	Sargassum muticum	Seaweed-Seaweed *Fucus vesiculosus*	Epibiota	Men (TC)
(Staehr et al. 2000)	*Sargasum muticum*	Seaweed-Mixed	Seaweeds	Men (BACI)
(Piazzi et al. 2001)	*Caulerpa racemosa*	Seagrass-*Posidonia oceanica* matte	Seaweeds	Men (BACI)
	Caulerpa racemosa	Seaweed-Mixed	Seaweeds	Men (BACI)
(Piazzi et al. 2002)	*Acrothamnion preissii*	Seagrass-*Posidonia oceanica*	Seaweeds	Men (CI)
	Wamersleyella setacea	Seagrass-*Posidonia oceanica*	Seaweeds	Men (CI)
(Forrest and Taylor 2003)	*Undaria pinnatifida*	Seaweed-Mixed	Seaweeds, mobile and sessile fauna	Men (BACI)
(Piazzi et al. 2003)	*Caulerpa racemosa*	Seaweed-Mixed	Seaweeds	Men (CI)
	Caulerpa taxifolia	Seaweed-Mixed	Seaweeds	Men (CI)

(Piazzi and Cinelli 2003)	*Caulerpa racemosa*	Seagrass-*Posidonia oceanica* matte	Seaweeds	Men (CI)
	Red turf: *A. preissii, W. setacea*	Seagrass-*Posidonia oceanica* matte	Seaweeds	Men (CI)
	Red turf: *A. preissii, W. setacea*	Seagrass-*Posidonia oceanica*	Seaweeds	Men (CI)
	Caulerpa racemosa	Seaweed-Mixed	Seaweeds	Men (CI)
	Caulerpa taxifolia	Seaweed-Mixed	Seaweeds	Men (CI)
	Red turf: *A. preissii, W. setacea*	Seaweed-Mixed	Seaweeds	Men (CI)
(Balata et al. 2004)	*Caulerpa racemosa*	Seaweed-Mixed	Seaweeds	Men (CI)
	Caulerpa taxifolia	Seaweed-Mixed	Seaweeds	Men (CI)
(Britton-Simmons 2004)	*Sargassum muticum*	Seaweed-Mixed	Canopy seaweeds, mobile animals	Man (Rem)
(Casas et al. 2004)	*Undaria pinnatifida*	Seaweed-Mixed	Seaweeds	Man (Rem)
(Wernberg et al. 2004)	*Sargassum muticum*	Seaweed-Seaweed *Halidrys siliquosa*	Epibiota	Men (TC)
(Wikstrom and Kautsky 2004)	*Fucus evanescens*	Seaweed-*Fucus vesiculosus*	Epibiota	Men (TC)
(Eklöf et al. 2005)	*Eucheuma denticulatum*	Bare-Mudflat	Epibiota/infauna	Men (CI)
	Eucheuma denticulatum	Seagrass-Mixed	Epibiota/infauna	Men (CI)
(Falcao and de Szechy 2005)	*Caulerpa scalpeliformis*	Bare-Mudflat	Seaweeds	Men (CI)
	Caulerpa scalpeliformis	Seaweed-Mixed	Seaweeds	Men (CI)
(Piazzi et al. 2005)	*Caulerpa racemosa*	Seaweed-Mixed	Seaweeds	Man (Add)
(Sánchez and Fernández 2005)	*Sargassum muticum*	Seaweed-Mixed	Seaweeds	Man (Rem)
(Sánchez et al. 2005)	*Sargassum muticum*	Seaweed-Mixed	Seaweeds	Men (BA)

Table 1. contd....

Table 1. contd.

Reference (by year)	Alien	Invaded habitat	Impact on (focal response)	Method
(Valentine and Johnson 2005)	*Undaria pinnatifida*	Seaweed-Mixed	Seaweeds	Man (Rem)
(Buschbaum et al. 2006)	*Sargassum muticum*	Seaweed-Seaweed *Fucus vesiculosus*	Epibiota	Men (TC)
	Sargassum muticum	Seaweed-Seaweed *Halidrys siliquosa*	Epibiota	Men (TC)
(Ekloef et al. 2006)	*Eucheuma denticulatum*	Seagrass-Mixed	Epifauna	Man (Add)
(Piazzi and Ceccherelli 2006)	*Caulerpa racemosa*	Seaweed-Mixed	Seaweeds	Man (Rem)
(Schmidt and Scheibling 2006)	*Codium fragile tomentosoides*	Seaweed-*Laminaria longicruris/ digitata*	Epibiota	Men (TC)
(Strong et al. 2006)	*Sargassum muticum*	Bare-Mudflat	Infauna	Men (CI)
(Thomsen and McGlathery 2006)	*Gracilaria vermiculophylla*	Invertebrate-Oyster reef	Seaweeds, sessile animals	Man (Add)
(Thomsen et al. 2006a)	*Gracilaria vermiculophylla*	Seaweed-Mixed	Seaweeds	Men (CI)
(Thomsen et al. 2006b)	*Sargassum muticum*	Seaweed-Mixed	Seaweeds	Men (CI)
(Vincent et al. 2006)	Multiple combined	Various-Mudflat, drift alga, seagrass	Seaweeds	Men (CI)
(Wikström et al. 2006)	*Fucus evanescens*	Seaweed-*Ascophyllum nodosum*	Mobile animals/herbivores	Men (TC)
	Fucus evanescens	Seaweed-*Fucus vesiculosus*	Mobile animals/herbivores	Men (TC)
(York et al. 2006)	*Caulerpa taxifolia*	Seagrass-*Posidonia australis*	Fish	Men (TC)
	Caulerpa taxifolia	Seagrass-*Zostera muelleri*	Fish	Men (TC)
(Harries et al. 2007)	*Sargassum muticum*	Seaweed-Mixed	Epibiota	Men (CI)
(Schmidt and Scheibling 2007)	*Codium fragile tomentosoides*	Seaweed-Mixed	Mobile animals	Man (Rem)
	Codium fragile tomentosoides	Seaweed-*Laminaria longicruris/ digitata*	Mobile animals	Men (TC)

(Berkenbusch et al. 2007)	*Zostera japonica*	Mudflat with ghost shrimps	Infauna	Man (Add)
(Berkenbusch and Rowden 2007)	*Zostera japonica*	Mudflat with ghost shrimps	Infauna	Men (CI)
(Mineur et al. 2008)	Multiple combined	Seaweed-Mixed	Seaweeds	Men (CI)
(Piazzi and Balata 2008)	*Caulerpa racemosa*	Seaweed-Mixed	Sessile plants and animals	Men (CI)
(Gribben et al. 2009a)	*Caulerpa taxifolia*	Bare-Mudflat	Sessile plants and animals	Man (Rem)
	Caulerpa taxifolia	Bare-Mudflat	Sessile plants and animals	Men (CI)
(McKinnon et al. 2009)	*Caulerpa taxifolia*	Bare-Mudflat	Mobile animals	Men (TC)
	Caulerpa taxifolia	Seagrass-*Halophila ovalis*	Mobile animals	Men (TC)
	Caulerpa taxifolia	Seagrass-*Zostera muelleri*	Mobile animals	Men (TC)
(Olabarria et al. 2009)	*Sargassum muticum*	Seaweed-Mixed	Seaweeds	Man (Rem)
(Piazzi and Balata 2009)	*Caulerpa racemosa*	Seagrass-*Posidonia oceanica* matte	Seaweeds	Men (CI)
	Womersleyella setacea	Seagrass-*Posidonia oceanica* matte	Seaweeds	Men (CI)
	Caulerpa racemosa	Seaweed-Mixed	Seaweeds	Men (CI)
	Womersleyella setacea	Seaweed-Mixed	Seaweeds	Men (CI)
(Raffo et al. 2009)	*Undaria pinnatifada*	Seaweed-*Macrocystis*	Epibiota	Men (CI or TC)
	Undaria pinnatifada	Seaweed-*Macrocystis*	Epibiota	Men (CI)
(Bulleri et al. 2010)	*Caulerpa racemosa*	Seaweed-Mixed	Seaweeds (+few invertebrates)	Man (Rem)
(Gestoso et al. 2010)	*Sargasssum muticum*	Seaweed-*Bifurcaria bifurcata*	Epibiota	Men (TC)
(Lang and Buschbaum 2010)	*Sargassum muticum*	Invertebrate-Oyster reef	Sessile and endobenthic species	Man (Add)

Table 1. contd....

Table 1. contd.

Reference (by year)	Alien	Invaded habitat	Impact on (focal response)	Method
(Lutz et al. 2010)	*Codium fragile tomentosoides*	Seaweed-*Codium fragile tasmanicum/novae-zelandiae*	Epiphytes	Men (TC)
(Thomsen 2010)	*Gracilaria vermiculophylla*	Seaweed-Mixed	Mobile animals	Man (Add)
(Thomsen et al. 2010a)	*Gracilaria vermiculophylla*	Bare-Mudflat	Epibiota/infauna	Man (Rem)
	Gracilaria vermiculophylla	Invertebrate-*Diopatra cuprea* basophyte	Epibiota	Men (TC)
	Gracilaria vermiculophylla	Invertebrate-*Mytilus edulis* basophyte	Epibiota	Men (TC)
(Gennaro and Piazzi 2011)	*Caulerpa racemosa*	Seaweed-Mixed	Seaweeds	Man (Add)
(Irigoyen et al. 2011)	*Undaria pinnatifida*	Seaweed-Mixed	animal benthic communities	Man (Rem)
(Klein and Verlaque 2011)	*Caulerpa racemosa*	Seagrass-*Posidonia oceanica* matte	Seaweeds	Man (Rem)
(Pacciardi et al. 2011)	*Caulerpa racemosa*	Bare-Mudflat	Mobile animals	Man (Rem)
	Caulerpa racemosa	Bare-Mudflat	Mobile animals	Men (CI)
(White and Shurin 2011)	*Sargassum muticum*	Seaweed-Mixed	Seaweeds	Man (Rem)
(Gallucci et al. 2012)	*Caulerpa taxifolia*	Bare-Mudflat	Meiofauna in sediment	Men (TC)
	Caulerpa taxifolia	Seagrass-*Zostera muelleri*	Meiofauna in sediment	Men (TC)
(Gestoso et al. 2012)	*Sargassum muticum*	Seaweed-*Bifurcaria bifurcata*	Epibiota	Men (TC)
	Sargassum muticum	Seaweed-*Saccorhiza polyschides*	Epibiota	Men (TC)
(Guerra-Garcia et al. 2012)	*Asparagopsis armata*	Seaweed-Seaweed *Corallina elongata*	Crustaceans (Peracarids)	Men (TC)
(Janiak and Whitlatch 2012)	*Grateloupia turuturu*	Seaweed-*Chondrus crispus*	Epibiota	Men (TC)
(Krumhans and Scheibling 2012)	*Codium fragile tomentosoides*	Seaweed-*Saccharina latissima*	Animals (colonizing)	Men (TC)

(Guerra-García et al. 2012)	*Asparagopsis armata*	Seaweed-Mixed	Crustaceans (+ few other fauna)	Men (CI)
(Vázquez-Luis et al. 2012)	*Caulerpa racemosa*	Seaweed-Seaweed Artificial-mimic	Amphipods	Men (TC)
	Caulerpa racemosa	Seaweed-Seaweed Empty bag	Amphipods	Men (TC)
	Caulerpa racemosa	Seaweed-Seaweed *Halopteris scoparia*	Amphipods	Men (TC)
(Willette and Ambrose 2012)	*Halophila stipulacea*	Seagrass-*Syringodium filiforme*	Fish	Men (TC)

Man = manipulative experiment, Add = addition (of invader), Rem = removal (of invader), Men = Mensurative experiment, CI = compare Control *vs* Impacted (invaded) site(s), BA = compare Before *vs* After (invasion) at site(s), TC = Trait Comparison—here focus is more specifically on how the invader differs in an ecological trait compared to a native species; do invasive species provide poorer habitat for epibiota or is less grazing compared to a native species). Community data: 5 types of community impact data are typically listed here in the following order (1) multivariate graphic expression (e.g., MDS plots), (2) multivariate statistical test (e.g., anosim/permanova), (3) taxonomic richness (r), (4) taxonomic diversity (d, e.g., Shannon Wiener index), and (5) taxonomic evenness (e). nd = data not reported for this particular type of community impact.

manipulative addition and removal experiments (8 vs. 15 studies) and several types of mensurative experiments (Before-After comparison = 2, Control-Impact comparison = 24, Before-After-Control-Impact comparison = 4, Trait Comparison = 19), with several studies showing more than one type of impact data. Most studies were done in Europe, North America or Australasia; these were dominated by studies from Italy (13), Spain (10), the USA (8), Australia (6) and Denmark (5). Only 7 studies were conducted on different continents; 3 were from Argentina, 2 from Tanzania and 1 each from Brazil and the West Indies. We found analogue geographical data-clustering when we examined which oceans and latitudes research was done in. Most studies were conducted in the northeast Atlantic (40, with 18 being from the Mediterranean Sea), followed by the northwest Atlantic (7), the southwest Pacific (6), the northeast Pacific (5), the southwest Atlantic (4), the west Indian Ocean (2), and the Caribbean Sea and Southern Ocean (1 study each). Furthermore, > 50% of all studies were done in a narrow latitudinal band of the northern hemisphere (35–45° N), and no data were collected south of –43° or north of 61°. More studies were conducted in the subtidal than intertidal zone (44 vs. 21), and from rocky coasts (40) compared to estuaries (17) or open sandy coasts (12). Finally, we found that 42 studies evaluated community impacts on rocky substrata, 24 on sedimentary substrata (including impacts in seagrass beds) and 5 on various biogenic substrata (native oysters, mussels and tube-forming polychaetes).

Conclusions. This mini-review revealed that marine plant ecologists have over the last decade addressed research gaps highlighted in earlier reviews. We documented an exponential increase in community impact studies over recent years. By using this trend-line we forecasted that a whopping 250 new community impact papers could be published over the next decade. It is clearly important that this future research effort build on—instead of repeating—past research (see list in Table 1). Furthermore, a surprisingly high number of studies has quantified multivariate community effects and effects on taxonomic richness. Still, we hope that more studies in the future aim to report all 5 major types of community effects; that is, report impacts on richness, diversity, evenness, and sample similarity with 2D-graphical displays and multivariate inferential tests. We encourage detailed data reporting that assesses impacts on dominant taxa and for all treatment-combinations (in factorial experiments) and sites (in nested designs), to provide a solid baseline standard that can be compared across impact studies (extensive data reporting is easily possible through online appendixes). Finally we found strong spatial clustering, as most studies were conducted from Europe, North America and Australasia and in a narrow latitudinal zone. It is therefore important that more studies are done outside these regions to ensure globally representative data (Pyšek et al. 2008, Nunez and Pauchard 2010, Martin et al. 2012).

Summary and discussion

Understanding ecological interactions between invaders and native marine plants is important to manage coastal ecosystems. Today, hundreds of alien species affect native marine plant communities around the world, and hundreds of marine plants are themselves invaders that impact native coastal ecosystems. However, sandy coastlines, wave-exposed rocky coastlines, and tropical reefs, situated away from urban centers, harbors and other human activities, appear to be relatively little impacted by marine plant invaders.

We noted that more studies have focused on negative than positive effects, even though invasive marine plants could represent a model system to investigate positive impacts of invaders because many are habitat-formers (Thomsen et al. 2010a, Byers et al. 2012). Furthermore, more studies have focused on direct rather than indirect effects, despite the latter potentially being equally important. Many of our examples of indirect effects were based on information evaluated from different studies or were hypothetical, rather than being from specific tests of interactions involving 3–4 species. Community impact studies and path analysis can be important tools to address this research gap (Britton-Simmons 2004). Most of the reviewed studies were 'unidirectional', focusing on how aliens impact natives or how natives impact aliens. Overall, we recommend that future research addresses positive effects, indirect effects, and reciprocal effects between invaders and native organisms, to increase our understanding of long-term implications of species interactions between invasive and native species.

We note that none of the reviewed papers emphasized local or regional extinction events, suggesting that when negative interactions dominate, native species are only partially excluded by invaders (Briggs 2007, Byrnes et al. 2007, Briggs 2010). This is perhaps in contrast to invasion impacts documented from less open freshwater and terrestrial ecosystems (Ricciardi 2004, Clavero and García-Berthou 2005). Indeed, coastal marine systems appear to be relatively open to invasions (Inderjit et al. 2006, Williams and Smith 2007). If invader-driven extinctions are rare in marine systems, then taxonomic richness and global homogenization (Olden and Rooney 2006) should increase, as new alien species are integrated into new communities and as local communities slowly converge towards a global species pool. However, it is premature to conclude that invader-driven marine extinctions do not occur, because (a) species inventories are difficult to compile and local extinctions of rare species may have been overlooked, (b) 'interaction equilibriums' may not have been reached and coastal systems may face future invasion-driven extinctions debts that will be collected (Tilman et al. 1994, Kuussaari et al. 2009), and (c) extinctions may happen if invasion effects interact with co-occurring human stressors, such as overfishing, climate change, pollution and eutrophication (Stachowicz et al. 2002, Thuiller et al. 2007, Bradley et al. 2010, Cote and Green 2012).

Throughout this review we have highlighted that invasion impacts are context-dependent, and that impact changes in space and time depend on combinations of invader attributes and invaded system attributes (Strayer et al. 2006, Thomsen et al. 2011a). Impacts from marine alien plants may change from negative to positive along abiotic stress gradients (Bertness and Callaway 1994) or from direct to indirect pathways if additional invasive species enter a community (Strauss 1991, Wootton 1994, Menge 1995, Rodriguez 2006). One way to address context dependency is to use orthogonal and nested community impact experiments, because these designs can simultaneously quantify direct and indirect interactions, different impact magnitudes, and can highlight both 'winners' and 'losers' in a particular invasion and environmental context. For example, when *Caulerpa* species or *G. vermiculophylla* convert mudflats to meadows, sessile organisms win through habitat formation whereas sedimentary infauna lose from habitat destruction. Factorial and nested experiments also allow researchers to quantify context dependency on multivariate community structure and to estimate how much change to community structure is explained by the invader vs. other test-factors, and if invasion impacts and other stressors interact in a synergistic, antagonistic or additive fashion (Thomsen and McGlathery 2006, Piazzi and Balata 2008).

A growing number of studies aims to identify simple rules to predict invasion impacts and potentially be applied in risk assessments. Within an area, an impact of invasive species on a defined ecosystem property can be standardized as the product of areal occupancy of the invaders, its mean density, and its mean per-capita effect throughout this area (Parker et al. 1999). However, few studies have applied this formula to invaders because the formula components can be difficult to measure and the per-capita effects difficult to define (Thiele et al. 2010). Even so, from this relationship we expect that *G. vermiculophylla* has a higher 'global' invasion impact than, for example, the morphologically similar coarsely branched red algal invaders *G. salicornia* or *Soliera filiformis* because the invaded area and abundances of *G. vermiculophylla* are much larger. A similar simple impact rule states that both negative (Lang and Buschbaum 2010, White and Shurin 2011) and positive (Thomsen 2010) impacts of invaders are density-dependent (Thomsen et al. 2011a). It has also been suggested that historical invasion data can be used to predict future invasion impacts (Ricciardi 2003, Kulhanek et al. 2011). For example, impacts of freshwater carp and zebra mussels can be partly predicted from models incorporating their abundance and pre-impact conditions (Ricciardi 2003, Kulhanek et al. 2011) (we are not aware of studies that have applied this method to predict marine plant invasion impacts). Finally, a 'distinctiveness' impact rule suggests that functionally distinct invaders have larger impacts than invaders that are functional similar to native species (Ricciardi and Atkinson 2004, Thomsen et al. 2014a). However, there are probably exceptions to this rule, depending on whether impacts are compared within or between trophic/functional groups and depending on what response variables are measured. Thus, impacts from marine plants associated with competition for limited resources or genetic hybridization are likely to be larger when invaders and natives are relatively similar (not different). Furthermore, invaders and natives can be so ecologically different that they simply do not interact at all (i.e., impact = 0). In short, the distinctiveness rule is most useful within clearly set model boundaries. With a rapid increase in the number of invasion impact studies (Table 1), it is important to organize this information. Marine invasion impact studies can be sorted along test attributes associated with the invader and the native species of invaded systems, resource levels and abiotic conditions (Thomsen et al. 2011b), how these attributes interact (e.g., how distinct the invader is from the native species), and how general the attributes are (are they only important in a specific unique invasion or are the attributes universally relevant for all invasions?). Marine plant invasion impact studies could also be organized according to what responses are measured, for example if impacts are reported within or between trophic levels and functional groups, and what level of the ecological hierarchy impact is investigated. Marine plant invasion impacts within trophic and functional groups probably involve both competition and facilitation processes (with high impacts when invaders are similar to native species) but are likely dominated by consumer-resource and habitat-formation interactions on higher trophic levels and functional groups (with high impact when invaders are different from native species) (Ricciardi and Atkinson 2004, Thomsen et al. 2014a).

We conclude that invaders have significant impacts on native marine plants and that invasive plants can have dramatic impacts on native plants and animals. Impacts can be both positive and negative, depending on the attributes of the invader, the invaded system and the impacted organism. The positive and negative effects have been manifested through direct mechanisms, such as formation and modification of habitats, food provision, competition, and consumption, and indirect mechanisms such as keystone competition, habitat cascades and consumption cascades. There is still much to learn about invasion impacts, particularly about long-term and large-scale effects, possible local extirpations of rare species, and if local communities are homogenized following invasions. We hope that our review will stimulate more research on ecological interactions between invaders and marine plants, and that new research will build upon, rather than repeat, the rapidly growing invasion literature to provide better tools to predict, prevent, ameliorate and manage invasion impacts.

Acknowledgements

We gratefully acknowledge the continuing financial support of the New Zealand Ministry of Science and Innovation and the National Institute of Water and Atmospheric Research (contract C01X0501). MST was supported by the Marsden Fund of New Zealand and TW was supported by a Future Fellows grant from the Australian Research Council.

References

Albins, M. 2013. Effects of invasive Pacific red lionfish *Pterois volitans* versus a native predator on Bahamian coral-reef fish communities. Biological Invasions 15: 29–43. doi: 10.1007/s10530-012-0266-1.

Albins, M.A. and M.A. Hixon. 2008. Invasive Indo-Pacific lionfish *Pterois volitans* reduce recruitment of Atlantic coral-reef fishes. Marine Ecology Progress Series 367: 233–238.

Alcoverro, T. and S. Mariani. 2004. Patterns of fish and sea urchin grazing on tropical Indo-Pacific seagrass beds. Ecography 27: 361–365. doi: 10.1111/j.0906-7590.2004.03736.x.

Ambrose, W.G. and B.W. Nelson. 1982. Inhibition of giant kelp recruitment by an introduced brown alga. Botanica Marina 25: 265–267.

Argyro, M., A. Demetropoulos and M. Hadjichristophoro. 1999. Expansion of the macroalga *Caulerpa racemosa* and changes in softbottom macrofaunal assemblages in Moni Bay, Cyprus. Oceanologica Acta 22: 517–528.

Balata, D., L. Piazzi and F. Cinelli. 2004. A comparison among assemblages in areas invaded by *Caulerpa taxifolia* and *C. racemosa* on a subtidal Mediterranean rocky bottom. P.S.Z.N. Marine Ecology 25: 1–13.

Baldwin, J.R. and J.R. Lovvorn. 1994. Expansion of seagrass habitat by the exotic *Zostera japonica*, and its use by dabbling ducks and brant in Boundary Bay, British Colombia. Marine Ecology Progress Series 103: 119–127.

Bell, S.S. 1991. Amphipods as insect equivalents? An alternative view. Ecology 72: 350–354.

Bellan-Santini, D., P.M. Arnaud, G. Bellan and M. Verlaque. 1996. The influence of the introduced alga, *Caulerpa taxifolia*, on the biodiversity of the Mediterranean marine biota. J. Mar. Biol. Ass. U.K. 76: 235–237.

Berkenbusch, K. and A.A. Rowden. 2007. An examination of the spatial and temporal generality of the influence of ecosystem engineers on the composition of associated assemblages Aquatic Ecology 41: 129–147.

Berkenbusch, K., A.A. Rowden and T.E. Myers. 2007. Interactions between seagrasses and burrowing ghost shrimps and their influence on infaunal assemblages. Journal of Experimental Marine Biology and Ecology 341: 70–84.

Bertness, M. and R. Callaway. 1994. Positive interactions in communities. Trends in Ecology and Evolution 9: 191–193.

Boudouresque, C.F., R. Lemée, X. Mari and A. Meinesz. 1996. The invasive alga *Caulerpa taxifolia* is not a suitable diet for the sea urchin Paracentrotus lividus. Aquatic Botany 53: 245–250.

Bradley, B.A., D.S. Wilcove and M. Oppenheimer. 2010. Climate change increases risk of plant invasion in the Eastern United States. Biological Invasions 12: 1855–1872. doi: 10.1007/s10530-009-9597-y.

Briggs, J.C. 2007. Marine biogeography and ecology: invasions and introductions. Journal of Biogeography 34: 193–198.

Briggs, J.C. 2010. Marine biology: the role of accommodation in shaping marine biodiversity. Marine Biology 157: 2117–2126.

Britton-Simmons, K.H. 2004. Direct and indirect effects of the introduced alga *Sargassum muticum* on benthic, subtidal communities of Washington State, USA. Mar. Ecol. Prog. Ser. 277: 61–78.

Bulleri, F., L. Airoldi, G.M. Branca and M. Abbiati. 2006. Positive effects of the introduced green alga, *Codium fragile* ssp. *tomentosoides*, on recruitment and survival of mussels. Marine Biology 148: 1213–1220.

Bulleri, F., D. Balata, I. Bertocci, L. Tamburello and L. Benedetti-Cecchi. 2010. The seaweed *Caulerpa racemosa* on Mediterranean rocky reefs: from passenger to driver of ecological change. Ecology 91: 2205–2212.

Buschbaum, A., A.S. Chapman and B. Saier. 2006. How an introduced seaweed can affect epibiota diversity in different coastal systems. Marine Biology 148: 743–754.

Byers, J., S. Reichard, J.M. Randall, I.M. Parker, C.S. Smith, W.M. Lonsdale, A.E. Atkinson, T.R. Seastedt, M. Williamson, E. Chornesky and D. Hayes. 2002. Directing research to reduce the impact of nonindigenous species. Conservation Biology 16: 630–640.

Byers, J., J.T. Wright and P.E. Gribben. 2010. Variable direct and indirect effects of a habitat-modifying invasive species on mortality of native fauna. Ecology 91: 1787–1798.

Byers, J., P. Gribben, C. Yeager and E. Sotka. 2012. Impacts of an abundant introduced ecosystem engineer within mudflats of the southeastern US coast. Biological Invasions 14: 2587–2600. doi: 10.1007/s10530-012-0254-5.

Byrnes, J.E., P. Reynolds and J.J. Stachowicz. 2007. Invasions and extinctions reshape coastal marine food webs. PLOS one 3: 1–6.

Carlton, J.T. 1996. Biological invasions and cryptogenic species. Ecology 77: 1653–1655.

Casas, G., R. Scrosati and M.L. Piriz. 2004. The invasive kelp *Undaria pinnatifida* (Phaeophyceae, Laminariales) reduces native seaweed diversity in Nuevo Gulf (Patagonia, Argentina). Biological Invasions 6: 411–416.

Casini, M., J. Lövgren, J. Hjelm, M. Cardinale, J.-C. Molinero and G. Kornilovs. 2008. Multi-level trophic cascades in a heavily exploited open marine ecosystem. Proceedings of the Royal Society B: Biological Sciences 275: 1793–1801. doi: 10.1098/rspb.2007.1752.

Castilla, J. and P. Neill. 2009. Marine bioinvasions in the Southeastern Pacific: status, ecology, economic impacts, conservation and management. pp. 439–457. *In*: Rilov, G. and J. Crooks (eds.). Biological Invasions in Marine Ecosystems. Springer, Berlin Heidelberg.

Castilla, J.C., N.A. Lagos and M. Cerda. 2004. Marine ecosystem engineering by the alien ascidian *Pyura praeputialis* on a mid-intertidal rocky shore. Marine Ecology Progress Series 268: 119–130.

Cebrian, E., E. Ballesteros, C. Linares and F. Tomas. 2011. Do native herbivores provide resistance to Mediterranean marine bioinvasions? A seaweed example. Biological Invasions 13: 1397–1408. doi: 10.1007/s10530-010-9898-1.

Cebrian, E., C. Linares, C. Marschal and J. Garrabou. 2013. Exploring the effects of invasive algae on the persistence of gorgonian populations. Biological Invasions: 1–10. doi: 10.1007/s10530-012-0261-6.

Ceccherelli, G. and F. Cinelli. 1997. Short-term effects of nutrient enrichment of the sediment and interactions between the seagrass *Cymodocea nodosa* and the introduced green alga *Caulerpa taxifolia* in a Mediterranean bay. Journal of Experimental Marine Biology and Ecology 217: 165–177.

Ceccherelli, G. and N. Sechi. 2002. Nutrient availability in the sediment and the reciprocal effects between the native seagrass *Cymodocea nodosa* and the introduced green alga *Caulerpa taxifolia* in a Mediterranean bay. Hydrobiologia 474: 57–66.

Clavero, M. and E. García-Berthou. 2005. Invasive species are a leading cause of animal extinctions. Trends in Ecology & Evolution 20: 110.

Cote, I. and S.J. Green. 2012. Potential effects of climate change on a marine invasion: The importance of current context. Current Zoology 58: 1–8.

Daskalov, G.M., A.N. Grishin, S. Rodionov and V. Mihneva. 2007. Trophic cascades triggered by overfishing reveal possible mechanisms of ecosystem regime shifts. Proceedings of the National Academy of Sciences 104: 10518–10523. doi: 10.1073/pnas.0701100104.

Davis, R.C., F.T. Short and D.M. Burdick. 1998. Quantifying the effects of green crab damage to eelgrass transplants. Restoration ecology 6: 297–302.

Dayton, P.K. 1972. Towards an understanding of community resilience and the potential effects of enrichment to the benthos of McMurdo Sound, Antarctica. Proceedings of the Colloquium on Conservation Problems in Antartica: 81–96.

de Villele, X. and M. Verlaque. 1995. Changes and degradation in a *Posidonia oceanica* bed invaded by the introduced tropical alga *Caulerpa taxifolia* in the North Western Mediterranean. Botanica Marina 38: 79–87.

DeGraaf, J.D. and M.C. Tyrrell. 2004. Comparison of the feeding rates of two introduced crab species, *Carcinus maenas* and *Hemigrapsus sanguineus*, on the blue mussel, *Mytilus edulis*. Northeastern Naturalist 11: 163–166.

Delefosse, M. and E. Kristensen. 2012. Burial of *Zostera marina* seeds in sediment inhabited by three polychaetes: Laboratory and field studies. Journal of Sea Research 71: 41–49.

Demopoulos, A.W.J. and C.R. Smith. 2010. Invasive mangroves alter macrofaunal community structure and facilitate opportunistic exotics. Marine Ecology Progress Series 404: 51–67. doi: 10.3354/meps08483.

deRivera, C.E., G.M. Ruiz, A.H. Hines and P. Jivoff. 2005. Biotic resistance to invasion: native predator limits abundance and distribution of an introduced crab. Ecology 86: 3364–3376. doi: 10.1890/05-0479.

Eastwood, M.M., M.J. Donahue and A.E. Fowler. 2007. Reconstructing past biological invasions: niche shifts in response to invasive predators and competitors. Biological Invasions 9: 397–407.

Eklöf, J.S., M. de la Torre Castro, L. Adelsköld, N.S. Jiddawi and N. Kautsky. 2005. Differences in macrofaunal and seagrass assemblages in seagrass beds with and without seaweed farms. Estuarine, Coastal and Shelf Science 63: 385–396.

Ekloef, J.S., R. Henriksson and N. Kautsky. 2006. Effects of tropical open-water seaweed farming on seagrass ecosystem structure and function. Marine Ecology Progress Series 325: 73–84.

Engelen, A., N. Henriques, C. Monteiro and R. Santos. 2011. Mesograzers prefer mostly native seaweeds over the invasive brown seaweed *Sargassum muticum*. Hydrobiologia 669: 157–165. doi: 10.1007/s10750-011-0680-x.

Engelen, A., A. Primo, T. Cruz and R. Santos. 2013. Faunal differences between the invasive brown macroalga *Sargassum muticum* and competing native macroalgae. Biological Invasions 15: 171–183. doi: 10.1007/s10530-012-0276-z.

Estes, J.A. and J.F. Palmisano. 1974. Sea otters: Their role in structuring nearshore communities. Science 185: 1058–1060.

Falcao, C. and M.T.M. de Szechy. 2005. Changes in shallow phytobenthic assemblages in southeastern Brazil, following the replacement of *Sargassum vulgare* (Phaeophyta) by *Caulerpa scalpelliformis* (Chlorophyta). Botanica Marina 48: 208–217.

Forrest, B.M. and M.D. Taylor. 2003. Assessing invasion impact: survey design considerations and implications for management of an invasive marine plant. Biological Invasions 4: 375–386.

Francour, P., V.M. Harmelin, J.G. Harmelin and J. Duclerc. 1995. Impact of *Caulerpa taxifolia* colonization on the littoral ichtyofauna of north-western Mediterranean Sea: preliminary results. Hydrobiologia 301: 345–353.

Gallucci, F., P. Hutchings, P. Gribben and G. Fonseca. 2012. Habitat alteration and community-level effects of an invasive ecosystem engineer: a case study along the coast of NSW, Australia. Mar. Ecol. Prog. Ser.

Gennaro, P. and L. Piazzi. 2011. Synergism between two anthropic impacts: *Caulerpa racemosa* var. *cylindracea* invasion and seawater nutrient enrichment. Mar. Ecol. Prog. Ser. 427: 59–70.

Gestoso, I., C. Olabarria and J.S. Troncoso. 2010. Variability of epifaunal assemblages associated with native and invasive macroalgae. Marine and Freshwater Research 61: 724–731. doi: 10.1071/mf09251.

Gestoso, I., C. Olabarria and J. Troncoso. 2012. Effects of macroalgal identity on epifaunal assemblages: native species vs. the invasive species *Sargassum muticum*. Helgoland Marine Research 66: 159–166. doi: 10.1007/s10152-011-0257-0.

Gollan, J.R. and J.T. Wright. 2006. Limited grazing pressure by native herbivores on the invasive seaweed *Caulerpa taxifolia* in a temperateAustralian estuary. Marine and Freshwater Research 57: 685–694.

Gribben, P.E. and J.T. Wright. 2006. Sublethal effects on reproduction in native fauna: are females more vulnerable to biological invasion? Oecologia 149: 352–361.

Gribben, P.E., J. Byers, M. Clements, L.A. McKenzie, P.D. Steinberg and J.T. Wright. 2009a. Behavioural interactions between ecosystem engineers control community species richness. Ecology Letters 12: 1127–1136.

Gribben, P.E., J.T. Wright, W.A. O'Connor, M.A. Doblin, B. Eyre and P.D. Steinberg. 2009b. Reduced performance of native infauna following recruitment to a habitat-forming invasive marine alga. Oecologia 158: 733–745.

Guerra-García, J.M., M. Ros, D. Izquierdo and M.M. Soler-Hurtado. 2012. The invasive *Asparagopsis armata* versus the native *Corallina elongata*: Differences in associated peracarid assemblages. Journal of Experimental Marine Biology and Ecology: 121–128.

Hairston, N.G., F.E. Smith and L.S. Slobodkin. 1960. Community structure, population control, and competition. American Naturalist 94: 421–425.

Harries, D.B., S. Harrow, J.R. Wilson, J.M. Mair and D.W. Donnan. 2007. The establishment of the invasive alga *Sargassum muticum* on the west coast of Scotland: a preliminary assessment of community effects. J. Mar. Biol. Ass. U.K. 87: 1057–1067.

Harris, L.G. and A.C. Jones. 2005. Temperature, herbivory and epibiont acquisition as factors controlling the distribution and ecological role of an invasive seaweed. Biological Invasions 7: 913–924.

Hernández, C.A., R.D. Seitz, R.N. Lipcius, C.M. Bovery and D.M. Schulte. 2012. Habitat affects survival of translocated Bay scallops, *Argopecten irradians concentricus* (Say 1822), in Lower Chesapeake Bay. Estuaries and Coasts 35: 1340–1345. doi: 10.1007/s12237-012-9510-2.

Hewitt, C.L. and M.L. Campbell. 2010. The relative contribution of vectors to the introduction and translocation of marine invasive species. Australian Department of Agriculture Fisheries, and Forestry, Canberra: 56.

Hoeffle, H., M.S. Thomsen and M. Holmer. 2011. High mortality of *Zostera marina* under high temperature regimes but minor effects of the invasive macroalgae *Gracilaria vermiculophylla*. Estuarine, Coastal and Shelf Science 92: 35–46.

Hoeffle, H., T. Wernberg, M.S. Thomsen and M. Holmer. 2012. Drift algae, an invasive snail and elevated temperature reduces the ecological performance of a warm-temperate seagrass via additive effects. Marine Ecology Progress Series 450: 67–80. doi: 10.3354/meps09552.

Huston, M.A. 1994. Biological diversity: the coexistence of species on changing landscapes. Cambridge University Press: 681.

Inderjit, A., D. Chapman, M. Ranelletti and S. Shalini Kaushik. 2006. Invasive marine algae: An ecological perspective. The Botanical Review 72: 153–178.

Irigoyen, A.J., G. Trobbiani, M.P. Sgarlatta and M. Raffo. 2011. Effects of the alien algae *Undaria pinnatifida* (Phaeophyceae, Laminariales) on the diversity and abundance of benthic macrofauna in Golfo Nuevo (Patagonia, Argentina): potential implications for local food webs. Biological Invasions 13: 1521–1532.

Janiak, D.S. and R.B. Whitlatch. 2012. Epifaunal and algal assemblages associated with the native *Chondrus crispus* (Stackhouse) and the non-native *Grateloupia turuturu* (Yamada) in eastern Long Island Sound. Journal of Experimental Marine Biology and Ecology 413: 38–44. doi: 10.1016/j.jembe.2011.11.016.

Jaubert, J.M., J.R.M. Chisholm, D. Ducrot, H.T. Ripley, L. Roy and G.S. Passeron. 1999. No deleterious alterations in *Posidonia* beds in the Bay of Menton (France) eight years after *Caulerpa taxifolia* colonization. J. Phycol. 35: 1113–1119.

Jaubert, J.M., J.R.M. Chisholm, A. Minghelli-Roman, M. Marchioretti, J.H. Morrow and H.T. Ripley. 2003. Re-evaluation of the extent of *Caulerpa taxifolia* development in the northern Mediterranean using airborne spectrographic sensing. Marine Ecology-Progress Series 263: 75–82.

Johnston, C.A. and R.N. Lipcius. 2012. Exotic macroalga *Gracilaria vermiculophylla* provides superior nursery habitat for native blue crab in Chesapeake Bay. Marine Ecology Progress Series 467: 137–146. doi: 10.3354/meps09935.

Jones, C.G., J.H. Lawton and M. Shachak. 1994. Organisms as ecosystem engineers. Oikos 69: 373–386.

Klein, J.C. and M. Verlaque. 2011. Experimental removal of the invasive *Caulerpa racemosa* triggers partial assemblage recovery. Journal of the Marine Biological Association of the U.K. 91: 117–125.

Kristensen, E., T. Hansen, M. Delefosse, G.T. Banta and C.O. Quintana. 2011. Contrasting effects of the polychaetes *Marenzelleria viridis* and *Nereis diversicolor* on benthic metabolism and solute transport in sandy coastal sediment. Marine Ecology Progress Series 425: 125–139. doi: 10.3354/meps09007.

Krumhans, K.A. and R.E. Scheibling. 2012. Detrital subsidy from subtidal kelp beds is altered by the invasive green alga *Codium fragile* ssp. *fragile*. Mar. Ecol. Prog. Ser. 456: 73–85.

Kulhanek, S.A., A. Ricciardi and B. Leung. 2011. Is invasion history a useful tool for predicting the impacts of the world's worst aquatic invasive species? Ecological Applications 21: 189–202.

Kuussaari, M., R. Bommarco, R.K. Heikkinen, A. Helm, J. Krauss, R. Lindborg, E. Öckinger, M. Pärtel, J. Pino, R. Rodà, C. Stefanescu, T. Teder, M. Zobel and I. Steffan-Dewenter. 2009. Extinction debt: a challenge for biodiversity conservation. Trends in Ecology & Evolution 24: 564–571.

Lang, A.C. and C. Buschbaum. 2010. Facilitative effects of introduced Pacific oysters on native macroalgae are limited by a secondary invader, the seaweed *Sargassum muticum*. Journal of Sea Research 63: 119–128.

Levin, P.S., J.A. Coyer, R. Petrik and T.P. Good. 2002. Community-wide effects of nonindigenous species on temperater rocky reefs. Ecology 83: 3182–3193.

Levine, J. 1999. Indirect facilitation: evidence and predictions from a riparian community. Ecology 80: 1762–1769.

Levine, J.M., C.M. D'Antonio, J.S. Dukes, K. Grigulus, S. Lavorel and M. Vila. 2003. Mechanisms underlying the impacts of exotic plant invasions. Proceedings of the Royal Society of London Series B: Biological Sciences 270: 775–781.

Linares, C., E. Cebrian and R. Coma. 2012. Effects of turf algae on recruitment and juvenile survival of gorgonian corals. Mar. Ecol. Prog. Ser. 452: 81–88.

Lubchenco, J. 1978. Plant species diversity in a marine intertidal community: importance of herbivore food preference and algal competitive abilities. The American Naturalist 112: 23–39.

Lubchenco, J. 1983. *Littorina* and *Fucus*: effects of herbivores, substratum heterogeneity, and plant escapes during succession. Ecology 64: 1116–1123.

Lubchenco, J. and B.A. Menge. 1978. Community development and persistence in a low rocky intertidal zone. Ecological Monographs 59: 67–94.

Lutz, M., A. Davis and T. Minchinton. 2010. Non-indigenous macroalga hosts different epiphytic assemblages to conspecific natives in southeast Australia. Marine Biology 157: 1095–1103. doi: 10.1007/s00227-010-1391-y.

Martin, L.J., B. Blossey and E. Ellis. 2012. Mapping where ecologists work: biases in the global distribution of terrestrial ecological observations. Frontiers in Ecology and the Environment 10: 195–201. doi: 10.1890/110154.

McClanahan, T.R., A.T. Kamukuru, N.A. Muthiga, M. Yebio and D. Obura. 1996. Effect of sea urchin reductions on algae, coral, and fish populations. Conservation Biology 10: 136–154. doi: 10.1046/j.1523-1739.1996.10010136.x.

McDermid, K.J., M.C. Gregoritza and D.W. Freshwater. 2002. A new record of a second seagrass species from the Hawaiian archipelago: *Halophila decipiens* Ostenfeld Aquatic Botany 74: 257–262.

McKinnon, J.G., P.E. Gribben, A.R. Davis, D.F. Jolley and J.T. Wright. 2009. Differences in soft-sediment macrobenthic assemblages invaded by *Caulerpa taxifolia* compared to uninvaded habitats. Marine Ecology Progress Series 380: 59–71. doi: 10.3354/meps07926.

Menge, B.A. 1995. Indirect effects in marine rocky intertidal interaction webs: patterns and importance. Ecological Monographs 65: 21–74.

Mineur, F., M.P. Johnson and C.A. Maggs. 2008. Non-indigenous marine macroalgae in native communities: a case study in the British Isles. Journal of the Marine Biological Association of the United Kingdom 88: 693–698. doi: 10.1017/s0025315408001409.

Monteiro, C., A.H. Engelen and R.O. Santos. 2009. Macro- and mesoherbivores prefer native seaweeds over the invasive brown seaweed *Sargassum muticum*: a potential regulating role on invasions. Marine Biology 156: 2505–2515.

Neira, C., L.A. Levin, E.D. Grosholz and G. Mendoza. 2007. Influence of invasive *Spartina* growth stages on associated macrofaunal communities. Biological Invasions 9: 975–993. doi: 10.1007/s10530-007-9097-x.

Nejrup, L., M. Pedersen and J. Vinzent. 2012. Grazer avoidance may explain the invasiveness of the red alga *Gracilaria vermiculophylla* in Scandinavian waters. Marine Biology 159: 1703–1712. doi: 10.1007/s00227-012-1959-9.

Nejrup, L.B. and M.F. Pedersen. 2010. Growth and biomass development of the introduced red alga *Gracilaria vermiculophylla* is unaffected by nutrient limitation and grazing. Aquatic Biology 10: 249–259. doi: 10.3354/ab00281.

Norkko, J., D.C. Reed, K. Timmermann, A. Norkko, B.G. Gustafsson, E. Bonsdorff, C.P. Slomp, J. Carstensen and D.J. Conley. 2012. A welcome can of worms? Hypoxia mitigation by an invasive species. Global Change Biology 18: 422–434. doi: 10.1111/j.1365-2486.2011.02513.x.

Nunez, M.A. and A. Pauchard. 2010. Biological invasions in developing and developed countries: does one model fit all? Biological Invasions 12: 707–714. doi: 10.1007/s10530-009-9517-1.

Nyberg, C.D., M.S. Thomsen and I. Wallentinus. 2009. Flora and fauna associated with the introduced red alga *Gracilaria vermiculophylla*. European Journal of Phycology 44: 395–403. doi: 10.1080/09670260802592808.

O'Leary, J.K. and T.R. McClanahan. 2010. Trophic cascades result in large-scale coralline algae loss through differential grazer effects. Ecology 91: 3584–3597. doi: 10.1890/09-2059.1.

Olabarria, C., I.F. Rodil, M. Incera and J.S. Troncoso. 2009. Limited impact of *Sargassum muticum* on native algal assemblages from Rocky intertidal shores. Marine Environmental Research 67: 153–158.

Olden, J.D. and T.P. Rooney. 2006. On defining and quantifying biotic homogenization. Global Ecology and Biogeography 15: 113–120.

Pacciardi, L., N.M. de Biasi and L. Piazzi. 2011. Effects of *Caulerpa racemosa* invasion on soft-bottom assemblages in the Western Mediterranean Sea. Biological Invasions 13: 2677–2690.

Padilla, D.K. 2010. Context-dependent impacts of non-native ecosystem engineers, the Pacific oyster *Crassostrea gigas*. Integrative and Comparative Biology 50: 213–225.

Parker, I.M., D. Simberloff, W.M. Lonsdale, K. Goodell, M.J. Wonham, P.M. Kareiva, M.H. Williamson, B. Von Holle, P.B. Moyle, J.L. Byers and L. Goldwasser. 1999. Impact: toward a framework for understanding the ecological effects of invaders. Biological Invasions 1: 3–19.

Parks, J.R. 2006. Shorebird use of smooth cordgrass (*Spartina alterniflora*) meadows in Willapa Bay, Washington M.Sc. thesis, Environmental Studies, The Evergreen State College.

Pedersen, M.F., P.A. Stæhr, T. Wernberg and M. Thomsen. 2005. Biomass dynamics of exotic *Sargassum muticum* and native *Halidrys siliquosa* in Limfjorden, Denmark—implications of species replacements on turnover rates. Aquatic Botany 83: 31–47.

Piazzi, L. and F. Cinelli. 2003. Evaluation of benthic macroalgal invasion in a harbour area of the western Mediterranean Sea. European Journal of Phycology 38: 223–231.

Piazzi, L. and G. Ceccherelli. 2006. Persistence of biological invasion effects: Recovery of macroalgal assemblages after removal of *Caulerpa racemosa* var. *cylindracea*. Estuarine Coastal & Shelf Science 68: 455–461.

Piazzi, L. and D. Balata. 2008. The spread of *Caulerpa racemosa* var. *cylindracea* in the Mediterranean Sea: An example of how biological invasions can influence beta diversity. Marine Environmental Research 65: 50–61.

Piazzi, L. and D. Balata. 2009. Invasion of alien macroalgae in different Mediterranean habitats. Biological Invasions 11: 193–204. doi: 10.1007/s10530-008-9224-3.

Piazzi, L., G. Ceccherelli and F. Cinelli. 2001. Threat to macroalgal diversity: effects of the introduced green alga *Caulerpa racemosa* in the Mediterranean. Marine Ecology Progress Series 210: 149–159.

Piazzi, L., D. Balata and F. Cinelli. 2002. Epiphytic macroalgal assemblages of *Posidonia oceanica* rhizomes in the western Mediterranean. Eur. J. Phycol. 37: 69–76.

Piazzi, L., D. Balata and G. Ceccherelli. 2003. Co-occurrence of *Caulerpa taxifolia* and *C. racemosa* in the Mediterranean Sea: interspecific interactions and influence on native macroalgal assemblages. Cryptogamia Algologie 24: 233–243.

Piazzi, L., D. Balata, G. Ceccherelli and F. Cinellia. 2005. Interactive effect of sedimentation and *Caulerpa racemosa* var. *cylindracea* invasion on macroalgal assemblages in the Mediterranean Sea. Estuarine, Coastal and Shelf Science 64: 467–474.

Polte, P., A. Schanz and H. Asmus. 2005. The contribution of seagrass beds (*Zostera noltii*) to the function of tidal flats as a juvenile habitat for dominant, mobile epibenthos in the Wadden Sea. Marine Biology 147: 813–822. doi: 10.1007/s00227-005-1583-z.

Posey, M.H. 1988. Community changes associated with the spread of an introduced seagrass, *Zostera Japonica*. Ecology 69: 974–983.

Powell, K.I., J.M. Chase and T.M. Knight. 2011. A synthesis of plant invasion effects on biodiversity across spatial scales. American Journal of Botany 98: 539–548.

Pyšek, P., D.M. Richardson, J. Pergl, V. Jarosik, Z. Sixtova and E. Weber. 2008. Geographical and taxonomic biases in invasion ecology. Trends in Ecology & Evolution 23: 237–244.
Pyšek, P., V. Jarošík, Pergl J. Hulme, M. Hejda, U. Schaffner and M. Vilà. 2012. A global assessment of invasive plant impacts on resident species, communities and ecosystems: the interaction of impact measures, invading species' traits and environment. Global Change Biology 18: 1725–1737.
Raffo, M.P., M.C. Eyras and O.O. Iribarne. 2009. The invasion of *Undaria pinnatifida* to a *Macrocystis pyrifera* kelp in Patagonia (Argentina, south-west Atlantic). Journal of the Marine Biological Association of the U.K. 89: 1571–1580.
Relini, G.M.R. and G. Torchia. 1998. Fish biodiversity in a *Caulerpa taxifolia* meadow in the Ligurian Sea. Ital. J. Zool. 65(Suppl.): 465–470.
Reynolds, L.K., L.A. Carr and K.E. Boyer. 2012. A non-native amphipod consumes eelgrass inflorescences in San Francisco Bay. Marine Ecology Progress Series 451: 107–118. doi: 10.3354/meps09569.
Ricciardi, A. 2003. Predicting the impacts of an introduced species from its invasion history: an empirical approach to zebra mussel invasions. Freshwater Biology 48: 972–981.
Ricciardi, A. 2004. Assessing species invasions as a cause of extinction. Trends in Ecology and Evolution 19: 619.
Ricciardi, A. and S.K. Atkinson. 2004. Distinctiveness magnifies the impact of biological invaders in aquatic ecosystems. Ecology Letters 7: 781–784.
Riisgård, H.U., C.V. Madsen, C. Barth-Jensen and J.E. Purcell. 2012. Population dynamics and zooplankton-predation impact of the indigenous scyphozoan *Aurelia aurita* and the invasive ctenophore *Mnemiopsis leidyi* in Limfjorden (Denmark). Aquatic Invasions 7: 147–162.
Rius, M. and C.D. McQuaid. 2009. Facilitation and competition between invasive and indigenous mussels over a gradient of physical stress. Basic and Applied Ecology 10: 607–613.
Rodriguez, L.F. 2006. Can invasive species facilitate native species? Evidence of how, when, and why these impacts occur. Biological Invasions 8: 927–939.
Ruesink, J., J.-S. Hong, L. Wisehart, S. Hacker, B. Dumbauld, M. Hessing-Lewis and A. Trimble. 2010. Congener comparison of native (*Zostera marina*) and introduced (*Z. japonica*) eelgrass at multiple scales within a Pacific Northwest estuary. Biological Invasions 12: 1773–1789. doi: 10.1007/s10530-009-9588-z.
Russell, D.J., G.H. Balazs, R.C. Phillips and A.K.H. Kam. 2003. Discovery of the sea grass *Halophila decipiens* (Hydrocharitaceae) in the diet of the Hawaiian green turtle, *Chelonia mydas*. Pacific Science 57: 393–397.
Sala, E., Z. Kizilkaya, D. Yildirim and E. Ballesteros. 2011. Alien marine fishes deplete algal biomass in the eastern Mediterranean. PLOS one 6: e17356.
Sánchez, I. and C. Fernández. 2005. Impact of the invasive seaweed *Sargassum muticum* (Phaeophyta) on an intertidal macroalgal assemblage. Journal of Phycology 41: 923–930.
Sánchez, I., C. Fernández and J. Arrontes. 2005. Long-term changes in the structure of intertidal assemblages after invasion by *Sargassum muticum* (Phaeophyta). Journal of Phycology 41: 942–949.
Sand-Jensen, K. and J. Borum. 1991. Interactions among phytoplankton, periphyton, and macrophytes in temporate freshwaters and estuaries. Aquatic Botany 41: 137–175.
Saunders, M. and A. Metaxas. 2008. High recruitment of the introduced bryozoan *Membranipora membranacea* is associated with kelp bed defoliation in Nova Scotia, Canada. Marine Ecology Progress Series 369: 139–151. doi: 10.3354/meps07669.
Schaffelke, B. and C.L. Hewitt. 2007. Impacts of introduced seaweeds. Botanica Marina 50: 397–417.
Scheibling, R.E. and S.X. Anthony. 2001. Feeding, growth and reproduction of sea urchins (*Strongylocentrotus droebachiensis*) on single and mixed diets of kelp (*Laminaria* spp.) and the invasive alga *Codium fragile* spp. *tomentosoides*. Marine Biology 139: 139–146.
Scheibling, R.E., D.A. Lyons and C.B.T. Sumi. 2008. Grazing of the invasive alga *Codium fragile* ssp. *tomentosoides* by the common periwinkle *Littorina littorea*: Effects of thallus size, age and condition. Journal of Experimental Marine Biology and Ecology 355: 103–113. doi: 10.1016/j.jembe.2007.12.002.
Schiel, D.R. and G.A. Thompson. 2012. Demography and population biology of the invasive kelp *Undaria pinnatifida* on shallow reefs in southern New Zealand. Journal of Experimental Marine Biology and Ecology 434: 25–33.
Schmidt, A.L. and R.E. Scheibling. 2006. A comparison of epifauna and epiphytes on native kelps (*Laminaria* species) and an invasive alga (*Codium fragile* ssp. *tomentosoides*) in Nova Scotia, Canada. Bot. Mar. 49: 315–330.

Schmidt, A.L. and R.E. Scheibling. 2007. Effects of native and invasive macroalgal canopies on composition and abundance of mobile benthic macrofauna and turf-forming algae. J. Exp. Mar. Biol. Ecol. 341: 110–130.

Sellheim, K., J.J. Stachowicz and R.C. Coates. 2010. Effects of a nonnative habitat-forming species on mobile and sessile epifaunal communities. Marine Ecology-Progress Series 398: 69–80. doi: 10.3354/meps08341.

Short, F., T. Carruthers, W. Dennison and M. Waycott. 2007. Global seagrass distribution and diversity: A bioregional model. Journal of Experimental Marine Biology and Ecology 350: 3–20.

Shurin, J.B., E.T. Borer, E.W. Seabloom, K. Anderson, C.A. Blanchette, B. Broitman, S.D. Cooper and B.S. Halpern. 2005. A cross-ecosystem comparison of the strength of trophic cascades. Ecology Letters 5: 785–791.

Sjotun, K., S.F. Eggereide and T. Hoisaeter. 2007. Grazer-controlled recruitment of the introduced *Sargassum muticum* (Phaeophycae, Fucales) in northern Europe. Marine Ecology Progress Series 342: 127–138.

Soler-Hurtado, M.M. and J.M. Guerra-García. 2011. Study of the crustacean community associated to the invasive seaweed *Asparagopsis armata* Harvey, 1855 along the coast of the Iberian Peninsula. Zool. baetica 22: 33–49.

Sousa, R., J.L. Gutierrez and D.C. Aldridge. 2009. Non-indigenous invasive bivalves as ecosystem engineers. Biological Invasions 11: 2367–2385. doi: 10.1007/s10530-009-9422-7.

South, P.M., S. Lilley, L.W. Tait, T. Alestra, M.J.H. Hickford, M.S. Thomsen and D.R. Schiel. in press. Transient effects of an invasive kelp on the community structure and primary productivity of an intertidal assemblage. Marine and Freshwater Research.

Stachowicz, J.J., J.R. Terwin, R.B. Whitlatch and R.W. Osman. 2002. Linking climate change and biological invasions: Ocean warming facilitates nonindigenous species invasions. PNAS (USA) 99: 15497–15500.

Staehr, P.A., M.F. Pedersen, M.S. Thomsen, T. Wernberg and D. Krause-Jensen. 2000. Invasion of *Sargassum muticum* in Limfjorden (Denmark) and its possible impact on the indigenous macroalgal community. Marine Ecology Progress Series 207: 79–88. doi: 10.3354/meps207079.

Strauss, S.Y. 1991. Indirect effects in community ecology: their definition, study and importance. Trends in Ecology and Evolution 6: 206–210.

Strayer, D.L., V.T. Eviner, J.M. Jeschke and M.L. Pace. 2006. Understanding the long-term effects of species invasions. Trends in Ecology & Evolution 21: 645–651.

Strong, J.A., M.J. Dring and C.A. Maggs. 2006. Colonisation and modification of soft substratum habitats by the invasive macroalga *Sargassum muticum*. Marine Ecology Progress Series 321: 87–97.

Therriault, T.W. and L.M. Herborg. 2008. A qualitative biological risk assessment for vase tunicate *Ciona intestinalis* in Canadian waters: using expert knowledge. ICES Journal of Marine Science: Journal du Conseil 65: 781–787.

Thibaut, T. and A. Meinesz. 2000. Are the Mediterranean ascoglossan molluscs *Oxynoe olivacea* and *Lobiger serradifalci* suitable agents for a biological control against the invading tropical alga *Caulerpa taxifolia*? Comptes Rendus de l'Académie des Sciences—Series III—Sciences de la Vie 323: 477–488.

Thiele, J., J. Kollmann, B. Markussen and A. Otte. 2010. Impact assessment revisited: improving the theoretical basis for management of invasive alien species. Biological Invasions 12: 2025–2035. doi: 10.1007/s10530-009-9605-2.

Thomsen, M.S. 2004. Species, thallus size and substrate determine macroalgal break forces and break places in a low-energy soft-bottom lagoon. Aquatic Botany 80: 153–161. doi: 10.1016/j.aquabot.2004.08.002.

Thomsen, M.S. 2010. Experimental evidence for positive effects of invasive seaweed on native invertebrates via habitat-formation in a seagrass bed. Aquatic Invasions 5: 341–346.

Thomsen, M.S. and K. McGlathery. 2005. Facilitation of macroalgae by the sedimentary tube forming polychaete *Diopatra cuprea*. Estuarine, Coastal and Shelf Science 62: 63–73.

Thomsen, M.S. and K. McGlathery. 2006. Effects of accumulations of sediments and drift algae on recruitment of sessile organisms associated with oyster reefs. Journal of Experimental Marine Biology and Ecology 328: 22–34. doi: 10.1016/j.jembe.2005.06.016.

Thomsen, M.S. and K.J. McGlathery. 2007. Stress tolerance of the invasive macroalgae *Codium fragile* and *Gracilaria vermiculophylla* in a soft-bottom turbid lagoon. Biological Invasions 9: 499–513. doi: 10.1007/s10530-006-9043-3.

Thomsen, M.S., K. McGlathery and A.C. Tyler. 2006a. Macroalgal distribution patterns in a shallow, soft-bottom lagoon, with emphasis on the nonnative *Gracilaria vermiculophylla* and *Codium fragile*. Estuaries and Coasts 29: 470–478.

Thomsen, M.S., T. Wernberg, P.A. Stæhr and M.F. Pedersen. 2006b. Spatio-temporal distribution patterns of the invasive macroalga *Sargassum muticum* within a Danish *Sargassum*-bed. Helgol. Mar. Res. 60: 50–58.

Thomsen, M.S., P. Stæhr, C.D. Nyberg, D. Krause-Jensen, S. Schwærter and B. Silliman. 2007a. *Gracilaria vermiculophylla* in northern Europe, with focus on Denmark, and what to expect in the future. Aquatic Invasions 2: 83–94.

Thomsen, M.S., T. Wernberg, P. Staehr, D. Krause-Jensen, N. Risgaard-Petersen and B.R. Silliman. 2007b. Alien macroalgae in Denmark—a broad-scale national perspective. Marine Biology Research 3: 61–72. doi: 10.1080/17451000701213413.

Thomsen, M.S., P. Adam and B. Silliman. 2009a. Anthropogenic threats to Australasian coastal salt marshes. pp. 361–390. *In*: Silliman, B.R., M.D. Bertness and D. Strong (eds.). Anthropogenic Modification of North American Salt Marshes University of California Press, 432 pp, New York.

Thomsen, M.S., T. Wernberg, F. Tuya and B.R. Silliman. 2009b. Evidence for impacts of non-indigenous macroalgae: a meta-analysis of experimental field studies. Journal of Phycology 45: 812–819.

Thomsen, M.S., T. Wernberg, A.H. Altieri, F. Tuya, D. Gulbransen, K.J. McGlathery, M. Holmer and B.R. Silliman. 2010a. Habitat cascades: The conceptual context and global relevance of facilitation cascades via habitat formation and modification. Integrative and Comparative Biology 50: 158–175. doi: 10.1093/icb/icq042.

Thomsen, M.S., T. Wernberg, F. Tuya and B.R. Silliman. 2010b. Ecological performance and possible origin of a ubiquitous but under-studied gastropod. Estuarine Coastal and Shelf Science 87: 501–509. doi: 10.1016/j.ecss.2010.02.014.

Thomsen, M.S., J.D. Olden, T. Wernberg, J.N. Griffin and B.R. Silliman. 2011a. A broad framework to organize and compare ecological invasion impacts. Environmental Research 111: 899–908.

Thomsen, M.S., T. Wernberg, J.D. Olden, J.N. Griffin and B.R. Silliman. 2011b. A framework to study the context-dependent impacts of marine invasions. Journal of Experimental Marine Biology and Ecology 400: 322–327.

Thomsen, M.S., T. Wernberg, A.H. Engelen, F. Tuya, M.A. Vanderklift, M. Holmer, K.J. McGlathery, F. Arenas, J. Kotta and B.R. Silliman. 2012. A meta-analysis of seaweed impacts on seagrasses: generalities and knowledge gaps. PLOS one 7: e28595.

Thomsen, M.S., J.E. Byers, D.R. Schiel, J.F. Bruno, J.D. Olden, T. Wernberg and B.R. Silliman. 2014a. Impacts of marine invaders on biodiversity depend on trophic position and functional similarity. Marine Ecology Progress Series 495: 39–47. doi: 10.3354/meps10566.

Thomsen, M.S., T. Wernberg, J.D. Olden, J.E. Byers, J.F. Bruno, B.R. Silliman and D.R. Schiel. 2014b. Forty years of experiments on invasive species: are biases limiting our understanding of impacts? Neobiota 22: 1–22.

Thornber, C.S., B.P. Kinlan, M.H. Graham and J.J. Stachowicz. 2004. Population ecology of the invasive kelp *Undaria pinnatifida* in California: environmental and biological controls on demography. Marine Ecology Progress Series 268: 69–80.

Thuiller, W., D.M. Richardson and G.F. Midgley. 2007. Will climate change promote alien plant invasions? *In*: Nentwig, W. (ed.). Biological Invasions, Ecological Studies. Springer-Verlag, Berlin Heidelberg 193: 197–211.

Tilman, D., R.M. May, C.L. Lehman and M.A. Nowak. 1994. Habitat destruction and the extinction debt. Nature 371: 65–66.

Tomas, F., E. Cebrian and E. Ballesteros. 2011. Differential herbivory of invasive algae by native fish in the Mediterranean Sea. Estuarine, Coastal and Shelf Science 92: 27–34.

Tronstad, M., R.O. Hall, T.M. Koel and K.G. Gerow. 2010. Introduced lake trout produced a four-level trophic cascade in Yellowstone Lake. Transactions of the American Fisheries Society 139: 1536–1550.

Trowbridge, C.D. 2002. Local elimination of *Codium fragile* ssp. *tomentosoides*: indirect evidence of sacoglossan herbivory. J. Mar. Biol. Ass. U.K. 82: 1029–1030.

Trowbridge, C.D. and C.D. Todd. 2001. Host-plant change in marine specialist herbivores: ascoglossan sea slugs on introduced Macroalgae. Ecological Monographs 71: 219–243.

Trussell, G., P. Ewanchuck and M.D. Bertness. 2002. Field evidence of trait-mediated indirect interactions in a rocky intertidal food web. Ecology Letters 5: 241–245.

Trussell, G., P. Ewanchuk, M.D. Bertness and B.R. Silliman. 2004. Trophic cascades in rocky shore tide pools: distinguishing lethal and non-lethal effects. Oecologia 139: 427–432.

Tsai, C., S. Yang, A.C. Trimble and J.L. Ruesink. 2010. Interactions betweem two introduced species: *Zostera japonica* (dwarf eelgrass) facilitates itself and reduces condition of *Ruditapes philippinarum* (Manila clam) on intertidal mudflats. Marine Biology 157: 1929–1936.

Valentine, J.P. and C.R. Johnson. 2005. Persistence of the exotic kelp *Undaria pinnatifida* does not depend on sea urchin grazing. Marine Ecology Progress Series 285: 43–55.

Vázquez-Luis, M., J.A. Borg, P. Sanchez-Jerez and J.T. Bayle-Sempere. 2012. Habitat colonisation by amphipods: comparison between native and alien algae. Journal of Experimental Marine Biology and Ecology 432–433: 162–170.

Viejo, R.M. 1997. The effects of colonization by *Sargassum muticum* on tidepool macroalgal assemblages. Journal of the Marine Biological Association of the United Kingdom 77: 325–340.

Viejo, R.M. 1999. Mobile epifauna inhabiting the invasive *Sargassum muticum* and two local seaweeds in northern Spain. Aquatic Botany 64: 131–149.

Vilà, M., J.L. Espinar, M. Hejda, P.E. Hulme, V. Jarošík, J.L. Maron, J. Pergl, U. Schaffner, Y. Sun and P. Pyšek. 2011. Ecological impacts of invasive alien plants: a meta-analysis of their effects on species, communities and ecosystems. Ecology Letters.

Vincent, C., D. Mouillot, M. Lauret, T.D. Chi, M. Troussellier and C. Aliaume. 2006. Contribution of exotic species, environmental factors and spatial components to the macrophyte assemblages in a Mediterranean lagoon (Thau lagoon, Southern France). Ecological Modelling 193: 119–131.

Wallentinus, I. and C.D. Nyberg. 2007. Introduced marine organisms as habitat modifiers. Marine Pollution Bulletin 55: 323–332.

Wernberg, T., M.S. Thomsen, P.A. Stæhr and M.F. Pedersen. 2001. Comparative phenology of *Sargassum muticum* and *Halidrys siliquosa* (Phaeophyceae: Fucales) in Limfjorden, Denmark. Botanica Marina 44: 31–39. doi: 10.1515/BOT.2001.005.

Wernberg, T., M.S. Thomsen, P.A. Staerh and M.F. Pedersen. 2004. Epibiota communities of the introduced and indigenous macroalgal relatives *Sargassum muticum* and *Halidrys siliquosa* in Limfjorden (Denmark). Helgol. Mar. Res. 58: 154–161.

White, E.M., J.C. Wilson and A.R. Clarke. 2006. Biotic indirect effects: a neglected concept in invasion biology. Diversity and Distributions 12: 443–455.

White, L.F. and J.B. Shurin. 2011. Density dependent effects of an exotic marine macroalga on native community structure. Journal of Experimental Marine Biology and Ecology 405: 111–119.

Wikstrom, S.A. and L. Kautsky. 2004. Invasion of a habitat-forming seaweed: effects on associated biota. Biological Invasions 6: 141–150.

Wikström, S.A., M.B. Steinarsdóttir, L. Kautsky and H. Pavia. 2006. Increased chemical resistance explains low herbivore colonization of introduced seaweed. Oecologia 148: 593–601.

Willette, D.A. and R.F. Ambrose. 2009. The distribution and expansion of the invasive seagrass *Halophila stipulacea* in Dominica, West Indies, with a preliminary report from St. Lucia. Aquatic Botany 91: 137–142.

Willette, D.A. and R.F. Ambrose. 2012. Effects of the invasive seagrass *Halophila stipulacea* on the native seagrass, *Syringodium filiforme*, and associated fish and epibiota communities in the Eastern Caribbean. Aquatic Botany 103: 74–82.

Williams, S.L. 2007. Introduced species in seagrass ecosystems: Status and concerns. Journal of Experimental Marine Biology and Ecology 350: 89–110.

Williams, S.L. and J.E. Smith. 2007. A global review of the distribution, taxonomy, and Impacts of Introduced Seaweeds. Annual Review of Ecology, Evolution, and Systematics 38: 327–359.

Wootton, J.T. 1994. The nature and concequences of indirect effects in ecological communities. Annual Review of Ecology and Systematics 25: 443–466.

Wright, J.T. and P.E. Gribben. 2008. Predicting the impact of an invasive seaweed on the fitness of native fauna. Journal of Applied Ecology 45: 1540–1549.

Wright, J.T., L.A. McKenzie and P.E. Gribben. 2007. A decline in the abundance and condition of a native bivalve associated with *Caulerpa taxifolia* invasion. Marine and Freshwater Research 58: 263–272.

Wu, Y.T., C.H. Wang, X.D. Zhang, B. Zhao, L.F. Jiang, J.K. Chen and B. Li. 2009. Effects of saltmarsh invasion by *Spartina alterniflora* on arthropod community structure and diets. Biological Invasions 11: 635–649. doi: 10.1007/s10530-008-9279-1.

York, P.H., D.J. Booth, T.M. Glasby and B.C. Pease. 2006. Fish assemblages in habitats dominated by *Caulerpa taxifolia* and native seagrasses in south-eastern Australia. Marine Ecology Progress Series 312: 223–234.

11

Physical Threats to Macrophytes as Ecosystem Engineers

Walker D.I.[1,]* and *Bellgrove A.*[2]

Introduction

In pristine systems various species of marine macrophytes (seagrasses, fucoid and laminarian kelps, intertidal fucoids, etc.) act as autogenic ecosystem engineers (case 3, Jones et al. 1994), dominating space, creating complex, three-dimensional habitats and modifying the flow of resources to other species. Whilst there are still extensive knowledge gaps in understanding the conditions under which species act as engineers or not (Bertness et al. 1999, Bruno and Bertness 2001, Crain and Bertness 2006, Jones et al. 1997, Wright and Jones 2006), generally there are commonalities in the ways they modulate their environments (Fig. 1).

Macrophytes are reliant on shallow water habitats with sufficient light. Anything that reduces light availability for photosynthesis can reduce primary productivity and hence the quality of the habitat provisioned by macrophytes. Canopy effects involve complex direct and indirect interactions and reductions in canopy have been shown to alter the effectiveness of marine macrophytes as ecosystem engineers (Benedetti-Cecchi et al. 2001, Pocklington 2012, Reed and Foster 1984, Schiel and Lilley 2007). Whilst there are differences between photosynthetic communities on soft and hard substrata, the processes that affect photosynthesis influence macrophytes in the same way, ultimately reducing or removing canopy (Fig. 1).

[1] Oceans Institute and School of Plant Biology, University of Western Australia, Crawley, Western Australia 6009, Australia.

[2] Centre for Integrative Ecology, School of Life and Environmental Sciences, Deakin University, Warrnambool Campus, PO Box 423, Warrnambool, VIC 3280.

* Corresponding author.

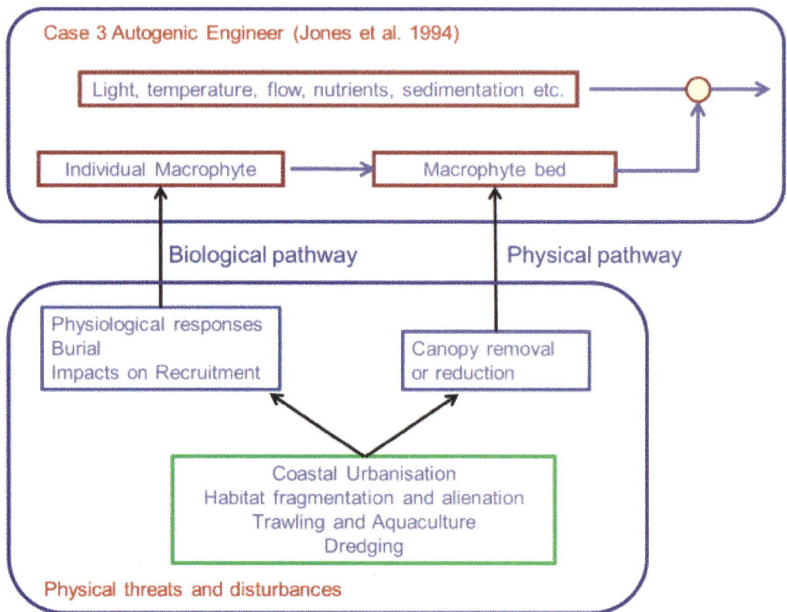

Figure 1. Marine macrophytes can act as autogenic ecosystem engineers, modulating the flow of various resources for other species. However, physical threats and disturbances can alter the engineering role by (1) impacting on individual plants via biological pathways that reduce health, size and/or density of plants, indirectly affecting the quality of the macrophyte bed or by (2) directly affecting the habitat created by macrophytes through physical removal or reduction of the canopy (and in some cases, below-ground biomass).

There are a range of anthropogenic threats to which marine macrophytes may be vulnerable. In this chapter, we focus on the threats of (i) coastal urbanization, (ii) habitat fragmentation and alienation, (iii) trawling and aquaculture, and (iv) dredging and the impacts of the physical disturbances caused by these threats on marine macrophytes. We examine the effects on biological processes that act on individual plants as well as the physical processes acting on the whole bed, or portions thereof. We review what is known about how these impacts can have flow-on affects to other species associated with the marine macrophytes. For the purposes of this chapter, we restrict our definition of marine macrophytes to include seagrasses and macroalgae (seaweeds) that have been shown, or have the potential, to act as ecosystem engineers. We use the term plant to refer to both seagrasses (angiosperms) and macroalgae (protists).

Coastal urbanisation

Greater than 60% of the world's population live in coastal zones, which occupy < 15% of the Earth's land surface, and coastal living is on the increase (reviewed by Airoldi and Beck 2007). Increasing urbanization of the coast places pressure on the associated ecosystems through a range of disturbances. Invariably there is commonality in the nature of disturbances to coastal ecosystems associated with human activities.

Physical disturbance through construction and land reclamation; discharge of domestic sewage (and industrial) effluent into the coastal marine environment; trampling through increased human visitation to shores; and sedimentation driven by a range of human activities (such as dredging; see below) all have the effect of reducing or completely removing macrophytes beds via both physical pathways and biological pathways, compromising the health or regeneration of individual macrophytes within beds (Fig. 1).

Construction and reclamation

Construction of marinas, port facilities, canal estates and coastal protection structures that are associated with increased coastal urbanization, both fragment and alienate macrophyte habitats, as well as change the distribution and orientation of available rocky-reef habitat (Airoldi and Beck 2007). The construction of hard structures may increase connectivity between populations and the potential for new invasions by providing 'stepping stones' in areas where natural reefs are sparse (Airoldi et al. 2005, Thompson et al. 2002). Construction can also result in the complete removal of macrophytes from some areas (Gros (1978; in French) cited in Thibaut et al. 2005). Coastal protection structures generally increase the amount of vertical surface area; potentially increasing the proportion of sessile invertebrates and encrusting coralline algae at the expense of foliose macrophytes (Vaselli et al. 2008). In southern Brazil, Rio Grande do Sul has 630-km long coastline of which < 1% is rocky reef (Esteves et al. 2003). Whilst the coastline is relatively undeveloped, increasing coastal development is resulting in alterations to the shoreline with > 31% of the coast impacted. Amongst the impacts are pollution from domestic effluents, closure of natural washouts and creeks, and shore armouring (Esteves et al. 2003).

Sewage outfalls

With increasing coastal development, comes the problem of dealing with increasing volumes of domestic sewage effluent. A common solution for many coastal cities and towns is to discharge sewage effluent (treated or untreated) into the near-shore marine environment. There is extensive evidence that sewage-effluent discharge alters the floristic composition on reefs proximate to ocean outfalls worldwide (Brown et al. 1990, Cormaci and Furnari 1999, Hirose 1978); with extensive declines and local-extinctions of perennial macrophytes (particularly canopy-forming fucoids) and increases in ephemeral and turf-forming species (Borowitzka 1972b, Brown et al. 1990, Fairweather 1990, Littler and Murray 1975, Littler and Murray 1978). Whilst there is laboratory-based evidence that the nitrogen concentrations in sewage effluent (and reduced salinity of effluent discharged in freshwater) can negatively affect the early development of fucoids (Adams et al. 2008, Doblin and Clayton 1995, Kevekordes 2000, Ogawa 1984), yet benefit the growth and competitive ability of ephemeral taxa such as *Ulva* (Rosenberg and Ramus 1981, Rosenberg and Ramus 1984, Steen 2004), there is little field-based evidence of the direct effects of sewage effluent on marine macrophytes. Clearly the rapidity at which canopy loss and local extinction has been shown to occur (Manning 1979) cannot be driven by impacts on recruitment alone of these long-lived perennial fucoids. To better manage the impacts of sewage effluent on marine macrophytes (and their associated communities) we must better understand the cause of their demise.

In addition to the floristic changes noted above, biodiversity typically declines in response to sewage-effluent discharge (Borowitzka 1972b, Fairweather 1990, Roberts et al. 1998, Rodriguez-Prieto and Polo 1996). This is likely a function of both direct toxicity effects (Adams et al. 2008, Kevekordes 2000) and indirect effects of canopy loss/reduction on flora and fauna (Bellgrove et al. in review, Littler and Murray 1975, Wootton 1991). The shifts in macroalgal dominance from canopy-forming macrophytes to ephemeral and turfing taxa around ocean outfalls result in reduced vertical stratification and therefore a change to the modulation of abiotic stresses such as temperature, light, etc. (Fig. 1 Physical pathway). We know from experimental canopy removals of large macrophytes around the world, that the resultant, more structurally simple, assemblages lack the buffering capacity afforded by macrophyte canopies (Benedetti-Cecchi and Cinelli 1992a, Bertocci et al. 2010, Jenkins et al. 1999, Jenkins et al. 2004, Schiel 2006).

There is extensive debate in the literature on top-down vs. bottom up control of epiphytes under conditions of eutrophication. The bottom-up effect of nutrient loading is well known to stimulate algal growth (Walker and McComb 1992) and is thought to be the cause of the excessive filamentous algae in the Baltic Sea (Bonsdorff et al. 1997, Elmgren and Larsson 2001), possibly by inhibiting settlement of grazers. Heck and Valentine 2006 summarise the debate very well, stating that if there is a proliferation of algae then grazers must be absent. For foundation species of macrophytes, the issue is one of reduction in their role as ecosystem engineers, and the flow-on effects on the associated organisms, whether the effect is top-down or bottom up.

Coralline algal turfs often proliferate at intermediate pollution levels (Bellgrove et al. 1997, Bellgrove et al. 2010, Borowitzka 1972a, Diez et al. 2009, Diez et al. 1999, May 1985), and there is accumulating evidence that coralline turfs and macrophyte canopies may represent alternative stable states (Airoldi 1998, Bellgrove et al. 2010, Benedetti-Cecchi and Cinelli 1992b, Bulleri et al. 2002, Mangialajo et al. 2008). Coralline turfs can themselves act as autogenic engineers, but the associated assemblages are different to those of canopy-forming macrophytes (Chapman et al. 2005, Kelaher et al. 2001, Mangialajo et al. 2008).

Trampling

The effects of trampling on terrestrial plants can be stark, with denuded tracks forming in areas of frequent animal and human traffic. Visitors to coastal parks will be aware of the common management strategy to restrict human traffic to defined areas to reduce the denudation of dunes. But as the visitor walks through the dunes and down onto the reef or seagrass bed, they may continue to place trampling pressure on the marine flora and fauna, but the effect may be less visually dramatic due to more random movements over the littoral and upper sublittoral zones (Goodsell and Underwood 2008). Impacts of human trampling on marine benthos can be significant, with a general reduction in the cover and biomass of erect algae and seagrasses, an increased cover of algal turfs and the concomitant indirect effects on the abundances and diversity of associated fauna (Milazzo et al. 2002a).

The impacts of human visitation and trampling of intertidal reefs have been relatively well studied. Foliose algae are susceptible to trampling and are often slow to recover. Fucoid algae in particular are affected by low intensity trampling; as little as a single step can reduce the individual biomass of the Australasian fucoid, *Hormosira banksii*, by 20% (Povey and Keough 1991). Several studies have reported increasing fucoid canopy loss with increasing trampling intensity (Araújo et al. 2009, Goodsell and Underwood 2008, Keough and Quinn 1998, Milazzo et al. 2002b, Povey and Keough 1991, Schiel and Taylor 1999), but the spatial extent of the trampling (number of trampled paths) can have a greater impact than the intensity (number of passages on a path) (Goodsell and Underwood 2008). Moreover, the magnitude of the effect of trampling and the period of canopy and community recovery, are temporally and spatially variable even within species (Keough and Quinn 1998, Milazzo et al. 2002b, Schiel and Taylor 1999).

Fucoids and some foliose red algae appear particularly susceptible to trampling as their discoid holdfasts and low attachment strength (McKenzie and Bellgrove 2009) mean whole fronds are easily dislodged proliferating large reductions in canopy cover/biomass (Araújo et al. 2009, Brosnan and Crumrine 1994, Goodsell and Underwood 2008, Keough and Quinn 1998, Povey and Keough 1991). In contrast, turfing algae that spread vegetatively with multiple attachment points to the substratum, are more robust to trampling and tend to break at the tips rather than be dislodged completely, and thus recover more quickly (Brosnan and Crumrine 1994, Povey and Keough 1991). This difference in susceptibility of trampling between fucoids and other foliose species compared with turfing species has resulted, in some cases, in a switch from a complex canopy-algal dominated system to an alternate low-profile, turf-dominated state (Araújo et al. 2009, Brosnan and Crumrine 1994, Milazzo et al. 2002b). Similar results have been seen for experimental manipulations of fucoid-canopy cover. However, other studies have shown that coralline turfs may suffer from increased UV exposure with the reduction of fucoid canopies through trampling (Keough and Quinn 1998, Schiel and Taylor 1999). Where turfs are able to withstand trampling and proliferate in the absence of dominate canopy algae, they may inhibit the recruitment and recovery of fucoid algae (Bellgrove et al. 2010).

The impacts of human trampling on seagrass meadows have been rarely studied, although shallow wading is a common activity in coastal waters (Eckrich and Holmquist 2000). Seagrass meadows may also be visited regularly for commercial and recreational bait harvesting (Pillay et al. 2010, Skilleter et al. 2006). Unlike terrestrial angiosperms, seagrasses lack supportive tissue and an evolutionary history of trampling by large herbivores, suggesting they may be vulnerable to human trampling (Eckrich and Holmquist 2000). Disturbance caused by trampling and bait pumping (which involves extraction and subsequent deposition of sediments to harvest bait species) has been attributed to the decline of *Zostera capensis* (with cascading effects on faunal composition and abundance) in South Africa (Pillay et al. 2010, Wynberg and Branch 1997). Similarly, in an experimental test of trampling intensity on *Thalassia testudinum* beds in Puerto Rico, above- and below-ground biomass were reduced with increasing trampling intensity, and had not recovered after 7 mo (Eckrich and Holmquist 2000). Importantly rhizome loss was greater in areas with softer sediments than firmer substrata (Eckrich and Holmquist 2000). Fish were unaffected by the trampling and seagrass loss, but shrimp abundances declined in intense trampling treatments (Eckrich and Holmquist 2000). Thus, while there are limited experimental studies of the effects of trampling in seagrass habitats, the effects are similar to those in reef systems with a general reduction in macrophyte biomass, increasing with trampling intensity. Understanding the complexities of the impacts warrants further attention.

Habitat fragmentation and alienation

The development of the coastline, particularly related to increased population pressure in coastal areas, leads to alienation and fragmentation of habitats available for macroalgae on the coast. Worldwide, population growth occurs along coastlines, where the competition for use of coastal space becomes intense. Coastal development may take the form of construction of ports, marinas and groynes; housing developments may alienate coastal terrestrial communities impacting on coastal water quality, whereas canal estates, where convoluted rock walls surround a coastal waterbody, have greater direct impact on the marine environment. All these developments have potential consequences for macroalgal habitats and hence for biodiversity. Some developments result in direct destruction of macroalgal communities, where construction destroys existing reef environments. The construction of ports and marinas often alienates existing macroalgal habitats, as well as fragmenting the remaining macroalgal distributions. Subsequent dredging and sediment infill may reduce the water quality and result in further losses of macroalgae. Canal estates are increasingly common, as the demand for real estate with water frontage increases. Such developments could be considered an increasing potential macroalgal habitat, and as providing fish 'nursery grounds' (Morton 1992). Water quality within such developments is often poor (Pers. obs. 1996) due to inadequate design, resulting in limited flushing capacity (Morton 1992, Cook et al. 2008), as well as being built off estuaries with already eutrophic conditions. Such canal estate developments have been subsequently been banned in some places (Scotland, the states of Victoria and New South Wales in Australia) as their impacts on coastal processes become clearer.

New South Wales has a coastal policy (2007) regulating development of canal estates as they can pose serious water quality threats, threaten the integrity of coastal wetlands and fisheries habitats, exacerbate flooding and potentially disturb acid sulphate soils. The flooding consequences for canal estates of sea level rise are economic consequences that most local councils are not prepared to accept as a risk.

Indirect changes in hydrodynamics and sedimentation may result from coastal development, which have the capacity to alter light availability as well as having direct effects on the environments in which macroalgae occur. Groynes alter sediment transport in the nearshore zone. Trowbridge (1996) demonstrated that the demography of *Codium setchellii* Gardner was heavily influenced by sand burial and scour resulting in *C. setchellii* being scarce on much of its geographical range. Consequences for macroalgae involve smothering or scour (Kendrick 1991, Airoldi et al. 1995).

Trawling and aquaculture

Trawling effects on soft bottoms

Bottom trawling, an industrial fishing method that drags large, heavy nets across the seafloor, stirs up huge, billowing plumes of sediment on shallow seafloors that can be seen from space (Fig. 2). Bottom trawling has been shown to kill vast numbers of marine macrophytes, as well as corals, sponges, fishes and other animals, especially in deeper water. Satellite images show that spreading clouds of mud remain suspended in the sea long after the trawler has passed (Gianni 2004).

In shallow waters effects are more damaging as these shallow waters have sufficient light for photosynthesis, and where seagrass beds are present, trawling can leave large scars increasing habitat fragmentation and making beds vulnerable to erosion and complete removal (Walker and Kendrick 1995).

Neckles et al. (2005) describe disturbance of the eelgrass *Zostera marina* by commercial mussel *Mytilus edulis* harvesting in Maine from dragging impacts, a less widespread and industrial form of fishing that still does substantial damage to seagrass beds in the north east USA. Dragging had disturbed 10% of the eelgrass cover in Maquoit Bay, with dragged sites ranging from 3.4 to 31.8 ha in size. Dragging removed above- and below-ground plant material from the majority of the bottom in the disturbed sites. These affected sites would require a mean of 10.6 yr for recovery

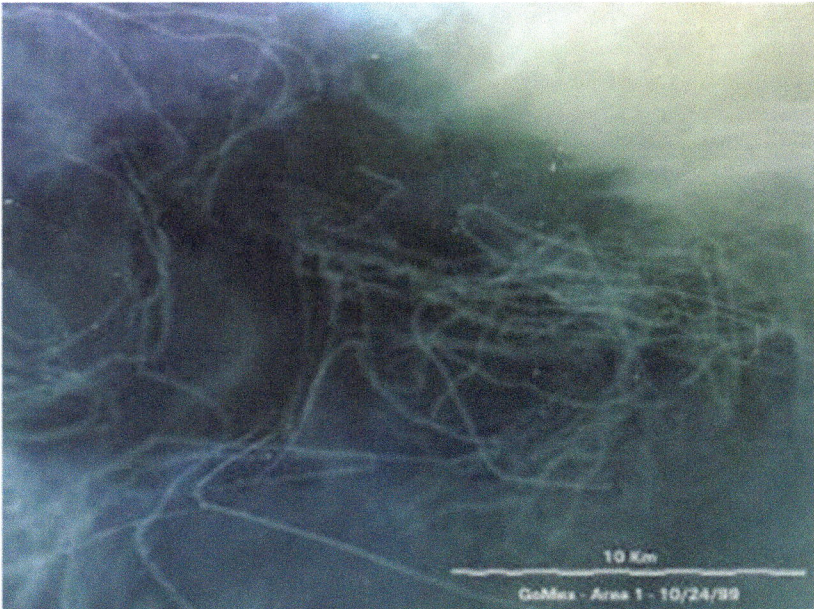

Figure 2. Gulf of Mexico, near Louisiana coast. Individual vessels can be seen as bright spots at end of sediment trails. Other bright spots are fixed oil and gas production platforms. One sediment trail can be traced for 27 km. Assuming a standard trawling speed of 2.5 knots, sediment from this trawl is visibly persistent for nearly 6 hours. Water depth < 20 m. Large, indistinct bright blue patches at lower left and upper right are cloud/haze (Credit: Landsat) http://www.sciencedaily.com/releases/2008/02/080215121207.htm.

of eelgrass shoot density. Where larger seagrasses such as *Posidonia* are present, recovery may take many decades (Walker et al. 2006) for these macrophytes to resume their role as ecosystem engineers.

Fishing effects on macroalgal habitats

Whilst most large-scale, industrial, bottom trawling and dredging tends to occur away from rocky reefs to protect the nets and maximise efficiency of the catch (e.g., Hinz et al. 2011), commercial and recreational harvesting can affect macroalgal beds and the associated communities. In the Mediterranean, historical and current illegal harvesting of the endolithic date mussel *Lithophaga lithophaga* is highly destructive to rocky reefs through the use of hammers and chisels, pneumatic hammers and/or explosives to remove the mussels from the rocks. For example, on the Ionian coast, SE Italy, 37% of rocky reef has been impacted (Fanelli et al. 1994). Through the destructive fishing process, macrophytes are removed (Fanelli et al. 1994, Guidetti 2011, Guidetti and Boero 2004), refugia for juvenile urchins are created (Guidetti 2011) and urchin barrens invariably persist (Guidetti 2011, Guidetti and Boero 2004). The loss of macroalgal canopy has community-scale effects, including reductions to fish recruitment and adult abundances (Guidetti and Boero 2004).

Overfishing of top predators can threaten macroalgal habitats in similar ways. In many parts of the world, overfishing of otters and finfish has driven trophic cascades in kelp forests. Generally sea urchins, relieved from predation pressure, proliferate in numbers and decimate kelp forests to create barrens with greatly altered community structure (Steneck et al. 2002).

Wild harvesting of canopy-forming macroalgae for use as food or extraction of alginates can affect the engineering role of the macroalgae through reductions in canopy cover (Fig. 1) and habitat fragmentation. Impacts may be great and long-lasting where harvesting is either too intense or too frequent to allow full community recovery. For example, Norwegian *Laminaria hyperborea* communities may take > 10 y to recover fully (Birkett et al. 1998), but harvesting is conducted on a 5-y cycle (Christie et al. 1998). In Central Chile, a long history of intensely harvesting the fucoid kelp *Durvillaea antarctica* (Cochayuyo) has led to a disappearance of adult plants from accessible sites in the midlittoral, semi-exposed and exposed habitats (Moreno 2001). When harvesting pressure is lifted in marine reserves, *D. antarctica* is able to recolonise these areas and the community composition shifts (Moreno 2001).

Effects of aquaculture on marine macrophytes

Aquaculture can directly and indirectly affect marine macrophytes by (1) wild harvest of algal species as feed material; (2) nutrient plumes from uneaten food and metabolic waste of caged species; (3) chemical pollution associated with disease and pest control in aquaculture cages; (4) sedimentation from detrital waste (Birkett et al. 1998, Feng et al. 2004). The impacts of these aquaculture artefacts will largely depend on the proximity of seagrass or macroalgal habitats to the aquaculture structures and the amount of water flow and mixing in the region. Because aquaculture structures are commonly placed in shallow and relatively sheltered areas, seagrasses may be more susceptible to the negative effects than macroalgae. Indeed, we found few published studies of the effects of aquaculture on macroalgae.

Soft bottoms are more commonly affected by aquaculture. Effects of aquaculture (Buschmann et al. 1996, Diaz-Almela et al. 2008, Holmer et al. 2008, Holmer and Frederiksen 2007, Terlizzi et al. 2010, Wisehart et al. 2007) in reducing primary productivity and reproduction in seagrasses (Diaz-Almela et al. 2008) and the effects on sediments from contamination associated with over feeding, disease control (Holmer et al. 2008, Holmer and Frederiksen 2007) and changes to invertebrate diversities and densities through canopy decline (Terlizzi et al. 2010, Wisehart et al. 2007) have been well-described. These effects indicate how vulnerable these communities are to the combined effects of aquaculture.

Dredging and sedimentation

Large scale dredging has major impacts on shallow coastal ecosystems (Erftemeijer and Lewis 2006) through light reduction and increased fine sedimentation through re-supension. Sedimentation can cause abrasion, anoxia, and reductions in light and has been shown to affect assemblage structure, usually driving a switch to less complex assemblages, but effects can be variable (Airoldi 2003, Airoldi and Cinelli 1997, Airoldi et al. 1996, Airoldi and Virgilio 1998, Balata et al. 2007, Santelices et al. 2002). Typically rocky reefs affected by sedimentation are dominated by turfing and filamentous algal species with reduced canopy-forming species and biodiversity (Airoldi and Cinelli 1997, Atalah and Crowe 2010, Balata et al. 2007). Reductions in canopy-forming species may be driven, at least in part, by recruitment failure, as sedimentation can preclude zygote attachment (Schiel et al. 2006).

Dredging and reclamation of marine environments, either for extraction of sediments or as part of coastal engineering or construction, can remove seagrasses. Land reclamation directly eliminates seagrass habitat and results in hardening of the shoreline, further eliminating seagrass habitat, as seen throughout Tokyo Bay (Japan) (Erftemeijer and Lewis 2006). Groynes or jetties alter sediment transport in the nearshore zone. Dredging removes seagrass habitat as well as the underlying sediment, leaving bare sand at greater depth, resulting in changes to ecosystem engineering roles of seagrasses (Ralph et al. 2006).

Cumulative impacts

Marine macrophytes can mediate the effects of disturbances on associated organisms (Bertocci et al. 2010) if they themselves can survive the disturbance. All of the physical threats to macrophytes reduce their roles as ecosystem engineers, but most research has been on single influences not on the combined effects.

Studies of combined effects are still not common but there is no doubt that such changes are cumulative. Eutrophication and sediment deposition are widely recognized as major problems for the functioning of coastal systems, and they are expected to increase during the next decades. Atalah and Crowe (2010) showed that nutrients caused an increase in the cover of green filamentous algae, magnified by the removal of grazers and that top down control (by grazers) is more important than bottom up factors (nutrients) in controlling macroalgal assemblage structure. The combined effect of grazer loss and nutrients was larger than the sum of their individual effects.

Bertolucci et al. (2010) also examined the combined effects of the loss of key functional species (canopy-forming macroalgae) and mechanical disturbance on macroalgal intertidal assemblages at 2 sites along the rocky coast of northern Portugal. They demonstrated the canopy buffering effects of the macroalgae, and concluded that current rates of loss of canopy-forming species in urban areas may be expected to exacerbate the effects of predicted climate change, including modifications in intensity and temporal patterns of storms.

When these effects are viewed in the light of potential global climate change, macrophytes are severely under threat from global scale changes!

Acknowledgements

Brinda Pandiyan is thanked for assistance with cataloging the literature used in this review.

References

Adams, M.S., J.L. Stauber, M.T. Binet, R. Molloy and D. Gregory. 2008. Toxicity of a secondary-treated sewage effluent to marine biota in Bass Strait, Australia: Development of action trigger values for a toxicity monitoring program. Marine Pollution Bulletin 57(6-12): 587–598.

Airoldi, L. 1998. Roles of disturbance, sediment stress, and substratum retention on spatial dominance in algal turf. Ecology 79(8): 2759–2770. doi: 10.2307/176515.

Airoldi, L. 2003. The effects of sedimentation on rocky coast assemblages. Oceanography and Marine Biology: an Annual Review 41: 161–236.

Airoldi, L. and F. Cinelli. 1997. Effects of sedimentation on subtidal macroalgal assemblages: an experimental study from a mediterranean rocky shore. Journal of Experimental Marine Biology and Ecology 215: 269–288.

Airoldi, L. and M. Virgilio. 1998. Responses of turf-forming algae to spatial variations in the deposition of sediments. Marine Ecology Progress Series 165: 271–282.

Airoldi, L. and M.W. Beck. 2007. Loss, status and trends for coastal marine habitats of Europe. *In*: Gibson, R.N., R.J.A. Atkinson and J.D.M. Gordon (eds.). Oceanography and Marine Biology, Vol. 45. Oceanography and Marine Biology 45: 345–405.

Airoldi, L., M. Fabiano and F. Cinelli. 1996. Sediment deposition and movement over a turf assemblage in a shallow rocky coastal area of the Ligurian Sea. Marine Ecology Progress Series 133: 241–251.

Airoldi, L., M. Abbiati, M.W. Beck, S.J. Hawkins, P.R. Jonsson, D. Martin, P.S. Moschella, A. Sundelof, R.C. Thompson and P. Aberg. 2005. An ecological perspective on the deployment and design of low-crested and other hard coastal defence structures. Coastal Engineering 52(10-11): 1073–1087. doi: 10.1016/j.coastaleng.2005.09.007.

Araújo, R., S. Vaselli, M. Almeida, E. Serrão and I. Sousa-Pinto. 2009. Effects of disturbance on marginal populations: human trampling on *Ascophyllum nodosum* assemblages at its southern distribution limit. Marine Ecology Progress Series 378: 81–92.

Atalah, J. and T.P. Crowe. 2010. Combined effects of nutrient enrichment, sedimentation and grazer loss on rock pool assemblages. Journal of Experimental Marine Biology and Ecology 388(1-2): 51–57. doi: 10.1016/j.jembe.2010.03.005.

Balata, D., L. Piazzi and F. Cinelli. 2007. Increase of sedimentation in a subtidal system: Effects on the structure and diversity of macroalgal assemblages. Journal of Experimental Marine Biology and Ecology 351(1-2): 73–82. doi: 10.1016/j.jembe.2007.06.019.

Bellgrove, A., M.N. Clayton and G.P. Quinn. 1997. Effects of secondarily treated sewage effluent on intertidal macroalgal recruitment processes. Mar. Freshw. Res. 48(2): 137–146. doi: 10.1071/mf96011.

Bellgrove, A., P.F. McKenzie, J.L. McKenzie and B.J. Sfilgoj. 2010. Restoration of the habitat-forming fucoid alga *Hormosira banksii* at effluent-affected sites: competitive exclusion by coralline turfs. Marine Ecology Progress Series 419: 47–56. doi: 10.3354/meps08843.

Benedetti-Cecchi, L. and F. Cinelli. 1992a. Canopy removal experiments in *Cystoseira*-dominated rockpools from the Western coast of the Mediterranean (Ligurian Sea). Journal of Experimental Marine Biology and Ecology 155: 69–83.

Benedetti-Cecchi, L. and F. Cinelli. 1992b. Effects of canopy cover, herbivores and substratum type on patterns of *Cystoseira* spp. settlement and recruitment in littoral rockpools. Mar. Ecol.-Prog. Ser. 90(2): 183–191. doi: 10.3354/meps090183.

Benedetti-Cecchi, L., F. Pannacciulli, F. Bulleri, P.S. Moschella, L. Airoldi, G. Relini and F. Cinelli. 2001. Predicting the consequences of anthropogenic disturbance: large-scale effects of loss of canopy algae on rocky shores. Marine Ecology Progress Series 214: 137–150.

Bertness, M.D., G.H. Leonard, J.M. Levine, P.R. Schmidt and A.O. Ingraham. 1999. Testing the relative contribution of positive and negative interactions in rocky intertidal communities. Ecology 80(8): 2711–2726.

Bertocci, I., F. Arenas, M. Matias, S. Vaselli, R. Araújo, H. Abreu, R. Pereira, R. Vieira and I. Sousa-Pinto. 2010. Canopy-forming species mediate the effects of disturbance on macroalgal assemblages on Portuguese rocky shores. Marine Ecology Progress Series 414: 107–116. doi: 10.3354/meps08729.

Birkett, D.A., C.A. Maggs, M.J. Dring, P.J.S. Boaden and S.R. Jenkins. 1998. Infralittoral Reef Biotopes with Kelp Species An overview of dynamic and sensitivity characteristics for conservation management of marine SACs. vol. 7. Scottish Association of Marine Science (UK Marine SACs Project), 174.

Borowitzka, M. 1972a. Intertidal algal species diversity and effect of pollution. Aust. J. Marine Freshwater Res. 23(2): 73–84.

Borowitzka, M.A. 1972b. Intertidal algal species diversity and the effect of pollution. Australian Journal of Marine and Freshwater Research 23: 73–84.

Brosnan, D.M. and L.L. Crumrine. 1994. Effects of human trampling on marine rocky shore communities. Journal of Experimental Marine Biology and Ecology 177(1): 79–97. doi: 10.1016/0022-0981(94)90145-7.

Brown, V.B., S.A. Davies and R.N. Synnot. 1990. Long-term monitoring of the effects of treated sewage effluent on the intertidal macroalgae community near Cape Schanck, Victoria, Australia. Botanica Marina 33: 85–98.

Bruno, J. and M.D. Bertness. 2001. Habitat modification and facilitation in benthic marine communities. pp. 201–218. *In*: Bertness, M.D., S.D. Gaines and M.E. Hay (eds.). Marine Community Ecology. Sinauer Associates, Inc., Sunderland, Massachusetts.

Bulleri, F., L. Benedetti-Cecchi, S. Acunto, F. Cinelli and S. Hawkins. 2002. The influence of canopy algae on vertical patterns of distribution of low-shore assemblages on rocky coasts in the northwest Mediterranean. J. Exp. Mar. Biol. Ecol. 267: 89–106.

Buschmann, A.H., D.A. Lopez and A. Medina. 1996. A review of the environmental effects and alternative production strategies of marine aquaculture in Chile. Aquacultural Engineering 15(6): 397–421. doi: 10.1016/s0144-8609(96)01006-0.

Chapman, M.G., J. People and D. Blockley. 2005. Intertidal assemblages associated with natural *Corallina* turf and invasive mussel beds. Biodiversity and Conservation 14: 1761–1776.

Christie, H., S. Fredriksen and E. Rinde. 1998. Regrowth of kelp and colonization of epiphyte and fauna community after kelp trawling at the coast of Norway. Hydrobiologia 375-76: 49–58. doi:10.1023/A:1017021325189.

Cormaci, M. and G. Furnari. 1999. Changes of the benthic algal flora of the Tremiti Islands (southern Adriatic) Italy. Hydrobiologia 398/399: 75–79.

Crain, C.M. and M.D. Bertness. 2006. Ecosystem engineering across environmental gradients: implications for conservation and management. Bioscience 56(3): 211–218.

Diaz-Almela, E., N. Marba, E. Alvarez, R. Santiago, M. Holmer, A. Grau, S. Mirto, R. Danovaro, A. Petrou, M. Argyrou, I. Karakassis and C.M. Duarte. 2008. Benthic input rates predict seagrass (*Posidonia oceanica*) fish farm-induced decline. Marine Pollution Bulletin 56(7): 1332–1342. doi: 10.1016/j. Marpolbul.2008.03.022.

Diez, I., A. Secilla, A. Santolaria and J.M. Gorostiaga. 1999. Phytobenthic intertidal community structure along an environmental pollution gradient. Mar. Pollut. Bull. 38(6): 463–472. doi: 10.1016/s0025-326x(98)90161-8.

Diez, I., A. Santolaria, A. Secilla and J. Maria Gorostiaga. 2009. Recovery stages over long-term monitoring of the intertidal vegetation in the 'Abra de Bilbao' area and on the adjacent coast (N. Spain). Eur. J. Phycol. 44(1): 1–14. doi: Pii 90850113810.1080/09670260802158642.

Doblin, M. and M.N. Clayton. 1995. Effects of secondarily treated sewage effluent on the early life history stages of two species of brown macroalgae: *Hormosira banksii* and *Durvillaea potatorum*. Marine Biology 122: 689–698.

Eckrich, C.E. and J.G. Holmquist. 2000. Trampling in a seagrass assemblage: direct effects, response of associated fauna, and the role of substrate characteristics. Marine Ecology Progress Series 201: 199–209. doi: 10.3354/meps201199.

Erftemeijer, P.L.A. and R.R.R. Lewis, III. 2006. Environmental impacts of dredging on seagrasses: A review. Marine Pollution Bulletin 52(12): 1553–1572. doi: 10.1016/j.marpolbul.2006.09.006.

Esteves, L.S., A.R.P. da Silva, T.B. Arejano, M.A.G. Pivel and M.P. Vranjac. 2003. Coastal development and human impacts along the Rio Grande do Sul beaches, Brazil. Journal of Coastal Research: 548–556.

Fairweather, P.G. 1990. Sewage and the biota on seashores: assessment of impact in relation to natural variability. Environ Monit Assess 14: 197–210.

Fanelli, G., S. Piraino, G. Belmonte, S. Geraci and F. Boero. 1994. Human predation along Apulian rocky coasts (SE Italy)—desertification caused by *Lithophaga lithophaga* (Mollusca) fisheries. Marine Ecology Progress Series 110(1): 1–8. doi: Doi 10.3354/Meps110001.

Feng, Y.Y., L.C. Hou, N.X. Ping, T.D. Ling and C.I. Kyo. 2004. Development of mariculture and its impacts in Chinese coastal waters. Reviews in Fish Biology and Fisheries 14(1): 1–10. doi: 10.1007/s11160-004-3539-7.

Gianni, M. 2004. High Seas Bottom Trawl Fisheries and their Impacts on the Biodiversity of Vulnerable Deep-Sea Ecosystems: Options for International Action. Executive Summary. Gland, Switzerland.

Goodsell, P.J. and A.J. Underwood. 2008. Complexity and idiosyncrasy in the responses of algae to disturbance in mono- and multi-species assemblages. Oecologia 157(3): 509–519.

Guidetti, P. 2011. The destructive date-mussel fishery and the persistence of barrens in Mediterranean rocky reefs. Marine Pollution Bulletin 62(4): 691–695. doi: 10.1016/j.marpolbul.2011.01.029.

Guidetti, P. and F. Boero. 2004. Desertification of Mediterranean rocky reefs caused by date-mussel, *Lithophaga lithophaga* (Mollusca : Bivalvia), fishery: effects on adult and juvenile abundance of a temperate fish. Marine Pollution Bulletin 48(9-10): 978–982. doi: 10.1016/j.marpolbul.2003.12.006.

Hinz, H., D. Tarrant, A. Ridgeway, M.J. Kaiser and J.G. Hiddink. 2011. Effects of scallop dredging on temperate reef fauna. Marine Ecology Progress Series 432: 91–102. doi: 10.3354/meps09166.

Hirose, H. 1978. Composition of benthic marine algae in relation to pollution in the Seto Inland Sea, Japan. Proceedings of the International Seaweed Symposium 9: 173–179.

Holmer, M. and M.S. Frederiksen. 2007. Stimulation of sulfate reduction rates in Mediterranean fish farm sediments inhabited by the seagrass *Posidonia oceanica*. Biogeochemistry 85(2): 169–184. doi: 10.1007/s10533-007-9127-x.

Holmer, M., M. Argyrou, T. Dalsgaard, R. Danovaro, E. Diaz-Almela, C.M. Duarte, M. Frederiksen, A. Grau, I. Karakassis, N. Marba, S. Mirto, M. Perez, A. Pusceddu and M. Tsapakis. 2008. Effects of fish farm waste on *Posidonia oceanica* meadows: Synthesis and provision of monitoring and management tools. Marine Pollution Bulletin 56(9): 1618–1629. doi: 10.1016/j.marpolbul.2008.05.020.

Jenkins, S.R., S.J. Hawkins and T.A. Norton. 1999. Direct and indirect effects of a macroalgal canopy and limpet grazing in structuring a sheltered inter-tidal community. Marine Ecology Progress Series 188: 81–92.

Jenkins, S.R., T.A. Norton and S.J. Hawkins. 2004. Long term effects of *Ascophyllum nodosum* canopy removal on mid shore community structure. Journal of the Marine Biological Association of the United Kingdom 84: 327–329.

Jones, C.G., J.H. Lawton and M. Shachak. 1994. Organisms as ecosystem engineers. Oikos 69: 373–386.

Jones, C.G., J.H. Lawton and M. Shachak. 1997. Positive and negative effects of organisms as physical ecosystem engineers. Ecology 78(7): 1946–1957.

Kelaher, B.P., M.G. Chapman and A.J. Underwood. 2001. Spatial patterns of diverse macrofaunal assemblages in coralline turf and their associations with environmental variables. Journal of the Marine Biological Association of the United Kingdom 81(6): 917–930.

Keough, M.J. and G.P. Quinn. 1998. Effects of periodic disturbances from trampling on rocky intertidal algal beds. Ecological Applications 8(1): 141–161.

Kevekordes, K. 2000. The effects of secondary-treated sewage effluent and reduced salinity on specific events in the early life stages of *Hormosira banksii* (Phaeophyceae). Eur. J. Phycol. 35: 365–371.

Littler, M.M. and S.N. Murray. 1975. Impact of sewage on the distribution, abundance and community structure of rocky intertidal macro-organisms. Marine Biology 30: 277–291.

Littler, M.M. and S.N. Murray. 1978. Influence of domestic wastes on energetic pathways in rocky intertidal communites. Journal of Applied Ecology 15: 583–595.

Mangialajo, L., M. Chiantore and R. Cattaneo-Vietti. 2008. Loss of fucoid algae along a gradient of urbanisation, and structure of benthic assemblages. Marine Ecology Progress Series 358: 63–74. doi: 10.3354/meps07400.

Manning, P.F. 1979. The biological effects of the discharge of secondarily treated sewage effluent at Boags Rocks, Gunnamatta Beach, Victoria. MSc, La Trobe University.

May, V. 1985. Observations on algal floras close to two sewage outlets. Cunninghamia 1(3): 385–394.

McKenzie, P.F. and A. Bellgrove. 2009. Dislodgment and attachment strength of the intertidal macroalga, *Hormosira banksii* (Fucales, Phaeophyceae). Phycologia 48(5): 335–343.

Milazzo, M., R. Chemello, F. Badalamenti, R. Camarda and S. Riggio. 2002a. The impact of human recreational activities in marine protected areas: what lessons should be learnt in the Mediterranean Sea? Marine Ecology-Pubblicazioni Della Stazione Zoologica Di Napoli I 23: 280–290. doi: 10.1111/j.1439-0485.2002.tb00026.x.

Milazzo, M., R. Chemello, F. Badalamenti and S. Riggio. 2002b. Short-term effect of human trampling on the upper infralittoral macroalgae of Ustica Island MPA (western Mediterranean, Italy). Journal of the Marine Biological Association of the United Kingdom 82(5): 745–748. doi: 10.1017/S0025315402006112.

Moreno, C.A. 2001. Community patterns generated by human harvesting on Chilean shores: A review. Aquat. Conserv. 11(1): 19–30. doi: 10.1002/aqc.430.

Neckles, H.A., F.T. Short, S. Barker and B.S. Kopp. 2005. Disturbance of eelgrass *Zostera marina* by commercial mussel *Mytilus edulis* harvesting in Maine: dragging impacts and habitat recovery. Marine Ecology-Progress Series 285: 57–73. doi: 10.3354/meps285057.

Ogawa, H. 1984. Effects of treated municipal wastewater on early development of sargassaceous plants. Hydrobiologia 116/117: 389–392.

Pillay, D., G.M. Branch, C.L. Griffiths, C. Williams and A. Prinsloo. 2010. Ecosystem change in a South African marine reserve (1960–2009): role of seagrass loss and anthropogenic disturbance. Marine Ecology Progress Series 415: 35–48. doi: 10.3354/meps08733.

Pocklington, J.B. 2012. The Ecological Role of Canopy-forming Fucoid Algae on Temperate Intertidal Rocky Shores. The University of Melbourne.

Povey, A. and M.J. Keough. 1991. Effects of trampling on plant and animal populations on rocky shores. Oikos 61: 355–368.

Ralph, P.J., D. Tomasko, K. Moore, S. Seddon and C.M.O. Macinnis-Ng. 2006. Human impacts on seagrasses: Eutrophication, sedimentation, and contamination. Seagrasses: Biology, Ecology and Conservation: 567–593. doi: 10.1007/978-1-4020-2983-7_24.

Reed, D.C. and M.S. Foster. 1984. The effects of canopy shading on algal recruitment and growth in a giant kelp forest. Ecology 65: 937–948.

Roberts, D.E., A. Smith, P. Ajani and A.R. Davis. 1998. Rapid changes in encrusting marine assemblages exposed to anthropogenic point-source pollution: a 'Beyond BACI' approach. Marine Ecology Progress Series 163: 213–224. doi: 10.3354/Meps163213.

Rodriguez-Prieto, C. and L. Polo. 1996. Effects of sewage pollution in the structure and dynamics of the community of *Cystoseira mediterranea* (Fucales, Phaeophyceae). Scientia Marina 60(2-3): 253–263.

Rosenberg, G. and J. Ramus. 1981. Ecological growth strategies in the seaweeds *Gracelaria foliifera* (Rhodophyceae) and *Ulva* sp. (Chlorophyceae): the rate and timing of growth. Botanica Marina 24: 583–589.

Rosenberg, G. and J. Ramus. 1984. Uptake of inorganic nitrogen and seaweed surface area : volume ratios. Aquatic Botany 19: 65–72.

Santelices, B., D. Aedo and A. Hoffmann. 2002. Banks of microscopic forms and survival to darkness of propagules and microscopic stages of macroalgae. Revista Chilena de Historia Natural 75: 547–555.

Schiel, D.R. 2006. Rivets or bolts? When single species count in the function of temperate rocky reef communities. Journal of Experimental Marine Biology and Ecology 338: 233–252.

Schiel, D.R. and D.I. Taylor. 1999. Effects of trampling on a rocky intertidal algal assemblage in southern New Zealand. Journal of Experimental Marine Biology and Ecology 235: 213–235.

Schiel, D.R. and S.A. Lilley. 2007. Gradients of disturbance to an algal canopy and the modification of an intertidal community. Marine Ecology Progress Series 339: 1–11. doi: 10.3354/meps339001.

Schiel, D.R., S.A. Wood, R.A. Dunmore and D.I. Taylor. 2006. Sediment on rocky intertidal reefs: Effects on early post-settlement stages of habitat-forming seaweeds. Journal of Experimental Marine Biology and Ecology 331(2): 158–172. doi: 10.1016/j.jembe.2005.10.015.

Skilleter, G.A., B. Cameron, Y. Zharikov, D. Boland and D.P. McPhee. 2006. Effects of physical disturbance on infaunal and epifaunal assemblages in subtropical, intertidal seagrass beds. Marine Ecology-Progress Series 308: 61–78. doi: 10.3354/meps308061.

Steen, H. 2004. Interspecific competition between *Enteromorpha* (Ulvales : Chlorophyceae) and *Fucus* (Fucales : Phaeophyceae) germlings: effects of nutrient concentration, temperature, and settlement density. Marine Ecology Progress Series 278: 89–101.

Steneck, R.S., M.H. Graham, B.J. Bourque, D. Corbett, J.M. Erlandson, J.A. Estes and M.J. Tegner. 2002. Kelp forest ecosystems: biodiversity, stability, resilience and future. Environmental Conservation 29(4): 436–459.

Terlizzi, A., G. De Falco, S. Felline, D. Fiorentino, M.C. Gambi and G. Cancemi. 2010. Effects of marine cage aquaculture on macrofauna assemblages associated with *Posidonia oceanica* meadows. Italian Journal of Zoology 77(3): 362–371. doi: Pii 921329889 10.1080/11250000903464075.

Thibaut, T., S. Pinedo, X. Torras and E. Ballesteros. 2005. Long-term decline of the populations of Fucales (*Cystoseira* spp. and *Sargassum* spp.) in the Alberes coast (France, North-western Mediterranean). Marine Pollution Bulletin 50(12): 1472–1489. doi: 10.1016/j.marpolbul.2005.06.014.

Thompson, R.C., T.P. Crowe and S.J. Hawkins. 2002. Rocky intertidal communities: past environmental changes, present status and predictions for the next 25 years. Environmental Conservation 29(2): 168–191. doi: 10.1017/s0376892902000115.

Walker, D.I., G.A. Kendrick and A.J. McComb. 2006. Decline and recovery of seagrass ecosystems—the dynamics of change. pp. 551–566. *In*: Larkum, A.W.D., R.J. Orth and C.M. Duarte (eds.). Seagrasses: Biology, Ecology and their Conservation. Springer Verlag Berlin.

Wisehart, L.M., B.R. Dumbauld, J.L. Ruesink and S.D. Hacker. 2007. Importance of eelgrass early life history stages in response to oyster aquaculture disturbance. Marine Ecology-Progress Series 344: 71–80. doi: 10.3354/meps06942.

Wootton, J.T. 1991. Direct and indirect effects of nutrients on intertidal community structure—variable consequences of seabird guano. Journal of Experimental Marine Biology and Ecology 151(2): 139–153. doi: 10.1016/0022-0981(91)90121-C.

Wright, J.P. and C.G. Jones. 2006. The concept of organisms as ecosystem engineers ten years on: progress, limitations, and challenges. Bioscience 56(3): 203–209.

Wynberg, R.P. and G.M. Branch. 1997. Trampling associated with bait-collection for sandprawns Callianassa kraussi Stebbing: effects on the biota of an intertidal sandflat. Environmental Conservation 24(2): 139–148. doi: 10.1017/s0376892997000209.

Index